Springer Texts in Business and Economics

For further volumes:
http://www.springer.com/series/10099

Illustration by Susan Resko

As you walk about town real estate investment opportunities abound!

G. Jason Goddard · Bill Marcum

Real Estate Investment

A Value Based Approach

Prof. G. Jason Goddard
Wells Fargo
Investor Real Estate
Winston-Salem, NC, USA

Prof. Bill Marcum
Wake Forest University
Wake Forest Road
Winston-Salem, NC, USA

Additional material to this book can be downloaded from http://extra.springer.com.

ISSN 2192-4333
ISBN 978-3-642-23526-9
DOI 10.1007/978-3-642-23527-6
Springer Heidelberg Dordrecht London New York

e-ISSN 2192-4341
e-ISBN 978-3-642-23527-6

Library of Congress Control Number: 2012935227

© Springer-Verlag Berlin Heidelberg 2012
This work is subject to copyright. All rights are reserved, whether the whole or part of the material is concerned, specifically the rights of translation, reprinting, reuse of illustrations, recitation, broadcasting, reproduction on microfilm or in any other way, and storage in data banks. Duplication of this publication or parts thereof is permitted only under the provisions of the German Copyright Law of September 9, 1965, in its current version, and permission for use must always be obtained from Springer. Violations are liable to prosecution under the German Copyright Law.

The use of general descriptive names, registered names, trademarks, etc. in this publication does not imply, even in the absence of a specific statement, that such names are exempt from the relevant protective laws and regulations and therefore free for general use.

Printed on acid-free paper

Springer is part of Springer Science+Business Media (www.springer.com)

Preface

Our inspiration for writing this book is our students. After teaching real estate finance courses for the last few years, students have complained that the text books that were being used were not up to the standards of the rest of the course. The authors contend that there is a gap in the existing publications for real estate finance. Currently there are two types of real estate books on the market: the first is highly academic with little practical relevance, the second is highly practical with little of the academic rigor that is required to truly understand the fundamentals of making a sound real estate investment decision.

The first type of existing real estate publication is a lengthy textbook that reads like a dictionary of terms with mathematical problems for use in a college setting. These books are typically very lengthy, and educators have to pick and choose the chapters that they plan to cover in a given semester. Covering the entire book is usually not possible during a traditional 16 week semester. Additionally, these books are usually written with residential real estate as the primary objective. Commercial real estate is often left for the chapters in the second half of the book. For classes geared toward commercial real estate, the text books available in the marketplace today do not allow for a smooth transition from topic to topic, as students are required to read chapters from various parts of a given text (or multiple texts) in an effort to achieve an adequate supplement to the material presented in class.

The second type of real estate book is written for audiences other than the academic market. The best of these books typically discuss topics such as *"the five things that every investor should know"* or *"this is how **you** can make money in real estate"*, or other such topics. These books often contain perspectives of experienced practitioners in the real estate industry that assume a similar level of expertise in the field in order to benefit from reading the book.

Given what has occurred in the real estate industry over the last few years, a paradigm shift is required in terms of real estate education. Rather than focusing purely on the mathematics or on the war stories of experience, there is a need in the literature for a book that attempts to fill the void between academically oriented text books on real estate finance and the various "practical" real estate books on the market that collectively do not provide the necessary ingredients for a successful investment venture in today's vibrant economic climate. Understanding real estate should be one part mathematical, another part definitional, and both must be accompanied by a firm grasp of how market trends (both micro and macro) affect the value of an investment opportunity.

The result of our identification of this "gap" in the existing investment real estate literature is "Real Estate Investment". The 12 chapters, along with the case studies, should serve as an adequate text for undergraduate and masters level courses in commercial real estate. Given the mix of academic and practical material, beginners and more experienced investors seeking to understand the lending perspective and to re-familiarize themselves with investment theory will benefit from studying this book. This book is also ideally suited for corporate training in investment real estate for small financial institutions, appraisal firms, and real estate investment companies. For instructors who adopt this as a course text, additional materials are available on the Springer website. These items such as lecture slides, case study and problem solutions would be of interest to the general reader as well. The website will also contain an excel based discounted cash flow worksheet that will aid in solving the case studies in this book, as well as providing a head start to those inspired to become investors after being exposed to the material in this book.

The authors would like to thank Barbara Fess, Economics Editor at Springer, for providing us with sound guidance and a due date far enough into the future where we had the time necessary to complete this project, but not too much time where we might have fallen behind schedule. We would also like to thank the readers of the RMA Journal who offered constructive feedback for the two articles published there that appear in part in this book. Finally, we would like to thank Edward G. Clarke and Leila Goddard who read portions of this book and offered welcome feedback.

As rock icon Joey Ramone used to say, "either you are in it, or you are *out* of it!" It is now time for you to get into it, and we hope that the messages conveyed in the chapters that follow inspire our students to excel and our general readership to prosper. We believe that real estate investment opportunities which exhibit the quality, quantity, and durability of income as defined in this book are more likely to provide the sound cash flows now and in the future than more speculative investments which require a lot more luck than skill.

Sincerely,
G. Jason Goddard and Bill Marcum

Contents

1	**Real Estate Investing After the Fall**	1
1.1	General Introduction to Real Estate Investing	1
	1.1.1 Speculative Investment and Value Investment	2
1.2	The Various Market Participants	3
1.3	Typical Investment Options	5
	1.3.1 Framework for Investment Property Selection	7
1.4	Property Rights and Title Assurance	7
	1.4.1 The Importance of Property Rights	7
	1.4.2 Title Assurance	10
1.5	Real Estate Market Cycle	14
	1.5.1 Global Linkages	14
	1.5.2 Comparative Advantage	15
	1.5.3 Supply and Demand Analysis	16
	1.5.4 Reasons for New Construction	19
	References	24
2	**Real Estate Finance: Loan Documentation and Payment Structures**	25
2.1	The Investment Motive	25
2.2	Documentation of Real Estate Conveyance and Indebtedness	27
	2.2.1 Mortgage	27
	2.2.2 Note	29
	2.2.3 Guaranty Agreement	34
2.3	The Various Forms of Mortgage Amortizations	35
	2.3.1 Types of Repayment Structures	36
	2.3.2 Fixed Rate Repayment Structures	37
	2.3.3 Variable Rate Repayment Structures	40
2.4	Pretend and Extend or Bankruptcy?	41
	2.4.1 Foreclosure and Bottom Fishing	42
	2.4.2 Bankruptcy	43
	References	48

Contents

3 Finance and Real Estate Valuation 49
- 3.1 Return to Fundamentals 49
- 3.2 Determining the Investor's Yield 50
  - 3.2.1 Basics of Investment Yield 50
- 3.3 Holding Period and Investment Strategy 52
- 3.4 The Concept of Compound Interest 53
- 3.5 Net Present Value (NPV) 55
- 3.6 Internal Rate of Return (IRR) 58
  - 3.6.1 Partitioning the Internal Rate of Return 59
  - 3.6.2 Weaknesses in the Internal Rate of Return Model 60
- 3.7 Valuing Real Estate Versus Homogenous Assets 61
- References 65

4 Real Estate Valuation 67
- 4.1 The Appraisal Process 67
  - 4.1.1 Beginning the Appraisal Process 68
- 4.2 The Three Approaches to Value 69
  - 4.2.1 The Sales Approach to Value 69
  - 4.2.2 The Cost Approach to Value 71
  - 4.2.3 The Income Approach to Value 74
- 4.3 Introduction to Cap Rates 80
  - 4.3.1 Market Extraction Method 81
  - 4.3.2 Market Survey Method 82
  - 4.3.3 Lender's Yield Method 83
  - 4.3.4 Band of Investments Technique 84
  - 4.3.5 Ellwood, Akerson, and Archeological Finance 86
  - 4.3.6 Conclusion 89
- References 92

5 The Anatomy of a Lease 95
- 5.1 The Lease and Its Impact on Cash Flow 95
  - 5.1.1 How the Price of Rent is Determined 96
  - 5.1.2 How Rent is Calculated 97
  - 5.1.3 Expense Sharing 100
  - 5.1.4 Impact of the Lease on Cash Flow 101
- 5.2 Various Leases Available in the Market 102
  - 5.2.1 Ground Leases 103
  - 5.2.2 Various Forms of Traditional Leases 104
- 5.3 Components of a Lease 105
  - 5.3.1 Setting the Boundaries 106
  - 5.3.2 Other Lease Contents 106
- 5.4 Lease Rollover Risk 108
- References 117

6	**Risk Analysis**		119
	6.1 Risk in Real Estate Investment		120
		6.1.1 Business Risk	120
		6.1.2 Management Risk	120
		6.1.3 Liquidity Risk	122
		6.1.4 Legislative Risk	122
		6.1.5 Inflation Risk	123
		6.1.6 Interest Rate Risk	124
		6.1.7 Environmental Risk	128
		6.1.8 Financial Risk	130
	6.2 Leverage Effects		131
	6.3 Statistics and Risk		133
	6.4 Partitioning the IRR		135
	6.5 Due Diligence Analysis		135
	6.6 Conclusion		136
	References		140
7	**Taxation in Investment Real Estate**		141
	7.1 Calculation of Property Taxes		141
	7.2 Effects of Interest Expense and Non-Cash Expenses on Taxable Income		142
	7.3 Introduction of After Tax Internal Rate of Return (ATIRR)		145
	7.4 Various Forms of Property Ownership		148
		7.4.1 The Global View	148
		7.4.2 Choose or Lose	149
	7.5 The Good, the Bad and 1031 Exchange		152
		7.5.1 Types of 1031 Exchanges	152
		7.5.2 Exchange Economics	154
		7.5.3 Exchanges at High Noon	155
		7.5.4 The Good: A Fistful of Dollars	156
		7.5.5 The Bad: Unintended Consequences (For a Few Dollars More)	157
	References		162
8	**Investing in Residential Apartment Projects**		163
	8.1 Property Life Cycle Pyramid		164
	8.2 Types of Residential Apartment Investment Projects		165
	8.3 Valuing Residential Apartment Projects		167
	8.4 Jasmine Court Apartments Case		169
	8.5 Mixed Use Properties as an Investment Alternative		172
	8.6 Off-Campus Student Housing as an Investment Alternative		173
		8.6.1 Location, Location, Location	173
		8.6.2 If You Build It, Will They Come?	174

		8.6.3	Market Analysis in Student Housing	175
		8.6.4	Class Project: Off-Campus Student Housing Occupancy Study	176
	References			181
9	**Investing in Retail and Office Property**			183
	9.1	Subtle Differences Between the Property Types		184
		9.1.1	Destination Versus Spontaneous Orientation	184
		9.1.2	Projected Demand for Space	185
	9.2	Classifying Retail Properties		185
		9.2.1	Outparcels and Single Tenant Properties	185
		9.2.2	Multi-Tenant Retail Properties	186
	9.3	Evaluating Retail Property Projects		188
		9.3.1	Retail Property Location Considerations	188
		9.3.2	Retail Tenant Considerations	188
	9.4	Classifying Office Investment Properties		190
		9.4.1	Office Property Delineation by Size and Class	190
		9.4.2	Office Property Categorization by Use and Design	191
	9.5	Evaluating Office Properties		192
	9.6	Office Condominiums as an Investment Alternative		193
	9.7	Case Studies in Retail and Office Property		193
	9.8	Conclusion		195
	References			202
10	**Investing in Warehouse and Industrial Property**			205
	10.1	Industrial and Warehouse Investment Properties		206
	10.2	Evaluating Industrial and Warehouse Property		208
		10.2.1	Location Considerations	208
		10.2.2	Projecting Demand for Industrial Space	208
	10.3	Outsourcing and Its Effect on Commercial Real Estate		209
	10.4	Self-Storage Facilities as an Investment Alternative		211
		10.4.1	Brief History of the Self-Storage Industry	211
		10.4.2	Delineation by Size and Class	213
		10.4.3	Projecting Demand for Self Storage Space	213
		10.4.4	Know Thy Customer	214
		10.4.5	The Ayatollah of Climate Controllah	216
	10.5	Conclusion		217
	References			223
11	**Securitization of Real Estate Assets**			225
	11.1	Origins of Securitization: The Development of the Secondary Mortgage Market		226
	11.2	The Agencies and Their Function		229
	11.3	The Structure of Agency Guaranteed MBS		231
		11.3.1	Mortgage Conformability	232
		11.3.2	Residential Mortgage Prepayments	233
		11.3.3	The Problem with Prepayments	233

11.4	CMOs and REMICs		235
	11.4.1	Types of CMOs or REMICs	235
11.5	Securitization Without Agency Guarantees		238
11.6	Replacing an Agency Guarantee with Credit Enhancements		239
	11.6.1	External Credit Enhancements	240
	11.6.2	Internal Credit Enhancements	241
	11.6.3	Structural Credit Enhancements	242
11.7	Commercial Mortgage-Backed Securities		242
	11.7.1	A Peculiar Arrangement: Servicing the Loans of a CMBS	243
	11.7.2	Struggling to Recover: The CMBS Market	244
	11.7.3	Reviving the CMBS Market Through Quantity, Quality and Durability (QQD)	246
11.8	Financial Markets and the Problems of Securitization		247
	11.8.1	Meltdown	247
	11.8.2	Who's to Blame	249
	11.8.3	Originate to Distribute	249
	11.8.4	The Rating Agencies	250
11.9	The Upside of Securitization		250
References			252

12 Real Estate Investment Trusts (REITs) 253

12.1	The History of REITs		253
	12.1.1	Origins of REITs	254
	12.1.2	What is a REIT?	254
12.2	Various Forms of REITs		256
	12.2.1	Equity REITs	256
	12.2.2	Mortgage REITs	257
	12.2.3	Hybrid REITs	258
	12.2.4	Mutual Fund REITs	259
12.3	REIT Investment Strategy and Portfolio Diversification		260
	12.3.1	REIT Quantity Strategies	261
	12.3.2	REIT Quality Strategies	261
	12.3.3	REIT Durability Strategies	262
	12.3.4	REIT Portfolio Diversification	262
12.4	REIT Valuation Techniques		263
	12.4.1	Gordon Dividend Growth Model	263
	12.4.2	FFO Multiple	264
	12.4.3	Net Asset Value (NAV)	265
	12.4.4	REIT Valuation Issues	266
12.5	Internationalization of REIT Concept		267
12.6	The Sendoff!		270
References			271

Glossary 273

Index 293

About the Authors

G. Jason Goddard is currently Vice President at Wells Fargo, where he has been a commercial lender for over 15 years. Mr. Goddard is currently real estate risk advisor for income producing investment real estate loans in the business and community banking segments, and works in Winston-Salem, NC. He obtained his MBA from the Bryan School at the University of North Carolina at Greensboro. Mr. Goddard is currently adjunct instructor at Wake Forest University, UNCG, and the University of Applied Sciences in Ludwigshafen, Germany. He is also an assistant editor of the Journal of Asia-Pacific Business. Mr. Goddard teaches the investment real estate course at the Schools of Business at Wake Forest University each spring and fall semester. Mr. Goddard also teaches the subject annually at the RMA-ECU Commercial Real Estate Lending School at East Carolina University in Greenville, NC. Mr. Goddard is co-author of three books: *International Business: Theory and Practice, Second Edition* (M.E. Sharpe 2006), *Customer Relationship Management: a Global Perspective* (Gower 2008), and "*The Psychology of Marketing: Cross-Cultural Perspectives*" (Gower 2010).

Bill Marcum is Citibank Faculty Fellow and Associate Professor of Finance at Wake Forest University, Winston-Salem NC. Dr. Marcum obtained his Ph.D. in Finance from the University of North Carolina at Chapel Hill and holds a Masters of Economics from the University of North Carolina at Greensboro. He has published numerous articles in leading academic journals and has three times won the senior teaching award at the Wake Forest Calloway School of Business and Accountancy. His research interests include real estate capital markets and investment, capital market efficiency and stock market predictability.

Real Estate Investing After the Fall

The spirit of property doubles a man's strength.
Voltaire

Contents

1.1	General Introduction to Real Estate Investing	1
	1.1.1 Speculative Investment and Value Investment	2
1.2	The Various Market Participants	3
1.3	Typical Investment Options	5
	1.3.1 Framework for Investment Property Selection	7
1.4	Property Rights and Title Assurance	7
	1.4.1 The Importance of Property Rights	7
	1.4.2 Title Assurance	10
1.5	Real Estate Market Cycle	14
	1.5.1 Global Linkages	14
	1.5.2 Comparative Advantage	15
	1.5.3 Supply and Demand Analysis	16
	1.5.4 Reasons for New Construction	19
References		24

1.1 General Introduction to Real Estate Investing

The topic of real estate investing has made worldwide headlines in recent years in both good and bad ways. During the economic expansion years from 2002 to 2007, real estate investment received much good press from everyone from government officials encouraging real estate ownership, to central banks encouraging bank lending via low interest rates, to various financial institutions offering ever more risky loan options, to investors who were seeking as much of a loan as they could possibly obtain while interest rates were low and while lending appetite was strong. We like to call this period of time the *Yes Era* of commercial and investment banking. Everyone was a winner, everyone got a trophy, and everyone it seems, got a loan.

During these good times, there were numerous television programs encouraging "investment flipping" primarily for single family residences purchased for investment. For every prime-time television program that was created in order to cater to the new found investment class of real estate speculators, there were seminars held in major cities offering attendees knowledge of "the few steps" that were needed in order to profit from real estate investment. Most of these steps concerned purchasing the video or book of the seminar host, and then ultimately buying a property with no money down in the hopes of a profit upon the eventual resale. In a strong property market, strategies such as these seemed infallible, as the profits almost seemed guaranteed after only some cosmetic improvements had been made, with the logic of what goes up must continue going up forever.

Then came the crash, and the bad news began rolling in. Stories of people being left holding numerous property investments with no buyers in sight started nightly news casts, internet discussion boards, and various other forms of news media. As it turned out, what goes up inevitably comes back down to earth. Investors have increasingly been reminded that not only do investment prices come back to earth following the collapse of over-valued real estate markets, values can actually plummet far below what was paid for the investment. Financial institutions have also been reminded of this fact, as the investment banking industry was negatively and significantly affected by the collapse of Lehman Brothers in October 2008, and the commercial banking industry will take years to approach the successes of the "Yes Era," if ever.

What does all of this mean? Contrary to what you may believe when beginning to read this book, the period after the collapse of an investment bubble is a great time to invest. For those who bought property at too high a price, or for those who bought a property for speculative purposes, the time after the bubble burst has been painful. If these properties were financed with debt, their commercial bank may now be asking them to pay down the balance of their loan in order to continue the financing arrangement. For those investors who did not partake in the rampant speculation of the "Yes Era", or for those investors who did participate but who also have the capacity to purchase more property today, the time is ripe for good deals. As will be discussed later in this book, one problem is that now is such a good time to purchase property that very few people are selling. As has been the case in other financial market corrections, sales will pick back up again, of this there is no question. What concerns the authors of this book is how investors in real estate can sufficiently learn from the mistakes of the recent past in order to not make the same mistakes.

1.1.1 Speculative Investment and Value Investment

One of the key lessons from the last bubble and market correction in real estate markets has been that speculative loans are by definition much more risky than loans supported by the income that the property produces over the holding period. In many of the "investment flipping" scenarios, the purchase of the investment was made relative to the sales price at the time of purchase as compared with the sales price at the time of eventual resale. Since during the holding period of the investment, no income accrues to the owner, the investment is speculative. This differs from a value investment, where the owner receives net cash flows from the property, typically on a monthly basis. The net cash flow is derived from the following formula:

> Monthly rent received from third party tenants
> − Property operating expenses paid by owner
> − Debt service paid on mortgage
> = Net cash flow to owner/investor

The formula shows that the value investment includes income each month (or year), while the speculative investment either does not produce monthly rent, or does not receive enough monthly rent to adequately cover the property operating expenses and debt service on the mortgage. Both speculative and value Investment will have a net cash flow once the property is sold, but only the value investment provides a net cash flow to the owner during the holding period.

For example, consider two different investment options. Both are purchased for $1 million, with $300,000 in equity. Both are sold at the end of 5 years for $1.2 million with a net cash flow after debt service of $400,000. The first investment produces no income for years 1–4, and the only income produced in year 5 is the resale of the property. The second investment produces $25,000 in net cash flow for each year of the holding period as is shown in Table 1.1.

The first investment would be considered speculative as the only means for a positive return on investment is the eventual resale of the property. The second investment produces positive income in the years when the investment is owned (years 1–5) and would be considered a value investment.

An over-dependence on speculative investments led to the inevitable market correction at the end of the Yes Era. It is the aim of this book to provide instruction on how to best evaluate investment real estate properties from a value investment standpoint.

1.2 The Various Market Participants

While the focus of this book is on evaluating income-producing property for investment, it is helpful to introduce the market participants in order to frame the discussion that follows. An aim of this book is to discuss the general parameters to consider from both the investor as well as the lender standpoint.

Table 1.1 Comparison of competing investments from an equity cash flow perspective

Period	0	1	2	3	4	5
Investment one	(300,000)	–	–	–	–	400,000
Investment two	(300,000)	25,000	25,000	25,000	25,000	425,000

In terms of the typical market participants in commercial real estate, the following is noted in Fig. 1.1:
- Investors
- Lenders
- Brokers
- Appraisers
- Attorneys
- Government

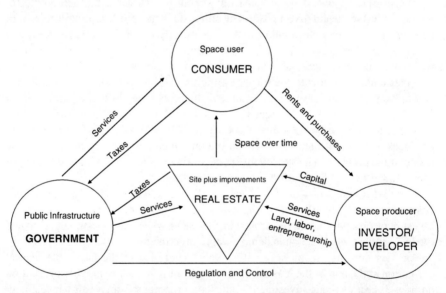

Fig. 1.1 Conceptual model of the real estate investment environment (Phyrr and Cooper (1982, p. 5))

Investors in the market can be individuals, partnerships, corporations, or those corporate entities specifically designed for real estate investments (Real Estate Investment Trusts (REITs) which will be discussed in greater detail in Chap. 12). The common theme for each is that all are seeking to utilize the available market knowledge and financial tools in order to acquire profitable real estate investments.

Lenders are typically commercial banks, but could also be financial institutions such as investment banks, pension funds, and insurance companies. The common theme is that these entities provide a portion of the purchase price to the investor in the form of debt. These various forms of debt will be discussed in greater detail in Chap. 2.

Brokers, appraisers, and attorneys all play pivotal roles in the acquisition and divestiture of investment real estate. Brokers act as an intermediary between the seller of the property and any prospective buyer. Commercial real estate brokers prepare "broker packages" that discuss the market characteristics of the subject property, highlight financial metrics, and provide demographic information concerning the general area. Brokers can also be utilized in finding replacement tenants in an investment property.

Appraisers are useful for prospective investors as they can be a source of market information. From an investment real estate perspective, market vacancy rates and market lease rates for comparable properties are very important in the due diligence process prior to purchase. Appraisers can provide this information to investors for a fee at the outset of the selection process, and they can provide lenders and investors with an appraisal, which consists of their opinion on the value of the subject property. Appraisers will evaluate the highest and best use of the property, and will then typically perform three different approaches to valuation. As will be discussed in greater detail in Chap. 4, the three approaches to value are the sales approach, the cost approach, and the income approach. Knowledge of all three approaches to value is important, but the primary focus of this book is on various income approach valuation techniques.

Attorneys help facilitate the proper conveyance of real estate from the seller to the buyer. We will discuss the role of the attorney in greater detail in this chapter when we discuss title assurance and property rights. For now, it should be noted that the attorney is necessary in order to complete the legal transfer of ownership from one party to another, and they are available to comment on any issues associated with the conveyance of the property.

The last participant in the investment real estate market is the government. As witnessed over the last few years, national governments have taken a much more active role in the management of the banking industry following the collapse of financial markets after the "Yes Era" lending Exuberance. At a more local level, government entities exert influence via the determination of appropriate zoning, via determinations on the amount of building which is permitted in a given area for a given purpose, as well as price controls and other regulations concerning the fair and equitable treatment of tenants, especially as related to fair housing laws.

1.3 Typical Investment Options

The focus of this book is on *value* investments within the following asset classes:
- Residential Apartments
- Retail
- Office
- Warehouse and Industrial

Because the first investment for most people is their home as a primary residence, it follows that many people (i.e., the typical investor) should understand the market demographics associated with residential housing better than other non-residential investment property options. The run up in real estate during the "Yes Era" began as a residential housing phenomenon. Given the knowledge associated with buying a primary residence, many investors take the first step into the investment real estate arena via the acquisition of a single family residence for investment purposes. Once a single family residence has been obtained as a successful investment, an investor may then purchase a second single family residence. Once investment success is achieved in multiple single family residences, a not too uncommon paradigm is to then sell those individual residences in order to purchase an apartment complex. Unlike single family residences, which do not have much available information concerning market rent rates and vacancies, information on apartment complexes is readily available in most developed country markets (Tiwari and White 2010). An apartment property is here defined as a residential property containing at least five third party tenants. The tenants are deemed to be third party as they have no relation to the owners of the property (i.e. they are different individuals where leases are negotiated under normal market conditions). Apartment properties make up the first primary investment real estate option considered in this book. Apartment properties will be discussed in greater detail in Chap. 8.

The second investment real estate property type we discuss is retail. This form of investment property can be either single tenant or multi-tenant. For example, an investor may be interested in purchasing a free-standing retail property containing a nationally known single tenant. Retail properties also consist of multi-tenant "strip centers," that may have a well-known company as an anchor tenant. An anchor tenant is defined as a company which serves as the primary draw for the property, and which takes up the largest percentage of the occupied space in the subject property. Retail properties are also known as shopping centers. Depending on the size and types of anchor tenants, these properties can be further classified as neighborhood centers, community centers, regional centers, and super regional centers (Phyrr and Cooper 1982). Retail properties are discussed in greater detail in Chap. 9.

The third investment real estate described is office buildings properties. Office properties can take many forms and can serve third party tenants as diverse as doctors, lawyers, CPA's, dentists, banks, and many other types of businesses. Office properties can be single tenant or multi-tenant, and ownership can be of the complete structure or of individual units (known as condos). Office properties will be discussed in greater detail in Chap. 9.

The fourth investment real estate vehicle discussed is warehouse and industrial property. Industrial tenants are involved in activities related to the production, storage, and distribution of tangible economic goods (Phyrr and Cooper 1982). Production facilities involve the manufacture of tangible economic goods, while warehouse facilities involve the storage of those wares. A subset of warehouse would be self-storage facilities. Warehouse and industrial properties are discussed in greater detail in Chap. 10.

1.3.1 Framework for Investment Property Selection

Now that we have discussed the difference between a speculative investment and a value investment, and that we have introduced the "Four Food Groups" of real estate investment property (i.e., apartment, retail, office, industrial), it is time to introduce a framework for analyzing the strengths and weaknesses of the various investments. Later in this book, we will discuss specific financial metrics such as the internal rate of return, net present value, and capitalization rates, but for now the following "QQD" framework is instructive:
- Quantity of the cash flows
- Quality of the cash flows
- Durability of the cash flows.

From our discussion in Sect. 1.1.1, the definition of a speculative investment should be clear. What probably is not clear is how to differentiate between competing value investments. What if you had three properties all producing the same net operating income (NOI), of say $150,000. Your first property consists of a ten unit apartment complex, which is 100% occupied with tenants on month-to-month leases. Your second property consists of a single tenant retail property (e.g., a drug store or grocery store) with a 25 year lease term. Your third property consists of a multi-tenant office building with average lease terms of 3 years. Because the NOI is equal for all three properties, the *quantity* of the cash flows is the same for the three investment alternatives. In terms of assessing the *quality* of the cash flows, an investor would need to evaluate the financial strength of the third party tenants. This is admittedly difficult for apartments as financial information for the tenants is not readily available to the investor. What is available is the history of payment performance of those tenants and the overall strength of the market. The *durability* of the cash flows involves the average lease term for the tenants relative to either the holding period of the investment (for an unlevered property) or the term of the loan (for a levered property). From a durability standpoint, the drug store or grocery store property clearly is the strongest of the three. Since apartment properties (along with self-storage properties, which are not part of this example) typically have lease terms of up to 1 year in length, how strong the durability of these properties is can be determined by the depth of the market in terms of replacement tenant potential.

1.4 Property Rights and Title Assurance

1.4.1 The Importance of Property Rights

The ability to convey ownership in real estate depends on the legal system in a given area. It is also important to research prior conveyance of a specific property to ensure that no other claimants exist. If a buyer purchases a property, they are in effect assuming outright ownership. If another party subsequently claims to own land that the new owner thought that they owned outright, legal disputes ensue.

The ability to effectively convey property ownership from one party (seller) to another (buyer) is now taken for granted in most developed economies. Today, prospective buyers can search by the property's address on internet sites, to determine current ownership, and they can also determine how much the current owner paid for the property, and if the property taxes are paid up to date. Most counties across the United States maintain websites that provide GIS maps of various properties in the specific county. While the data is more extensive for residential properties (they are sold most often on the basis of market comparables), basic ownership information for commercial properties also is typically available.

1.4.1.1 Origins of United States Property Rights System

Extensive public information for property ownership was not always available in the United States. As discussed by Peruvian economist Hernando De Soto, after the American Revolution, the property rights system contained many flaws (De Soto 2000). Prior to the American Revolution real estate ownership was based on the British system. In England, the use of written instruments to transfer title of real property began in 1677 with the passage of the statute of frauds. This law was enacted to prevent fraud and perjury and to allow obligations to be enforced without depending on the memory of witnesses. Prior to this, possession was the sole determinant of land ownership (Buchanan 1988). Given the differences in available land and population density of the new nation relative to England, there resulted more illegal owners of land than those with legal title. Those whose land claims were illegitimate were known as squatters. These land claimants assumed ownership by such means as "tomahawk rights", which meant that the land was claimed based on improvements made to the land while it was in their possession. George Washington referred to these illegal squatters as *banditti...skimming and disposing of the cream of the country at the expense of the many*. Thomas Jefferson urged that these now majority holders of land be somehow made legitimate. The solution was the concept of preemption, whereby a settler could buy land that they had improved before it was offered for sale to the general public. Jefferson realized that if the young country was to survive, that the majority of illegal land holders had to be brought into the legal system of land ownership.

1.4.1.2 Property Rights Issues Today in Developing Countries

The issue of legal land claims still has relevance today, as many developing countries find themselves with a majority of illegal land claimants. Other than governmental solvency issues, a lack of legal title to land does not allow for the individual to pledge the property as collateral for a small business loan. DeSoto refers to this as "dead capital", with the assumption that corrective measures in property rights could help spur economic development.

In a study in the Philippines, it was determined that 168 steps over a period of 25 years were required to "formalize" urban property. In Egypt, DeSoto determined that 71 steps in conjunction with 31 bureaucratic entities were required to access (legally) desert land for construction. In Haiti, it took 12 years to obtain a sales

1.4 Property Rights and Title Assurance

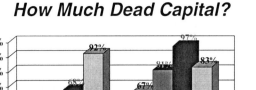

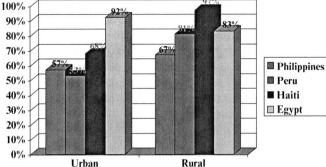

Fig. 1.2 Amount of dead capital in select developing countries (Adapted from De Soto (2000))

contract following a 5 year land lease. Over the time period 111 steps were involved to gain the sales contract. In DeSoto's home of Peru, it took 1 year to register a business, and five stages to form a legally owned home. The first stage alone would serve to discourage most citizens as there were 207 steps in the first stage! Based on Desoto's research, the level of dead capital in these developing countries was extraordinarily high as shown in Fig. 1.2.

Unfortunately, in the years since the publication of DeSoto's landmark book, there is still much work to be done in terms of adequate provision of property rights and transparency in general in many developing countries throughout Latin America, Eastern Europe, and parts of Asia (Tiwari and White 2010). These countries may be known as "developing countries" owing to their penchant for developing new ways of keeping legally titled land out of the hands of their citizens. The problem of property ownership has not been limited only to developing countries. After the reunification of Germany, many legal squabbles ensued between those parties who were currently occupying land and those who had fled East Germany after World War II but still had a claim on the land (Dieterich et al. 1993). The former communist and the developing countries appear to have been influenced by the tradition of French Liberty, where the aim of the system is (supposedly) the prevention of huge inequalities. A lasting legacy of English common law in the United States has been the Lockean perspective of English Liberty, where private property rights are deemed essential for a just society (Audi 2009).

1.4.1.3 Limits on Property Rights

In closing our discussion on property rights, it is helpful to discuss certain situations that limit the owner's use of the property. Property rights are obviously limited to uses which are legally permitted in a given area, and this is affected based on zoning ordinances in a given area. There can also be environmental restrictions for a specific use, such as if water or minerals exist on a given property. The state controls riparian (water) rights, mineral rights, and has the ability to control the use of a property from a pollution standpoint. Additionally, the government reserves eminent domain for situations when it is necessary to purchase a property held in private ownership for fair market value in order to accomplish something in the public interest. This public interest would typically involve highway expansion or some other public land usage. A final restriction of property rights concerns residential housing. Fair housing in the United States pertains to investments in one to four family dwellings and apartment complexes. The owner of the property may not discriminate against tenants on the basis of race, religion, national origin, handicap, or other such characteristics.

1.4.2 Title Assurance

1.4.2.1 The "Estate" in Real Estate

As should now be clear, property rights entail the right to control, occupy, develop, improve, exploit, pledge, lease, or sell a given property. The type of ownership conveyed reflects the use of the property. In the case of residential (i.e. owner occupied) properties, ownership is typically conveyed on a fee simple basis. This is the most complete form of ownership (known as a freehold estate), as the owner can sell, lease, divide up, or otherwise exploit the property as they see fit, as long as the intended act is legal. The term "fee" has evolved from the term "fief" given that the king in England would grant land in parcels called fiefs. "Simple" refers to the ownership being without encumbrances or restrictions (Buchanan 1988).

Sometimes zoning restrictions are said to infringe on a property owner's rights, especially when zoning restrictions limit an owner's ability to convert a given property to a specific use. Typically, zoning restrictions are based on common sense, but sometimes area demographics can change leading to the conventional wisdom of the past no longer applying in the current situation. For example, in many rural communities in the eastern part of North Carolina, land that has been for generations used for agricultural purposes may now be better suited for commercial development, especially when heirs to a given property no longer are interested in farming. If the local municipality can be convinced that it is in the economic interest of the community to change the zoning of a given tract of land, then the new commercial development can occur. This has been the case in Greenville, North Carolina with the conversion of former farm land in order to construct the Brody School of Medicine at East Carolina University, as well as in various private developments of off-campus student housing serving the needs of a growing university. In circumstances where the zoning is not changed, the fee simple owner of the property could in turn serve as a lessor and can in turn allow a farming operation access to the land for a fee. The farming operation is considered a lessee, as they are paying to have access to the land for a specific period of time. Alternatively, if the zoning is changed from agricultural use to commercial development, the lessor could seek financing for a commercial construction project (i.e. become

1.4 Property Rights and Title Assurance

a mortgagor). The lender (or mortgagee) would obtain a secured interest in the property via a mortgage. If the lessor defaults on the loan, this secured interest has value to the mortgagee. Unfortunately, this situation has proven to be all too common in the aftermath of the "Yes Era", given the rise in foreclosures in recent years in the United States (Goddard 2010; Wells Fargo Economics Group 2010).

Another form of ownership is one that only lasts as long as the life of the owner. In these so-called life estates, the ownership of the property reverts back to the grantor or their heirs. This form of ownership is primarily seen in residential properties, but it could prove to be an issue for any property type. Given the travails of investing in the stock market, real estate has become a core holding in most portfolios, and it often represents the largest asset owned in a given family. Thus life estates have practical application when the primary owner is determining how ownership should be determined after their death. The primary owner may grant ownership to a specific individual only until that individual dies. This could be the case in a divorce where an individual is concerned how the former spouse will bequeath the property after that individual's passing. The life estate thus allows for the temporary transfer of ownership, with the grantor ensuring that they can determine the eventual heirs of the property. From a lending perspective, this form of ownership is not as favorable as a fee simple estate, given the uncertainty with the eventual termination date of the life estate.

A third less common form of estate is the future estate. In this form of ownership, the property rights are not conveyed until some date in the future. A reversionary interest is created when a grantor conveys less than ownership to a grantee, and retains the right to reclaim full ownership of the property at a later date. During the period that the grantee has access to the property, the grantor maintains the right to sell or even to mortgage the property. Under this scenario, the grantee only has the right to occupy the property for a specified period of time. Another example of a future estate is when a grantor conveys the reversionary interest to a third party upon termination of the grantee estate. This interest is known as the remainder.

A final form of estate has particular importance in this book. Leasehold estates involve the conveyance of property rights for a specific period of time. This form of estate typically involves a lease between the owner of the property and a tenant (lessee) who occupies the property for a specified period of time. An estate for years involves a lease with an exact duration of tenancy. We will discuss leases more thoroughly in Chap. 5, but for purposes of illustration, an estate for years could be as long as 99 years (as is the case for many ground leases) or as short as 2 years. Some commercial tenants such as the drug store chain Walgreens in the United States sign leases for up to 75 years, although the tenant has the option of not renewing the lease after each 25 year period.

Another form of leasehold estate is an estate from year to year. This form of estate involves periodic tenancy, where the tenant can remain in the property until the tenant provides a notice of termination of tenancy. Some real estate investors prefer such month-to-month leases as they provide the owner with the ability to increase rents (in an escalating rent environment) or to replace an undesirable tenant if they so choose. Many commercial lenders would desire longer leases given the increased certainty of the cash flows when leases are of longer length. Sometimes a tenant occupies a property with the knowledge of the owner, but without an official lease. This situation is referred to as an estate at will. If the same tenant occupies the property without the owner's knowledge, this is known as an estate at sufferance. This relates back to our earlier discussion of squatter's rights in early America and of the current situation in many developing countries today.

1.4.2.2 Title Assurance Defined

One of the primary points in the preceding discussion on property rights concerned whether the seller of a property had the legal right to convey ownership of the subject property. Buyers and commercial lenders seek certainty regarding this issue, and the certainty comes in the form of both a legal opinion as to the quality of the title conveyed, and in the form of title insurance. As part of the legal opinion on the quality of title, an abstract of title is issued by an attorney specializing in real estate law. This abstract would disclose prior sales or prior claims on the subject property by other individuals, the present seller, or other various market participants. The historical verification of the lineage of ownership involves the deed. The deed is a written instrument used to convey the title of real estate from one entity to another. If the review of the title history shows that a current possessor of a property has insufficient documentation (i.e. a squatter), then adverse possession is affirmed. In this case the seller does not have sufficient proof that they own the property being sold. This would be the case if a squatter on a given property was trying to sell or pledge the subject property. If a title is rather deemed free from reasonable doubt and litigation, then the title is classified as marketable. If a marketable title exists, the transfer of ownership can legally occur. If there are any known encumbrances on the title, the attorney will take steps to clear the exceptions if at all possible.

1.4.2.3 Methods of Title Assurance

There are various title assurance methods available in the marketplace, and the cost of the title assurance rises with the amount of coverage. In a perfect world, a buyer (or a lender) of a given property would desire complete certainty that the title to the subject property is free from any and all encumbrances. As will be discussed below, complete certainty is rarely achieved due to the limits in terms of recorded history of property transfer in a given area, and owing to the various limits on property rights which are typically imposed by government entities.

The most complete form of title assurance is the general warranty deed. In this context, a grantor warrants that the title conveyed is free and clear of all encumbrances other than those specifically cited. The general warranty deed offers the buyer the most protection, and is accordingly the most expensive title assurance to obtain in the marketplace.

A second form of title assurance is the special warranty deed. In this context, a grantor is only assuring title for the period that they actually owned the property.

1.4 Property Rights and Title Assurance

Thus the grantor does not speak to encumbrances prior to their ownership. If a buyer is purchasing a property which has had only one longstanding owner, this form of title assurance may be acceptable, but most lenders will require the general warranty deed.

A third form of title assurance is the quit claim deed. This is similar to an "as is" vehicle purchase, as no warranties are made since the grantor is simply forfeiting their claim of rights in a given property. This provides the grantee (buyer) the least amount of protection when compared to the other two title assurance methods. This form of title assurance is utilized in divorce proceedings or in the case of partnership buyouts. In these situations, the aim may only be to remove a specific owner from property ownership while leaving other present owners intact.

Title insurance is available for whichever title assurance method is utilized. Title insurance involves a one time insurance payment to ensure that the policy holder is protected from loss due to title defects. Title insurance is typically required by lenders as it serves to virtually eliminate risk associated with title defects.

As mentioned previously, there may be encumbrances on a property from a title standpoint even under title insurance. One possible encumbrance could be a mechanics lien, which may have been placed on the property for unpaid services rendered by a contractor. This lien can be removed once the contractor has been paid. Another possible encumbrance could be the presence of an easement. An easement is a non-possessory interest in land by an entity other than the owner of the property. The easement could be for the purposes of a right-of-way where a governmental (or pseudo-governmental) entity must have access to the property for specific reasons. For example, a utility company will often have a right-of-way easement in order to access public utility lines that cross the subject property. Another example of an easement would be for a property located directly behind another property that has primary road access. In situations such as this, an easement may be required so that the owner of the property without primary road access has the ability to exit and enter their property. This requires an easement over a portion of the property that does have primary road access. In this case, the owner of the property with primary road access would have an easement (i.e. a title encumbrance) as the owner of the property without primary road access has typically paid a lump sum payment in order for perpetual use of this portion of the property. While the owner of the property that has primary road access does obtain a fee for allowing the easement to exist (in favor of the property behind their property), this is an example of something that encumbers the title of a given property. A final example of a typical encumbrance of a title would be a survey exception. In this case, the acreage surrounding the subject property is measured to ensure that no portion of the land parcel encroaches onto the land of an adjacent property. In some cases, survey exceptions would warrant not purchasing the property. In situations like these, the area of encroachment are vital to the use of the property, such as the building itself or perhaps a portion of the parking lot. In many cases, survey exceptions could be essentially harmless, such as a back portion of the property that does not affect the value or use of the property. In any event, these survey exceptions would show up as defects in the title policy.

1.5 Real Estate Market Cycle

1.5.1 Global Linkages

During the years before the fall in real estate, there were seemingly endless discussions in both popular and academic circles about the benefits of globalization. The many benefits of globalization include increasingly integrated world markets, and the continued improvement of market information (Ajami et al. 2006). While the benefits of integrated markets are still present "after the fall", the internationalization of real estate has primarily concerned large urban areas (Tiwari and White 2010). In these large urban areas, much of the same market fundamentals affect the supply and demand for real estate. In fact, the off shoring of office jobs from the developed world to developing nations such as India has helped give rise to a nascent retail sector in large urban centers of the developing world (Singh and Bose 2008). The market forces that have helped shape this integration of supply and demand for investment property in urban markets is still in effect. One of the lessons or perhaps reminders after the fall has been that markets do not move in unison. While globalization has increased the integration of world markets, even urban centers experience differences in the real estate market cycle. Additionally, smaller cities and more rural communities are not as connected as the large cities are to other large cities throughout the world. Thus these smaller communities experience much more differentiation in market cycles than is seen in the urban centers.

The profitability of an investment property is highly dependent on the region where the property is located. If you are investing in an office building in the Atlanta or Boston area, occupancy rates over the last 10 years could have fallen in your investment property due to the off shoring of service sector positions. Additionally, if you had been investing in retail properties in an area of overbuilding, vacancy rates may have risen to levels that impacted your ability to service the mortgage debt on the property. Thus, the specific location of a property is very important in determining the viability of an investment alternative. For major metro markets, third party information is available concerning market vacancy and rental rate information. In the United States, third party sources such as REIS and CB Richard Ellis, supply market information to investors for a fee. In European markets, Jones, Lang, and Lasalle is known for producing a quarterly "property clock" which discloses in which area of the market cycle a specific market (i.e. Frankfurt) falls relative to a specific property type (i.e. office). The property clock (as shown in Fig. 1.3) consists of a circle with four quadrants signifying whether the rental rates are accelerating, whether rental growth is slowing, whether rental rates are falling, or if rental rates are bottoming out. Various cities are shown as points along the circle depending on which quadrant their market is currently located.

Fig. 1.3 Jones Lang Lasalle's "Property Clock" diagram

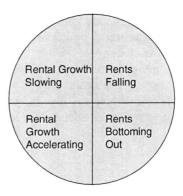

1.5.2 Comparative Advantage

Some geographic areas outperform others owing to the concept of comparative advantage developed by the economist David Ricardo. Seen in a real estate context, some locations will outperform other locations due to specific factors which are location determined. Sometimes these specific factors could be natural advantages. Areas near seaports would certainly accrue benefits to storage and warehouse facility owners. Areas near specific natural resources could also benefit owners of investment property by increasing the demand for space in a given market. Many property types would receive capitalization effects owing to proximity to tourist or beach destinations (retail, residential, hotel, etc.). Ceteris paribus, these natural advantages serve as a comparative advantage from one location relative to another.

Other comparative advantage determinants are not as easily quantified. One such determinant of comparative advantage in real estate markets is the quality of the workforce. Workforce quality is determined by the average education levels of the population, the availability of educational opportunities in an area, and on the availability of skill sets necessary to conduct business in an increasingly global and knowledge-intensive environment. A given market's strength in workforce quality can be determined based on the number and types of firms that are present in the market. If a subject market has a sufficient cluster of firms specializing in a high skill service or manufacturing operation, this bodes well for the overall education level and skill set level of the area.

Another component of a given market's comparative advantage is the proximity to major consumer markets. One of the first considerations when conducting a market feasibility study is the number of rooftops near the proposed location. If the proposed property is an office building, the consideration of rooftops involves commuting time to and from the location by possible customers and office employees. If the proposed property is retail oriented, commuting time, average daily drive-by traffic, and the ease of ingress and egress are very important. For apartments and industrial properties the prevalence of rooftops is helpful regarding both tenancy and employment potential.

Thus another important characteristic of comparative advantage in real estate location would be the proximity of major employment hubs. Most commercial broker packages highlight the strengths that a given market provides in this area. Most broker packages will disclose the primary employers in a given area and will compare population growth with regional and national averages. Evaluating the primary employers for a given area is known as economic base analysis, which involves the calculation of a location quotient as is shown below.

$$\frac{RE_j/RE_{tot}}{USE_j/USE_{tot}}$$

The formula above involves determining the amount of employment in a given industry (RE_j) relative to the total employment in the region (RE_{tot}). Additionally, the level of employment in the same sector nationally (USE_j) should be compared with overall national employment levels (USE_{tot}). If the quotient provides a value greater than 1.00×, then the employment is considered to be an economic driver in the area. If the quotient provides a value less than 1.00×, then the given sector is considered a supporting player in terms of the region's economic growth (Brueggman and Fisher 2010). Information on national and regional employment by sector can be obtained from most state and federal government economic statistics bureaus.

While this sort of analysis primarily concerns feasibility studies for proposed construction projects, this information is also helpful when deciding on various investment alternatives for existing properties. Regardless of whether a property is new or existing, future implications of employment changes will impact the profitability of the investment property. Once the base employers in a given market are better understood, market sensitivity analysis can be performed using the employment multiplier. The employment multiplier is found by dividing total employment by the base employment in a given market. Sensitivity analysis can be performed to better understand the effect of possible future changes in the base employment level.

1.5.3 Supply and Demand Analysis

Once the comparative advantage of a given market is determined, the next step is to predict the level of supply and demand for an investment property in a given market. Whether an investor is considering investing in an existing (stabilized) property, or in a property which is either new construction or is experiencing a current occupancy rate which is less than what is necessary to cover the annual expenses and mortgage debt (i.e. non-stabilized), the market equilibrium is very important. The market equilibrium is here defined as the currently experienced level of occupancy, lease rates, and property expenses per property type in the region. Where a current market is in terms of the availability of space will

1.5 Real Estate Market Cycle

determine whether lease rates in the area are increasing or decreasing. For example, consider a given market that has an overall occupancy rate of 80%. This implies that the amount of currently available space is 20% higher than the current level of demand. This could imply that the current vacant space is deemed unsuitable based on the age or condition of the space, or it could mean that the vacant space is relatively new and has not been on the market long enough to be occupied by willing tenants. In any event, vacancy exists in the market. Should the level of vacancy in the market suddenly decrease drastically, the price associated with renting the remaining space may increase as the space is all the more dear in the market. As the price of rental space rises, property developers will sense a profit opportunity, and new construction projects will ensue. The question at this point concerns whether the new and existing space will be sufficiently absorbed by the market participants. If new space is added to a market that has an occupancy rate of 90%, and the resulting occupancy becomes 85%, negative absorption is said to have occurred. If new construction continues unabated, the market vacancy rate could increase further. As the vacancy rate for a given property type increases, the rental rates in a given area drop. In perfect markets, the decrease in rental rates would signal that additional construction is not warranted. Unfortunately, during the "Yes era", much lending was done on a speculative basis (i.e. with no leases in place at the time of closing) which contributed to the recent financial crisis witnessed in numerous markets worldwide.

Thus, new construction and the desirability of existing space in a given market can help to determine the forward progression of lease rates. As the quantity of space demanded falls in a given market, so does the price of rental space. Property owners in a market with negative absorption may consider alternative uses for a given property. As mentioned earlier, this may involve a change in zoning for the site, and most certainly would involve tenant improvement expenses incurred by the owner in order to change the interior of the subject property.

Fig. 1.4 by Mueller (1999) serves as illustration of this concept. A dark line in Fig. 1.4 depicts the long term average occupancy rate for a given market. For purposes of illustration, let's assume that this is equal to 85%. When the current occupancy rates are less than 85%, rents are not increasing and in fact may be decreasing depending on the current market cycle quadrant for the given market. Once the current market occupancy rises above 85%, new construction may ensue. Rental rates will increase until supply equals demand. Rental rates may increase beyond the market equilibrium point, but eventually new construction will begin to diminish the rental rates for newly signed leases. In Fig. 1.5, the market quadrants are added to the viewpoint.

As shown in Fig. 1.5, the real estate market cycle can be broken up into four phases. The first phase is the recovery phase. This is where many markets find themselves today after the recent financial crisis. During the recovery phase, no new construction is planned and vacancy rates are beginning to show improvement. The initial consequence for rental space is that rates begin to bottom out. As the recovery of the market improves, vacancy rates drop and rental rates rise. These factors contribute to new construction being planned and eventually to the

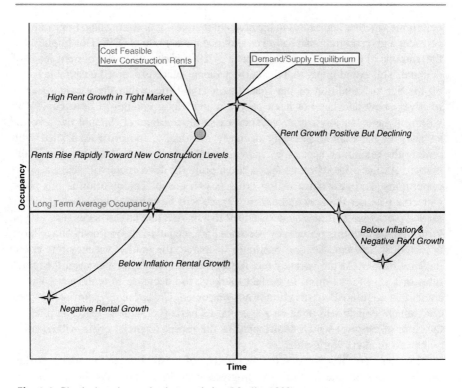

Fig. 1.4 Physical market cycle characteristics (Mueller 1999)

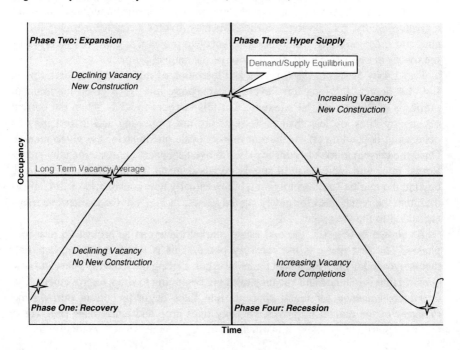

Fig. 1.5 Market cycle quadrants (Mueller 1999)

beginnings of the expansion phase. The expansion phase continues as long as rental rates continue to increase and the underlying economic conditions of the market remain positive. Eventually the demand and supply equilibrium point is achieved. Unfortunately for all concerned, a *loud siren* does not sound when the equilibrium point is achieved signaling that no additional supply is needed in the market. The equilibrium point of the market is typically determined after it has been realized. The absence of an equilibrium siren means that those projects that were completed after the equilibrium point was achieved will not experience occupancy and rental rates as were expected at the start of construction. This period of uncertainty is the third phase and is known as the hyper-supply phase. Towards the middle to end of this third phase, increased vacancy ensues as new construction continues with the resultant negative absorption. The fourth phase is known as the recession phase. In this phase of the market cycle, rental rates drop, vacancy rates increase, and new construction ceases. The length of the recession phase depends on the level of overbuilding (which implies over-lending) in the market, the severity of the economic downturn, and the strength of the market fundamentals as mentioned earlier in the chapter.

1.5.4 Reasons for New Construction

We have now reached the time for conclusions concerning the real estate market cycle. The factors that lead to new construction often depend on where a given location is in the real estate market cycle. As mentioned earlier in the chapter, one reason for new construction can be an increase in the total employment in an area or in the base employment in the area generally. As mentioned previously, low vacancy rates in a market can also be the source of new construction, given the increased profit potential for real estate development based on the increase in rental rates. As discussed in the early part of this chapter, a low level of interest rates (similar to what was experienced during the Yes Era) can also lead to an increase in new construction in a given market. If banks are providing more liberal lending terms than in years past as shown by the level of interest rates and the tolerance for risk in lending, then new construction that may not have been undertaken in the past may become viable.

The demand for retail space is affected by changes (either good or bad) in the economic conditions of an area. The supply of real estate can either be a reaction to changes in demand for a given market, or can be speculative in nature given lending parameters of banks and the level of interest rates. While banks and other lending institutions are less likely to approve speculative real estate projects in recessions, they are also more likely to approve these same projects in expansionary times. Since the determination of when a market moves into the recession phase is usually an ex post measure, lenders are more likely to approve speculative projects at the end of the expansionary cycle, which is exactly the wrong time to be approving such projects.

It is for this reason that the due diligence process for investment real estate does not end with the market analysis. Similar properties located within the same market can produce different results based on the quantity, quality, and durability of the cash flows, and based on the specific location of the property within the submarket. Regardless of the level of market analysis undertaken, nothing can replace visiting the subject property yourself in order to assess the overall quality of the location of the property being considered for investment.

Questions for Discussion
1. Describe the circumstances which led to the financial crisis as related to commercial real estate markets in 2008.
2. Explain the difference between a speculative investment and a value investment.
3. Who are the key players in the commercial real estate markets and what are the primary investment options discussed in Chap. 1?
4. Elaborate on the QQD framework. How is this helpful in differentiating between investment alternatives?
5. What can Colonial America teach developing countries today concerning property rights?
6. Explain the differences between freehold, leasehold, life, and future estates.
7. Describe the various methods of title assurance. Provide an example for each type to illustrate when the method would be appropriate.
8. Define comparative advantage in a real estate context. How might this be used to increase investment performance?
9. Explain the real estate market cycle as related to new construction, bank lending risk tolerance, and rental rates.
10. What stage of the real estate market cycle is your home market in currently? Does the stage differ by property type and if so why? Who could you ask to determine this information?

Mini-Case: Market Research Analysis Case Study

Congratulations! You have been hired as a senior loan approving officer at World Class Bank. You have received the following commentary concerning a current loan request on your desk for an off-campus student housing property in Greensboro, NC.

Area Analysis

Location

Greensboro, the county seat of Guilford County, is located in the Piedmont section of North Carolina. The Piedmont region consists of Greensboro, Winston-Salem, High Point, and seven counties. The main east/west, north/south highways that intersect within the Greensboro area include: Interstate 40 and 85, US 29, 70, 220, and 421. The Greensboro thoroughfare system is well planned and moves traffic efficiently.

Social Forces

Guilford County encompasses 651 square miles while Greensboro spans 109 square miles. Guilford County is part of an eight county Standard Metropolitan Statistical Area (MSA), with a 1990 population of 1,071,000, which was over 15% increased from the 1980 census. The 2000 population in Greensboro increased over 120% as compared with 1990. The 2010 population increased also by 120% over that of 2000. The city of Greensboro offers a variety of housing styles in a wide range of prices. Numerous apartment complexes are available throughout the city. According to a survey conducted in 2010, there were approximately 50,000 units in the city, with an average vacancy rate of 8%. There are currently 800 apartment units under construction in the city, with an additional 3,200 units proposed for development.

There is one county wide public school system. In addition to its public school system, Guilford County has many private schools. Greensboro hosts two public state universities, three private colleges, and several technical institutes. Greensboro also has a variety of parks and recreational facilities available, a Cultural Arts Center, over 500 churches, and four hospitals.

Economic Forces

The unemployment rate in Greensboro at the end of 2010 was 9.5%. This compares with unemployment rates at the same time in Winston-Salem of 8.7%, and 10.1% in High Point. The state of North Carolina's unemployment rate was 9.8%, with the national average at 9.5% at the end of 2010. The average household income according to the US Census Bureau is $53,000, with the median effective buying income for Greensboro of $37,000. The PTR Airport is located 8 miles west of downtown Greensboro and is a major business and private center. Federal Express opened a $230 million air cargo hub for the airport in June 2009. Job projections for the facility are 1,500 employees. The Piedmont Triad Partnership forecasts 20,000 new jobs could be created as a result of the Federal Express hub within 16 years. Thus, growth in the

Greensboro and Guilford County area is strong due to the many assets the area has to offer. The future of Greensboro appears good as development, job opportunities, and a strong real estate market continue.

Governmental Forces

An elected city council and county commissioners make up the management of both the Greensboro and Guilford County government system. Guilford County has an 11 member Board of Commissioners. The government systems of both the county and city work well together and have the reputation as being one of the more progressive in the state. The City of Greensboro has an adequate level of police, sheriff, and fire department employees. Services available to business and industry have been adequate in the past. Water and sewer service is provided by the City of Greensboro. AT&T provides telephone service, and several companies provide cellular phone services. Piedmont Natural Gas Company provides natural gas and Duke Power Company supplies the electric power to the area. The City of Greensboro provides city bus service.

Environmental Forces

City water and sewer are available within Greensboro with well and septic systems available in areas without city water and sewer. Due to the large amount of residential growth over the past several years, Greensboro has experienced problems with adequate water and sewer in the past. The City of Greensboro has negotiated with the cities of Winston-Salem, High Point, and Reidsville to purchase water for Greensboro. Greensboro also has a long-term water problem due to increased development and population. The recently completed Randleman Dam is the long-term solution to the water problems. Environmentally conscience decisions have been made by limiting development of neighboring watershed areas. Greensboro has an average annual temperature of 58° and an annual average rainfall of 42.5 in.

Greensboro is a pleasant place to live due to the pleasant climate, stable economy, a convenient location to recreational facilities, major highways, good quality of educational facilities, growing diversified industries, and the progressive government, the demand for residency in the area is felt to increase with future growth. These factors also effectively lure new industry locating to the southeast to the Greensboro area. Greensboro has experienced progressive economic development and business recruitment as growth and stability continues.

Neighborhood/Competitive Market Analysis

The subject property is located on the western side of the central quadrant of Greensboro, NC, within Greensboro city limits, across from South Aycock Street from the UNC-Greensboro campus. The neighborhood boundaries are considered to be Walker Avenue to the north, Aycock Street to the east, Spring Garden Street to the south, and Elam Avenue to the west. The neighborhood is an established residential area located adjacent to the UNC-G campus to the west. The commercial influences are located along the southern boundary of the neighborhood with single family and multi-family influences intermixed within the neighborhood. The existing homes within the neighborhood are older with many being rented to students at UNC-G. Many of the older and larger single family residences have been converted to duplex or triplex units. Some smaller single family residences have been razed making way for newer multi-tenant apartment buildings. This has happened within the last 3–6 years.

The major traffic arteries serving the neighborhood are Aycock Street, Spring Garden Street, and Freeman Mill Road. These roads provide convenient access to the surrounding area. The average daily traffic count is approximately 55,000 cars per day at the intersection of Aycock and Spring Garden, according to the NC DOT.

The UNC-G campus is expanding southward. The southwest portion of the campus is new. UNC-G has recently constructed and leased large on-campus apartment buildings along the Aycock and Spring Garden area, and these units are managed by the university. The target market for the subject property is UNC-G students. Along with on-campus housing, there is a need for off-campus housing. The historical enrollment of the university has increased from 12,731 students in 2000 to 18,478 in 2010. The freshman class represents 20% of the total students. Freshmen are encouraged, but not required to live on campus. All UNC-G students have the option of living on campus, with approximately 4,000 students living in the 23 residence halls on campus.

Minimal vacant land is available with development conforming to trends and zoning in the neighborhood. The future of the area appears good as growth and development continues.

Questions for Discussion
1. What additional information is vital in order to determine the supply and demand for student housing in the area?
2. How might the required additional information be compiled, and from whom, and over what time horizon is the data useful for student housing?
3. Is all of the information in this report helpful?
4. How can we determine when a market is saturated?
5. Are student housing investment properties recession proof?

References

Ajami, R., Khambata, D., Cool, K., & Goddard, G. J. (2006). *International business theory and practice* (2nd ed.). Armonk, NY: M.E. Sharpe.
Audi, R. (2009). *Business ethics and ethical business*. Oxford: Oxford University Press.
Brueggman, W. B., & Fisher, J. (2010). *Real estate finance & investments* (14th ed.). New York: MacGraw-Hill.
Buchanan, M. R. (1988). *Real estate finance* (2nd ed.). Washington, DC: American Bankers Association.
De Soto, H. (2000). *The mystery of capital: Why capitalism triumphs in the west and fails everywhere else*. New York: Basic Books.
Dieterich, H., Dransfeld, E., & Voβ, W. (1993). *Urban land & property markets in Germany*. London, UK: University College London (UCL) Press.
Goddard, G. J. (2010). The global housing boom: Aftermath of a global financial crisis. In G. Raab, R. A. Ajami, & G. J. Goddard (Eds.), *Psychology of marketing: Cross-cultural perspectives*. London: Gower House Publishing.
Mueller, G. R. (1999). Real estate rental growth rates at different points in the physical market cycle. *Journal of Real Estate Research, 18*(1), 131–150.
Phyrr, S. S., & Cooper, J. R. (1982). *Real estate investment: Strategy, analysis, decisions*. New York: John Wiley & Sons.
Singh, H., & Bose, S. K. (2008). My American cousin: A comparison between India and the US malls. *Journal of Asia-Pacific Business, 9*(4), 358–372.
Tiwari, P., & White, M. (2010). *International real estate economics*. London: Palgrave Macmillan.
Voltaire, F. M. A. (1929). Property. *Philosophical dictionary* (1764), Reprinted. New York: A.A. Knopf.
Wells Fargo Securities LLC Economics Group Special Report. (2010). Credit Quality Monitor: June 2010, Published on Wells Fargo Securities Economics Website, June 8, 2010.

Real Estate Finance: Loan Documentation and Payment Structures

A promise made is a debt unpaid.
Robert William Service

Contents

2.1 The Investment Motive 25
2.2 Documentation of Real Estate Conveyance and Indebtedness 27
    2.2.1 Mortgage 27
    2.2.2 Note 29
    2.2.3 Guaranty Agreement 34
2.3 The Various Forms of Mortgage Amortizations 35
    2.3.1 Types of Repayment Structures 36
    2.3.2 Fixed Rate Repayment Structures 37
    2.3.3 Variable Rate Repayment Structures 40
2.4 Pretend and Extend or Bankruptcy? 41
    2.4.1 Foreclosure and Bottom Fishing 42
    2.4.2 Bankruptcy 43
References 48

2.1 The Investment Motive

In Chap. 1, two of the market participants that were introduced were investors and lenders. In this chapter, these two market participants come to the forefront. When structuring loan documentation and amortizations for investment property indebtedness, the types of arrangements provided often depend on the motivation for investment by the investor, and as we will discuss in a subsequent section, the objectives of the lender.

As noted in Chap. 1, some investors are motivated to purchase investment property based on the hope that the value of the asset will appreciate over time. When making an investment decision, prospective owners determine how much to buy a property for relative to the projected income stream of the subject property.

The ratio of projected net income relative to sales price is referred to as the capitalization (or cap) rate. During the "yes era", cap rates fell to historical lows in many markets. Given the numerous years of solid economic performance in many markets worldwide, the perceived low risk associated with investment property helped to drive down cap rates. Additionally, the prevalence of low interest rates and the increased number of investors in the market also contributed to lower cap rates.

The prevalence of low interest rates in the market also helped to motivate buyers as the rate of return on an investment property was improved as the cost of debt decreased. The high level of liquidity in the markets also helped to increase financial institutions' risk tolerance, which led to longer amortization schedules for investment property than previously seen.

Amortization periods for residential (i.e. primary residence) properties have been offered at 30 years in length for many years, but only during the "Yes era" did investment property amortizations approach this level. Given the motivations of future sale at a higher price, longer loan amortization schedules, and low interest rates, the result was lower payments for the investors, which ceteris paribus increased their returns. If we revisit Table 1.1 from Chap. 1, and consider that the numbers in the table represent equity cash flows to the investor, then lower rates, longer amortizations, and lower payments would increase positive cash flows to the investor over the holding period of the investment.

As we are discussing interest rates, the fact that interest rate expense is a tax deductible expense in most developed countries serves to increase the motivation for investors, regardless of the level of interest rates in the marketplace (Economist 2010).

A final motivation for investing which will be discussed here is the diversification motive. Given how erratic the stock market has become over the last decade, real estate increasingly has become viewed as a means to reduce volatility in investment portfolios. In the aftermath of the financial crisis of 2008, many investors would more than likely question whether real estate was a help or a hindrance in this regard, although "after the fall" property valuations will hopefully be more predictable and in accordance with historical real estate property valuation increases.

Figure 2.1 exhibits the 1, 3, and 5 year returns as of 12/31/2009 for common stock, bond, and CPI benchmarks (Vanguard, 2009). As you can see from Fig. 2.1, there has been quite a bit of variance in the returns for the US REIT index when seen from the 1, 3, and 5 year return viewpoints. The MSCI US REIT Index was up 97% as of February 28, 2010, and was down 57% at the same time of the year for 2009. The large positive movement during 2009 for the REIT index increased the 5 year returns into positive territory by June 30, 2010. Given that the levels of returns seen in 2008 and 2009 are not sustainable, it would appear that real estate as an investment alternative will experience returns in the future more in line with historical averages that traditionally place real estate above bonds, but lower than growth oriented stocks from a return perspective. In Chap. 3, we will update this table to validate whether subsequent returns are more in line with historical averages.

2.2 Documentation of Real Estate Conveyance and Indebtedness

Stocks	1 Year	3 Years	5 Years
Russell 1000 Index (Large Caps)	-28.43%	-5.36%	0.79%
Russell 2000 Index (Small Caps)	27.17%	-6.07%	0.51%
Dow Jones Wilshire 5000 Index (Entire Market)	29.35%	-5.01%	1.09%
MSCI All Country World Index ex USA (International)	42.14%	-3.04%	6.30%
Bonds			
Barclays Aggregate Bond Index (Broad Taxable Market)	5.93%	6.04%	4.97%
Barclays Mutual Bond Index	12.91%	4.41%	4.32%
Citigroup 3-Month Treasury Bill Index	0.16%	2.22%	2.88%
CPI			
Consumer Price Index	2.72%	2.28%	2.56%
Real Estate Investment Trusts			
Vanguard REIT Index Fund (as of 06/30/2010)	54.95%	-8.41%	0.47%
MSCI US REIT Index (as of (06/30/2010)	55.23%	-8.54%	0.43%

Fig. 2.1 Comparative investment returns (Vanguard 2009, 2010)

2.2 Documentation of Real Estate Conveyance and Indebtedness

When a real estate investor has found the appropriate property for purchase, there is often a need for partnering with a financial institution in order to secure the funds required in order to purchase the investment property. Investors will seek lenders at competing financial institutions to bridge the gap between the cash going into the purchase by the investor and the purchase price of the subject property. There are three primary documents that we will discuss in this section:
- Mortgage
- Note
- Guaranty Agreement

2.2.1 Mortgage

The first primary document between a lender and an investor is the mortgage. The purpose of this document is to provide evidence that the investor (borrower) has pledged real property (the investment being purchased) to another party (the lender) as security for the loan. The mortgage document will describe the mortgagor (borrower) and the mortgagee (lender) and will also provide a legal description of the property being pledged as security for the loan. The mortgage document will provide evidence to the covenants of seisin and warranty, which validates that the owner of the property does in fact own the title conveyed, and that there are no significant issues concerning the title of the property as was discussed in Chap. 1.

If a property is owned jointly by individuals, the mortgage will also provide for a provision between spouses (or partners) such that the interest of one spouse is transferred to the other at death. This is known as dower rights, when the husband is the first to pass away, and curtesy rights when the wife is the first to pass away. It should be noted that the mortgage provides security only for the real property and does not document pledging of chattel, which is defined as personal property or trade fixtures installed by the tenant.

The mortgage instrument also contains a list of covenants, clauses, and other contractual arrangements between the borrower and the lender. Many of the clauses elaborate on things that the mortgagor must do in order to maintain the property to the satisfaction of the lender and various governmental authorities. For example, the mortgage will specify that the taxes and insurance for the property must be paid on time, that the condition of the property must be maintained, and that the property must be kept free from liens and other encumbrances. The mortgage may in fact specify that the borrower may not obtain a second lien of indebtedness on the property. This junior mortgage would be subordinate to the first lien on the property, but in the aftermath of the financial crisis, lenders have learned that second liens can cause trouble in a foreclosure situation. If junior mortgages are allowed, the mortgage document could contain a subordination clause, whereby any seller financed or other bank financed loan will become secondary to the mortgagee's loan.

Another typical clause in a mortgage document is the due on sale clause. If the property is sold during the term of the loan, this clause specifies that the loan must be paid out. Since the bank has credit qualified (hopefully!) the borrower, but has not done the same amount of due diligence on the eventual purchaser of the property, the due on sale clause does not allow for the loan to be assumed by a third party. The bank may very well qualify the purchaser for a loan to buy the property, but the pricing and other loan terms will need to be renegotiated based on the strength of the individual, the strength of the income stream from the subject property, and various other considerations.

Sometimes the purpose of the loan dictates that an open ended mortgage be provided in the mortgage document. If the underlying loan is for construction or is a revolving line of credit, there is a need for the borrower to be able to advance funds at a future date from when the loan is closed. For example, if a client is approved for construction financing for a loan up to $3 million, there may not be a need for all of the funds at loan closing. In fact, only a small amount may be needed upfront in order to purchase the land or to begin site work. Over the term of the construction period, the borrower will need to access the funds as the project moves to completion.

It should be noted that the mortgage document only includes land, any buildings on site, easements, natural resources, rents from real estate, and personal fixtures (not including fixtures owned by tenants as was mentioned earlier). Thus, a mortgage will typically contain an after acquired property clause. This provides that anything after the documents are executed that becomes part of the real estate is included in the collateral securing the bank loan.

2.2 Documentation of Real Estate Conveyance and Indebtedness

In some states in the United States, a deed of trust document is used in lieu of a mortgage. In this case, the property is transferred to a trustee (neutral party) by a borrower (trustor) in favor of a lender (beneficiary) and is re-conveyed upon payment in full. At loan closing, the seller conveys the subject property to the buyer, who simultaneously conveys the property to the trustee selected by the lender, which is usually a title insurance company. The trustees hold this legal title until the loan is paid in full, or when the loan goes into default. Thus, the deed of trust is a document that gives the lender the right to sell the property if the borrower cannot repay the loan. The deed of trust document allows for a cleaner foreclosure proceeding given the trustee intermediary. Currently deeds of trust are utilized in Alaska, Arizona, California, Mississippi, Missouri, Nevada, North Carolina, Virginia, and Washington DC. Approximately 15 states use both the deed of trust and the mortgage. The remaining states use the mortgage document.

Some readers may wonder why there is such a discrepancy between which states use one document versus another, and would certainly question the origins of the states who use both documents. In the aftermath of the "yes era", when regulation in financial markets is on the rise, it may be easier to recall that the banking industry in the United States has historically been highly regulated. During the antebellum period, before the United States had a single currency, some states were subject to different operating environments even within the same state. The ties of tradition are often quite difficult to overcome, which may shed some light on the regional differences in loan documentation to this day.

Whether the chosen method of documentation is a mortgage or a deed of trust, once a mortgage is recorded at the public place of record, it serves as notice to all that a lien has been recorded on the property. The effective date of the recording sets the placement of mortgage liens as first, second, or third.

2.2.2 Note

A second important document used in securing a loan against an investment property is the note. The note is a document which serves as evidence of debt between a borrower and the lender. Taken together with the mortgage document, the note and mortgage are evidence of an obligation to repay a loan and to pledge property as security for a loan. Should the borrower default on the loan, the typical course of action is to sue on the note, and to foreclose on the mortgage or deed of trust.

In order to illustrate the difference between a note and a mortgage, the following paragraphs will itemize common items that are included in a note. Since the note is evidence of indebtedness between a borrower and a lender, one of the first things included in a note is the amount borrowed. As mentioned in the discussion of future advance clauses in mortgages, the amount borrowed does not have to be advanced upfront. The more complicated the loan structure, the more likely that the note will be accompanied by a loan agreement, which memorializes all terms and agreements between borrower and lender. The note will also disclose the amount of interest

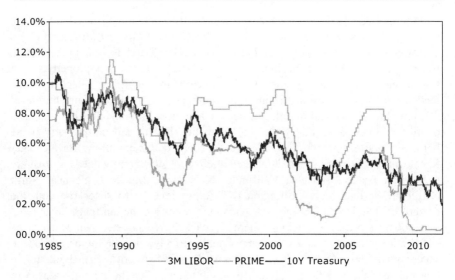

Fig. 2.2 The Long View of Variability: LIBOR, US Prime, and 10 Year US Treasuries (Bloomberg via Wells Fargo Securities 2011)

paid, the index for the variable rate, and whether an interest rate swap exists for the loan. The index for the variable rate could be based on the bank's prime rate, but more probably for investment real estate loans from a bank the index will be tied to US Treasuries or the LIBOR. What is important to the investor regarding the variable rate index chosen is its variability over time, and whether the index is seen as being influenced by the lender. For this second reason, variable rate indexes may more likely be tied to the US Treasury rates or LIBOR rather than prime as is shown in Figure 2.2.

As can be seen from the charts above, if the movement of the LIBOR and Prime rates are viewed in the long term context, LIBOR has experienced more variability than has prime. From an investor's perspective, the increased variability of the LIBOR lays evidence that the rate is not manipulated by financial institutions or government entities, or that it is not manipulated as much as the prime rate which is seen more like an orchestrated staircase in the figure above. The lack of manipulation implies that the lender does not have as much control of the rate, at least in the eyes of many investors.

The note will describe the amount of the monthly payment, the due date for the payments, and when the term of the loan matures. The loan term for typical investment real estate (or otherwise commercial) loans is typically not equal to the amortization of the loan. For example, if an individual obtains a personal mortgage for their primary residence, it is very likely that the term of the loan and the amortization of the loan are equal (i.e. both from 15 to 30 years). This is not the case for most commercial mortgage loans. If an investor receives a commercial loan based on a 20 year amortization, the term of the loan may only be 5 years. Thus, the bank is only committing to funding the loan for 5 years, and after 5 years a balloon payment occurs. This will require the customer to be re-qualified for the loan, and for the pricing to be updated to reflect the current market conditions and to reflect how well the borrower honored their loan commitment over the loan term.

The note will also itemize the calculation of any late fees or prepayment penalties. Typically, if a borrower fails to pay the loan payments for a sufficient period of time (say 3 months), the note will specify that the interest rate will increase by a significant pre-determined level (say 300 basis points) until the customer returns the loan to current status, or cures other noted deficiencies.

The note will also itemize any default provisions. A default provision is defined as any particular event that can trigger the balance of the loan to be due to the lender in its entirety prior to the stated maturity date. Common default provisions are failure to adequately maintain the condition of the property, the sale of the property securing the loan, any material changes to the management structure of the borrowing entity (if other than an individual), and any material changes in the financial condition of any person or entity associated with the loan (as borrower or as guarantor).

A primary default provision in a note for commercial real estate is the annual debt coverage ratio covenant. The note may specify that the borrower or subject collateral must produce an annual debt service coverage ratio of at least 1.25. In other words, the annual net operating income of the property must exceed the annual debt service on the loan by 25%. Net operating income is defined as the residual income before tax for the property after any operating expenses have been deducted. In order to calculate compliance with the debt service coverage ratio, the investor must supply annual financial statements to the bank. It is recommended that an investor also supply a current property rent roll which lists all of the current tenants in the subject property, the amount of square footage for each tenant, the annual lease rate paid per square foot for each tenant, and the expiration dates of the current leases. The rent roll can be utilized in conjunction with either a historical profit and loss statement or a pro-forma statement which projects the income and expense statement for the property for the current year. These statements can be used to project the probability that the property will meet the minimum debt service coverage for the current year, and is seen as a forward looking view of whether a default is imminent for the property. As will be discussed in mini-case one at the end of Chap. 3, the rent roll can tell the lender and the investor much about the future viability of the investment property.

Should the property fail to meet the minimum debt coverage ratio (DCR) provision, the bank may declare the loan to be in default, which means that the loan balance is due in full. Typically, a single missed DCR covenant will not result in this action, but it depends on the severity of the missed covenant. For example, if the calculated DCR is 1.20 relative to the required minimum of 1.25, the lender may simply waive the covenant violation. What is important in determining the proper course of action is viewing the property in forward-looking terms. If the DCR covenant is projected, based on current occupancy, to meet the minimum hurdle rate for the next fiscal year, covenant waiver may be the appropriate action. If the current year DCR is projected at far less than 1.25, a principle curtailment may be necessary in order to return the property to a performing asset on the books of the lender. If the DCR is projected at less than 1.00, then the customer may have difficulty in repaying the loan in the coming months. In situations such as these, more drastic action may prove necessary. The investor will be given a period of time to cure the default via additional collateral, refinance, or principal curtailment (also known as a capital call for the investors). If a suitable solution is not provided within the prescribed forbearance period, foreclosure may be imminent.

The concept of being in default has to do with the borrower failing to fulfill their contractual obligations, agreements, or duties. These duties could involve the property owner failing to pay the property taxes or insurance premiums for the subject property. Technical default can arise if the borrower fails to adequately maintain the condition of the property.

An area of particular interest in the aftermath of the recent financial crisis is the calculation of prepayment penalties. Prepayment penalty calculations range in complexity from a simple percentage of the loan balance to a more complicated calculation used to determine the breakage fee given the prepayment of the loan by a specific amount at a particular point in the amortization of the balance. How a prepayment penalty is calculated should be specified in the note or the swap agreement if an interest rate swap is arranged.

Let's assume that an investor has obtained a loan for $3 million secured with two commercial buildings. At the time the loan was made, the client locked into a fixed rate of 6% (through an interest rate swap or otherwise). Since the client locked in at 6%, they were forecasting that rates were to rise over the loan term, thus wanting to lock in their fixed rate prior to any increases. One of the side effects of government intervention in financial markets is that lowering interest rates is used as a tool for increasing aggregate demand in an economy following a recession. In the wake of "the Yes Era", it is therefore probable that interest rates became lower than the 6% fixed rate obtained by the client in this example.

If interest rates in the current market are less than the fixed rate obtained and the client wishes to break the rate commitment, a penalty will be assessed. This is a distinction of commercial mortgages relative to residential mortgages, as the primary costs associated with refinancing a residential mortgage are typically the closing costs associated with the refinance. In the absence of prepayment penalties for a fixed rate commitment, the client's willingness to refinance their loan as rates drop would only be hindered by the costs of the refinance compared with the benefits of the lower rate. Prepayment penalties offer the lender a chance to recoup a portion of the interest which is lost over the loan term should a customer prepay the note. Thus, fixed rates offered by a bank without prepayment penalties are typically at higher interest rates than those offered with a prepayment penalty of some sort.

The motivation for a lower rate may not be the only trigger for a prepayment penalty. In our example, the loan was secured with two properties. What if a prospective buyer gave our client an offer that he simply could not refuse to

2.2 Documentation of Real Estate Conveyance and Indebtedness 33

purchase one of the properties? In order to facilitate the sale, the collateral would need to be released by the lender, which would presumably come with a curtailment of outstanding principal of some amount. The prepayment of principal would trigger the prepayment penalty.

Another trigger for a prepayment may be the performance of the property. If occupancy rates in either of the properties securing the bank loan suffered relative to prior years, the lender may deem that the current value of the property has fallen to a point that requires the client to pay down the loan balance. This situation has unfortunately become all too common in the aftermath of the recent financial crisis. Those investors who did not keep an adequate amount of cash reserves on hand have had a harder time dealing with the decline in property values. One alternative to principal reduction in this case would be the pledging of additional unencumbered property as collateral for the bank loan.

The preceding discussion helps clarify the benefits of interest rate swaps. An interest rate swap allows the bank to remove interest rate risk from the balance sheet, and allows the client to obtain the fixed rate that they desire. Interest rate swaps introduce a third party as the bank acts like a middle man between the customer with the bank loan desiring a fixed rate and a counter party. The counter party is the party who is making the opposite bet on interest rate movements than the client who is borrowing the money from the bank. In the scenario outlined earlier, the counter party would benefit from market conditions with interest rates lower than 6%. If rates were to increase above the 6% level, the original bank client would be "in the money" and would receive an interest payment from the counterparty (via the bank's derivatives area) for any month where rates were above 6%. In the event that the original bank client wished to terminate an "in the money" interest rate swap, they would receive a breakage fee from the bank. An interesting observation has been that clients are often very astute when calculating breakage fees owed to them, but often claim ignorance as to the same breakage calculations when they have to pay the penalty!

Regardless of the type of prepayment penalty calculation in effect, having an interest rate swap allows the client who owns the property to benefit financially if their interest rate assumption is correct and rates rise after they lock in the fixed rate. In many cases, swaps are portable, so if a client wants to sell a property, they can move the swap to another property rather than pay the prepayment penalty. The bank is able to profit from facilitating the derivatives trade, with the down side coming when a loan balance is fully hedged (i.e. 100% of the loan balance is swapped) and the bank demands to be paid out on the loan. In situations such as these, where deterioration in the financial performance of the property has facilitated the bank's desire to have the loan paid out, the bank customer has to pay out the loan balance along with any breakage fee on the swap should no other alternatives exist.

During the "Yes Era", more banks moved from the concept of "hedging" of loans to matching the amount of the swap and the amount of the loan. In the aftermath of the crisis, many financial institutions (as well as many investors) may return to the traditional idea of hedging their bets and only swapping a portion of the loan balance.

2.2.3 Guaranty Agreement

The final area of loan documentation that we will discuss is the guaranty agreement. For most loans under $5 million, commercial lenders will typically require the personal endorsements of the principals involved. The purpose of the guaranty agreement is to document what is expected of the guarantors over the term of the loan. A guarantor is typically not the primary source of repayment for a commercial mortgage. The primary source of repayment is typically the net operating income from the subject property. The guarantors may have outside sources of income, personal liquidity, and net worth to serve as a secondary source of repayment in the event that the subject property does not produce enough net operating income to service the loan payment each month. A tertiary source of repayment for a lender is often the liquidation of the collateral in a foreclosure situation. The guaranty agreement is the agreement between the lender and those individuals or entities that are guaranteeing the loan.

Many real estate investors often form holding companies for investment real estate. Holding companies make it easier to separate the profits and associated tax liabilities for investment properties that are owned by different individuals. In situations where numerous owners are involved, pro-rata guarantees may be the preferred method of personal guarantee for the investors. This form of guarantee is considered a conditional guarantee as the individual is only obligated on the loan balance up to the percentage of ownership. Banks will often increase the guaranty to something similar to "125% pro-rata" as the financial strength of the various guarantors is seldom the same, and if the loan moves into a foreclosure situation early in the loan term, the legal fees associated with the liquidation of the property may exceed the amount of the loan balance that has been curtailed to that point.

Sometimes conditional (or limited) guarantors are not actively involved in the daily operations of the subject property. Limited guarantors are often participating in a real estate investment for the purposes of diversifying their investment portfolio, thus the lack of daily involvement in the management of the property. Given the various actors who may be guaranteeing loan exposure, an important feature of any guarantee agreement is the concept of consideration. Consideration of guaranty is a legal classification, and it involves something of value being exchanged by parties to an agreement, thus making the agreement legally binding (Fitch 1993). If a guarantor is not deemed to have received consideration (or benefit) from a transaction, it may be difficult to enforce that individual's guaranty in a foreclosure or collection scenario. The guaranty agreement declares that each individual guaranteeing the loan has received consideration for their guarantee.

The guaranty agreement further declares the individual's responsibility to provide the bank with annual financial statements and personal (or corporate) tax returns. The lender will receive and analyze the continued financial strength of the individuals typically on an annual basis. The guaranty agreement also provides declaration that the individual is not behind on their payment of taxes and that they are not party to any ongoing legal issues.

The guaranty agreement will disclose any specific covenants that the individual is expected to achieve during the term of the loan. For example, the lender may deem it necessary for the guarantor to maintain a certain minimum in terms of personal liquidity and net worth. As mentioned in the discussion on prepayment penalties, customers with low personal liquidity have trouble when the bank demands a curtailment of principal. If a guarantor is pledging collateral to the loan, it is listed in the guaranty agreement and the security pledge agreement if the collateral consists of marketable securities.

Similar to the note, the guaranty agreement lists default provisions specific to the guarantors on the loan. Should the guarantor fail to provide the lender with annual financial information in a timely manner, or should the credit reports deteriorate due to the failure of the guarantor to pay their creditors or tax authorities in a timely manner, the bank could declare the borrower's loan in default.

2.3 The Various Forms of Mortgage Amortizations

Now that we have introduced the various forms of documentation associated with commercial mortgages, we will elaborate on the various forms of mortgage amortizations. As mentioned in the discussion on notes, an investor's loan structure will be specified during the loan closing process. As alluded to in the beginning of this chapter, how the specific repayment on the loan is structured will depend on both the objectives of the lender and the investor. We will first discuss the various objectives and then will discuss some common forms of mortgage amortizations.

Sometimes lenders will prefer a quicker amortization of the principal balance on the loan. This can be achieved in numerous ways. The lender may offer a reduced loan amount, a shorter amortization schedule, or a repayment structure which recaptures principal quicker than a typical amortization. Sometimes the investor as borrower will also desire a quicker loan repayment. Some borrowers are debt averse or are simply averse to paying interest to the lender. This may prove to be a more typical scenario in the aftermath of the financial crisis. While interest rates are near historical lows when borrowing money, the rates achieved on savings accounts and certificates of deposit are even lower. A rational investor may compare the interest rate achieved on an investment product with that paid on the loan and decide that a quicker repayment is the preferred option.

Alternatively, some investors desire to lower their monthly debt obligations regardless of the implications for interest paid to the lender. The desire for a lower payment could be related to the planned resale of the property. This could involve a speculative investment as was discussed in Chap. 1, or it could involve a cash flow positive investment where the desired holding period for the investor is less than the amortization of the loan. In the second situation, a speculative resale plan also exists, but the investment is not purely speculative as compared with those plans where the holding period is very short and the property is not currently producing income sufficient to pay the monthly debt obligations. In cases of longer payback of principal, the lender may be agreeable to the structure if their yield is sufficient to compensate them for the risk.

A final determinant for loan structures involves underperforming properties. If the occupancy rate for a property has materially declined, or if there are other material changes in the property's operating performance, a lender may desire a repayment structure which increases the probability that the borrower can continue to pay the monthly loan payments, but does not reward a client for an underperforming property. For example, in a workout situation, the lender may be agreeable to providing some payment relief, but this would not involve the lowering of the contract interest rate per the note. The borrower's loan may be modified toward a longer amortization period, or perhaps to provide a few months of payments without reduction of principal, but this is typically achieved via the borrower agreeing to improve the bank's collateral position in some way.

As in most business situations, the payment structure agreed to by the borrower and lender reflect the economic circumstances at the time that the loan is closing, as well as the objectives of the parties involved.

2.3.1 Types of Repayment Structures

The development of fixed rate mortgage loans in the US began in a slow fashion. During the years prior to World War II, banks considered loans to be very risky. The terms on offer were very conservative, often requiring 50% equity and full loan repayment within 5 years (Brueggeman and Fisher 2010). In those days, only the very wealthy could afford loans, and the banks loaned the funds based on the outside sources of income, liquidity, and net worth of the borrower, as opposed to basing the loan on the income and value of the property. Over the years, given the vast expansion of the US economy, banks began to offer longer amortizations, require less equity, and generally opened up the credit markets to the average citizen. During the late 1990s and early twenty-first century, some banks may have taken the liberalization of lending terms too far, given the subprime mortgage crisis.

In the following section the following types of commercial mortgage repayment structures will be reviewed:
- Constant Payment Mortgage (CPM)
- Constant Amortization Mortgage (CAM)
- Graduated Payment Mortgage (GPM)
- Adjustable Rate Mortgage (ARM)
- Zero Amortization Mortgage (ZAM)

2.3 The Various Forms of Mortgage Amortizations

Payment Number	Payment Amount	Interest	Principal Reduction	Balance
				$1,000,000.00
1	6443.01	5,000.00	1443.01	$998,556.99
2	6443.01	4,992.78	1450.23	$997,106.76
3	6443.01	4,985.53	1457.48	$995,649.29
4	6443.01	4,978.25	1464.76	$994,184.53
5	6443.01	4,970.92	1472.09	$992,712.44
6	6443.01	4,963.56	1479.45	$991,232.99
7	6443.01	4,956.16	1486.85	$989,746.14
8	6443.01	4,948.73	1494.28	$988,251.87
9	6443.01	4,941.26	1501.75	$986,750.12
10	6443.01	4,933.75	1509.26	$985,240.86
11	6443.01	4,926.20	1516.81	$983,724.05
12	6443.01	4,918.62	1524.39	$982,199.66

Fig. 2.3 $1 million CPM loan payments for the first 12 months

2.3.2 Fixed Rate Repayment Structures

A constant payment mortgage (CPM) consists of a level principal and interest payment based on a fixed rate. This is the most common form of mortgage for investment real estate. It is known as a CPM, as the customer is paying the same, fixed, principal and interest payment each month for the term of the loan. For example, assume that a customer has obtained a loan for $1 million at a rate of 6% fixed, based on a 25 year amortization and a 3 year term. The term is less than the amortization as the bank will want to review the financial condition of the property, borrower, and guarantor at the end of the 3 year term in order to decide if they want to continue with the loan. The customer also has the ability to renegotiate the rate and terms at the end of the initial term. The CPM payment is based on the loan of $1 million being paid at a rate of 6% based on a 25 year amortization as is shown in Figure 2.3.

Assume that the loan was originated on January 15, 2012 and the first payment is due the following month. Each month the customer is responsible for paying $6,443.01 in principal and interest. After the first year, the loan balance has amortized down to $982,199.66.

The interest due in the first payment is $5,000. This is calculated based on the original loan amount of $1 million at a rate of 6% due in monthly installments. As you can see, the principal due is the remaining payment amount to equal the CPM of $6,443.01 for the month. While the composition of interest and principal changes each month based on the outstanding loan balance at the end of the prior month, the client must pay the same total payment amount each month under this repayment structure. This mortgage style is popular as the borrower has a set, certain payment for the length of the loan. Any amount of loan payment above the required $6,443.01 will reduce the outstanding principal balance on the loan.

The loan constant is a helpful tool regarding CPM payments. This is the interest factor that can be multiplied by the beginning loan amount to obtain the payment required to fully pay out the loan by the maturity date. In the example above, the loan constant can be found utilizing a financial calculator by inputting the following:

$$N = 300, \ I = 6\%, \ PV = -\$1, \ FV = \$0, \text{ and solving for PMT} = 0.006443$$

When the loan constant is multiplied by the original loan amount, the monthly payment is obtained. Additionally, if the full loan amount was input as the PV instead of the $-1 as shown above, then the loan payment of $6,443.01 would be calculated. A financial calculator can also be used to determine the outstanding loan balance after the first year by changing N to reflect 1 year being completed on the CPM. An illustration of the calculator inputs is shown below:

$$N = 300, \ I = 6\%, \ PV = \$1,000,000, \ FV = \$0, \text{ and solving for PMT}$$
$$= (\$6,443.01)$$

$$\text{Then change N to 288, and solve for PV} = \$982,199.61$$

A constant amortization mortgage (CAM) is another common form of commercial mortgage. This payment structure is also known as "straight line" repayment as the client is paying the same amount of principal each month, with interest varying depending on the outstanding balance. By the payment structure varying the interest expense paid each month but keeping the principal constant, the loan balance pays down quicker as compared to the CPM.

For example, the same $1 million loan at 6% interest is based on a 25 year amortization as it was previously. This time the payment structure is a CAM. Figure 2.4 shows the composition of principal and interest payments for the first year.

With the CAM, the monthly principal payment remains constant, and the amount of interest depends on the outstanding principal balance for the prior month. If you

Month	Beginning Principal Balance	Monthly Principal Payment	Monthly Interest Payment	Monthly Total Payment	New Principal Balance
1	$1,000,000.00	3,333.33	5,000.00	8,333.33	$996,666.67
2	$ 996,666.67	3,333.33	4,983.33	8,316.67	$993,333.33
3	$ 993,333.33	3,333.33	4,966.67	8,300.00	$990,000.00
4	$ 990,000.00	3,333.33	4,950.00	8,283.33	$986,666.67
5	$ 986,666.67	3,333.33	4,933.33	8,266.67	$983,333.33
6	$ 983,333.33	3,333.33	4,916.67	8,250.00	$980,000.00
7	$ 980,000.00	3,333.33	4,900.00	8,233.33	$976,666.67
8	$ 976,666.67	3,333.33	4,883.33	8,216.67	$973,333.33
9	$ 973,333.33	3,333.33	4,866.67	8,200.00	$970,000.00
10	$ 970,000.00	3,333.33	4,850.00	8,183.33	$966,666.67
11	$ 966,666.67	3,333.33	4,833.33	8,166.67	$963,333.33
12	$ 963,333.33	3,333.33	4,816.67	8,150.00	$960,000.00

Fig. 2.4 $1 million CAM loan payments for the first 12 months

2.3 The Various Forms of Mortgage Amortizations

Fig. 2.5 Payments at various interest rates

Year	Payment	Implied Interest
1	$5,500.00	4.40%
2	$5,775.00	4.88%
3	$6,063.75	5.37%
4	$6,366.94	5.88%
5	$6,685.28	6.39%

compare the remaining balance after 1 year for the CAM ($960M) with that of the CPM ($982M), you will see that the CAM method provides for a quicker reduction in the outstanding loan balance.

Borrowers who are debt averse may prefer the CAM given the quicker reduction in the loan balance. Lending institutions may also favor the CAM for situations where the recapture of principal more quickly is the preferred repayment structure.

Another common repayment structure is the graduated payment mortgage (GPM). While these types of mortgages are more common in residential real estate, the repayment structure can be applied in a commercial situation. The concept with a GPM is that lower payments relative to a CPM will be made earlier in the loan, and then payments rise in a stair-like fashion throughout the loan term. Assume for example that a college senior applies for a residential mortgage loan. Based on the low interest rates in the market, perhaps the student wishes to obtain a loan for a property in the town where their job will start after graduation. In this case, there is a significant difference between the borrower's income today and what is expected in the next year. A GPM would allow the student to pay a lower payment during their senior year, and then to have the payment step up once they begin working. This has particular significance for young professionals as well. As a newly employed person in a well-compensated profession such as the law or medical fields (or perhaps financial services!), even with a current job there is a distinct possibility that income will rise in the future, thus making the GPM a good prospect.

In commercial situations, a GPM could be favorable in situations where the leases for a given property increase over time at a rate higher than inflation. In cases where the income from a subject property has a step-up effect, a GPM could allow the bank to recapture principal in a similar manner to how the leases are structured.

The first step in creating a GPM is to start with what the loan payment would be under a CPM scenario. As in our example earlier, a $1 million loan for 25 years at 6% would produce a monthly payment under a CPM of $6,443.01. The lender and the customer could agree that this payment level at the present time is not possible, and they could instead require a payment of $5,500 monthly for the first year, with increases of 5% each year. This is shown in Fig. 2.5.

An issue with the GPM involves negative amortization. In situations when the pay rate on the loan is less than the contract rate of interest on the loan, the additional interest which is accruing but is not being paid will be added to principal. In other words, if a GPM is structured without a clean-up payment at the end of the year where the client pays all accrued interest for the year, the loan balance will increase during the years where the monthly payment is less than what is accruing on the loan balance. In Figure 2.5, the implied interest rate shows what the interest rate would be for a $1 million loan for 25 years at the GPM payment for a given year. Until the implied interest rate equals the contract rate of 6%, negative amortization will occur.

While the GPM can be a favorable structure when income levels of the borrower or the subject property are expected to increase over time, there is risk to the lender. If the projected increase in income does not occur (i.e. the individual loses job or the property loses tenants), then the payment will increase at a level higher than the increase in corresponding income. Additionally, if a final clean-up payment is not included in the GPM, the balance of the loan will increase for each of the first 4 years in our example.

2.3.3 Variable Rate Repayment Structures

Until now, we have dealt with fixed rate mortgage repayment structures. Obviously a borrower may prefer to have a variable rate on the mortgage. Sometimes the desired structure involves principal reduction over the loan term, and sometimes it does not. Some variable rate mortgages are similar to the CAM structure discussed previously, only the rate is not fixed. The concept for this type of structure is the same as for the CAM. The loan amount is divided by the number of months in the amortization, and the result is the monthly principal curtailment. The monthly interest payment would depend on the outstanding loan balance relative to the variable rate for the given month.

Earlier in this chapter we viewed some historical charts for typical variable rate indexes over the last 40 years (i.e. 1-month contract LIBOR and the US prime rate). A variable rate loan would be priced based on a spread over the index being employed. For example, assume that the investor obtained the same $1 million loan on a 25 year amortization. This time the rate was variable and the loan was priced at LIBOR + 300 basis points. If the 1-month contract LIBOR rate was 0.25% in a given month, then the variable rate would be 3.25% for that month. If LIBOR increased to 0.40% the next month, the rate on the loan would also increase to 3.40%.

Sometimes, variable rate loans maintain a fixed spread for a period of time, but then increase in subsequent years during the term. These mortgage payment structures are known as adjustable rate mortgages (ARM). ARMs were a contributing factor in the sub-prime mortgage crisis, as borrowers capitalized on historically low interest rates in order to obtain a larger mortgage than would have been possible if the loan payments had been fixed. The rates offered these clients were often lower (or teaser) rates early in the loan term, with payments increasing over time.

For example, assume that a borrower obtained the same $1 million loan for 25 years, this time opting for the ARM repayment structure. If the variable rate was based on 1-month contract LIBOR with a 300 basis point spread over the index, it is very possible that during 2004 the interest rate on the loan would have been near 4%. The ARM calls for the rate to reset at the beginning of each year over the

Fig. 2.6 Possible ARM movements

Amort. Left	ARM Year	Rate	Payment
25	1	4%	$5,278.37
24	2	5%	$5,827.07
23	3	7%	$6,969.97
22	4	8%	$7,556.15
21	5	9%	$8,147.01

loan term. Over the next 4 years, the variable rates could have increased as shown is in Fig. 2.6.

Based on the scenario outlined above, the borrower's monthly payment increased by over 50%. The probability of a similar increase in borrower income over this period was low, thus the ability to pay back the loan was hindered based on the repayment structure of the loan.

Since the onset of the recent financial crisis, this particular loan structure has become less popular. This is attributable to the bad press that this repayment structured received and also due to the historically low interest rates providing incentive for borrowers to seek fixed rate mortgages in recent years. The variable rate mortgage will not go away entirely but it is expected that the level of disclosure on the part of the lender will increase in the coming years.

A final repayment structure that we will discuss is the zero amortization mortgage (ZAM). As the name implies, this repayment structure does not reduce the principal on the loan during the loan term. This is also known as an interest only loan. Repayment structures of this type are typically associated with speculative loans, as the borrower does not desire to reduce the loan balance over the loan term as another source of repayment is planned. In situations where an investor has purchased a property with the aim of a quick resale (typically after improvements to the property have been made), the investor is essentially utilizing the bank loan to purchase the property, with the hope that the eventual resale of the property will be at a level to compensate the investor for the interest paid, for any improvements made to the property, and for a profit of some sort for the investor. In terms of financial calculator inputs, the ZAM repayment structure has the PV and FV as the same, as the loan balance has not decreased.

In the aftermath of the financial crisis, many lending institutions have made the attempt at reducing their exposure to loans structured in this manner. When borrowers have not had the ability to repay the loan in full, the next best option is to begin amortizing the loan balance utilizing one of the other repayment structures that we have covered in this section.

2.4 Pretend and Extend or Bankruptcy?

Sometimes borrowers find themselves in situations that cannot be remedied via typical mortgage structures. If a subject property has experienced a significant drop in occupancy rates or otherwise a decrease in net operating income, the commercial lender may ask the borrower to curtail the principal balance by an amount that is outside of the capability of the principals guaranteeing the loan (and outside of the normal loan amortization schedule). When the customer informs the bank of their inability to adhere to the wishes of the lender, two scenarios can unfold. The bank could inform the borrower that the loan should be paid out by a specified date. If the client is given 6 months to find an alternative lender, the client will then attempt to refinance the balance elsewhere. If at the end of this forbearance period, the loan is still in place, the lender can either commence foreclosure proceedings or can simply extend the loan for a longer period of time. The second option has been called "pretend and extend", as the lender is in effect pushing the loan maturity out into the future in the hopes that the situation improves. This approach is less likely to be practiced by internationally active banks, as they are subject to the Basel Accord, as administered by the Bank for International Settlements (BIS) in Basel Switzerland. This agreement, which has gone through numerous revisions, was enacted in the aftermath of the Asian Financial Crisis. Under this agreement, banks must hold capital in reserve in proportion to the overall level of risk in their loan portfolios. Given the desire to accurately reflect the current risk in a particular loan, "pretend and extend" may prove to be a thing of the past in coming years. Unfortunately, the riskiest of commercial real estate was approved during the "Yes Era", and many of these loans have yet to mature. As these loans mature in the coming years, foreclosure and bankruptcy may be the primary method of addressing the drop in property values.

2.4.1 Foreclosure and Bottom Fishing

If an investor cannot keep current on their loan payments with their lender, or if they cannot sell the property in its current state, there are numerous options available. The borrower may decide to transfer the equity in the mortgage to the mortgagee (lender). Under this scenario, title is transferred via a quit claim or warranty in a process known as voluntary conveyance. In this case, the borrower is freed from the obligation and the lender must then make arrangements with any other creditors that might have a claim on the subject property.

Another option is known as friendly foreclosure. In this case the borrower agrees to cooperate with the lender to achieve a quick foreclosure. This is similar to a prepackaged bankruptcy where the borrower agrees to terms with their creditors on the method and means of turning over their assets.

Once all other workout scenarios have been considered, the next option is to seek foreclosure against the mortgage. The foreclosure is a judicial process as the lender will sue on the debt, attach a judgment against the subject property (and anything else that they have legal claim against), and then facilitate the sale of the property.

In Chap. 3 we will discuss investment strategies in commercial real estate. One such strategy that bears mention here involves investors who seek properties in foreclosure. By honoring the desire to "buy low and sell high", investors who purchase distressed properties are akin to bottom fishers. The key is finding a property that is in good condition to allow for a return to profitability after purchase. Sometimes the property in foreclosure is cash flow positive. In situations like this, there are other factors that led to the problem situation.

The "Yes Era" saw a wave of foreclosures for various reasons. All had the underlying cause of negative cash flow, only the symptoms were different. The first form of foreclosure in an investment real estate perspective consists of properties that were "too speculative". These are indebted properties with raw land as collateral, or properties that had low occupancies at loan origination. Unless a "bottom fisher" has a realistic plan for improving the raw land, or unless they have interested tenants for the low occupancy properties, this type of foreclosed property may not prove out the "sell high" part of the investment plan.

The second form of investment property foreclosure is categorized as "over leveraged". These are properties with decent occupancy levels where the amount of bank financing was too high at loan origination, reflective of a more aggressive lending environment. Properties in this classification can realize profitable returns if they are purchased out of foreclosure at a reasonable price relative to the net operating income.

The third form of investment property foreclosure is categorized as "global concerns". These properties have arrived in foreclosure not owing to underlying cash flow weaknesses for the subject property, but due to larger issues associated with the ownership. The owners may have other properties which are causing the inability to meet their payments. For example, if an investor owns three properties of equal value and two of them are performing at a debt coverage ratio (DCR) of 1.20x, while the third has experienced vacancy issues and is only achieving a DCR of 0.50x, there could result the inability of the investor to make payments for the two otherwise performing properties. Additionally, the real estate asset could be performing well, but the investor's operating business (or source of wage income) could have decreased or been entirely eliminated, thus making foreclosure a likely event.

What should be evident from this discussion is that depending on the circumstances of the buyer and the seller, foreclosure properties can offer the potential to buy low and sell high. What is less evident is in which category a given property in foreclosure happens to be, as the circumstances of the seller are seldom known at a foreclosure sale or auction. Similar to commercial fishing on the high seas, the bottom fishers in real estate have been very busy of late, with overfishing of the prized assets almost assured (OECD 2010). The ability to find good deals in bad circumstances in the future will require a bit of luck, but also much skill.

2.4.2 Bankruptcy

Along with the possibility of foreclosure, the possibility of bankruptcy also awaits the speculative investor. In the United States, there are four primary types of bankruptcy. Since we just discussed fishing, one form of bankruptcy (Chap. 12) is for fisherman and farmers. The other three forms of bankruptcy will be briefly discussed as this chapter concludes.

Chapter 7 bankruptcy is known as straight bankruptcy. This is where all assets of the principal are liquidated to pay outstanding debts. Each creditor must petition the court in this proceeding in order to obtain funds to repay their obligations. This form of bankruptcy may not have anything to do with the investment property as it may just be that the principal in question was over leveraged. For this form of bankruptcy, if the debtor is not behind on their personal residence payments, the personal mortgage debt could be reaffirmed so that the individual does not lose their home. This is the worst form of bankruptcy as far as lenders are concerned. The appearance here is that the individual did not attempt to work out their debts and preferred to liquidate the assets in question. In this scenario, lenders are more likely to lose the principal that they have invested.

Chapter 11 bankruptcy is reserved for business owners. This form of bankruptcy involves reorganization with a plan to repay the debts. In this scenario, there is a court supervised plan for at least two-thirds of the debts which is endorsed by the majority of the creditors. Thus, there is some effort made to reach an agreeable repayment solution with the creditors. This form of bankruptcy, along with Chap. 7, can be filed only once every 6 years.

The final form of bankruptcy that we will briefly discuss is Chap. 13. This is the individual reorganization plan that is approved by the court. In Chap. 13 bankruptcy, the primary residence is protected. This is also known as a wage earner proceeding. This plan is available to individuals as long as they have approximately $1 million in secured debt and under approximately $350,000 in unsecured debt. For commercial real estate investors, if properties are held in the personal name these limits will soon be breached. Interestingly, there is only a 6 month waiting period for this form of bankruptcy. During the "Yes Era", some individuals may have filed numerous bankruptcies, as liberal lending policies encouraged banks to lend to individuals with bankruptcy in their past in an effort to make everyone a winner and to not discriminate against individuals who had filed bankruptcy in the past.

Questions for Discussion

1. Explain in your own words the difference between a note and a mortgage.
2. Describe how the debt coverage ratio is utilized by lenders to differentiate loans based on risk.
3. Describe how an adjustable rate mortgage could prove harmful to both a borrower and a lender in an increasing interest rate environment.
4. Differentiate between the various forms of bankruptcy. Which form of bankruptcy is most problematic from the perspective of the lender?
5. Elaborate on different situations when the CPM, CAM, and GPM would be the most preferred method of repayment structure for a loan. Be sure to include a discussion of the borrower's risk tolerance and lender's return in your answers.

2.4 Pretend and Extend or Bankruptcy?

Problems

1. A borrower has a choice between a fully amortizing CPM mortgage loan for $475,000 at 6% interest over 25 years versus the same loan amortized over 15 years.
   (a) What would be the monthly payment for each loan alternative?
   (b) What would be the initial six payments with a CAM assuming a 20 year amortization?

2. A fully amortizing loan is made for $620,000 at 5% interest for 25 years. Payments are made monthly. Calculate the following:
   (a) Monthly payment.
   (b) Interest and principle payment during month 1.
   (c) Total interest and principle paid over 25 years.
   (d) The outstanding balance if the loan is repaid at the end of year 12.
   (e) Total monthly interest and principle payments through year 12.
   (f) What is the breakdown of interest and principle during month 55?

3. A 30 year fully amortizing mortgage loan was made 10 years ago for $350,000 at 7% interest. The borrower would like to prepay the mortgage by $25,000.
   (a) Assuming no prepayment penalties, what would the new mortgage payment be?
   (b) Assuming that the loan maturity is shortened rather than lowering the payment, what would the new mortgage maturity be (in months remaining)?

4. A partially amortizing mortgage is made for $500,000 for a term of 10 years. The borrower and lender agree that a balance of $275,000 will remain and be repaid as a lump sum at that time.
   (a) If the interest rate is 6%, what are the monthly payments over the 10 year period?
   (b) If the borrower chooses to repay the balance at the end of year 5, what would the balance be at that time?
   (c) What would the balance of the loan be after 10 years if the loan payments had originally been based on a 20 year amortization?

5. A borrower and lender agree on a $2,000,000 loan at 6.5% interest. An amortization schedule of 20 years has been set, but the lender has scheduled the loan to mature after 5 years.
   (a) What will the balance be at the loan maturity in 5 years?
   (b) Why would the lender structure a loan in this manner?

6. An investor has agreed to a loan of $3.5 million on an office building at an interest rate of 8% with payments calculated using a 5.5% pay rate and a 25 year amortization. After the first 5 years, the payments are to be adjusted so that the loan can be amortized over the remaining 20 years.
   (a) What is the initial payment?
   (b) How much interest will accrue during the first year?
   (c) What will be the balance on the loan after 5 years?
   (d) What will be the monthly payments starting in year 6?

Mini-Case: Workout Loan Scenario

It is 6 o'clock in the evening on Friday of a very tough week. You are in the process of packing up your things to head out of the office when the phone rings. Hesitantly, you answer the phone. On the other line is a client of your bank in an area that you do not support. Your name was provided to him as a backup for a teammate who has already left the office for the day. The customer is in a panic as they have two loans with your bank which are going to mature on Monday. He has previously arranged for refinance for both loans with your bank, but now has a counter-proposal for you to consider.

The customer is located in Greenville, South Carolina, and his firm has two loans with your bank which are up for renewal currently. Both are secured with a medical office building located in Greenville. The first mortgage consists of a term loan with a current balance of $3,821,435, while the second mortgage consists of a line of credit. The line of credit has a total commitment of $1 million, with a current balance today of $974,000.

The client informs you that his current refinance offer from your bank requires him to pay down the balance on the line of credit to $800,000. The line of credit will then be renewed for 1 year on an interest only, variable rate basis. The line will be priced at 1 month contract LIBOR plus 300. The loan is to be renewed on a 3 year term based on a 15 year amortization. The rate is 6% fixed. The customer would like to instead pay down the term loan by $250,000, and keep the line of credit at $1 million for the next year.

The client says that he can save over $2,100 per month in loan payments on the first mortgage, and says that this would be a "win-win situation" as he would be saving money on his payments, and the bank would have less total outstanding exposure than under the current arrangement.

You inform the client that you will have to further research this issue, and promise to give him a call in the next hour with an answer to his question. Once you hang up with the client, your mind quickly thinks of how a timeline might materialize in order for this loan to close on Monday should you agree to the changes that the customer proposes. "If I agree to this, then I would have to inform the documentation preparation department of the changes before I leave the office today, so that the documents can be redrawn and provided to the client for the Monday closing."

As you are mentally going through the feasibility of this timeline, you are also researching the client's relationship with your bank. Upon reading the underwriting analysis for the client in the bank's customer database, you see that your teammate who had left for the day had the primary aim of getting the line of credit balance paid down when they structured the current refinance proposal. The line of credit balance has been frozen for the last 8 months, thus you doubt that the client has a lump sum source of repayment to pay the line balance down over the next year.

2.4 Pretend and Extend or Bankruptcy?

As you read through the information in the customer database, you see that the client who called you is the sole owner of the borrowing entity for both loans. He is also the sole guarantor for both loans. As you review his personal financial position, you see that he has $100,000 in personal liquidity, and a personal net worth of $2.5 million, not including assets held in the borrowing entity. The individual is contingently liable for total indebtedness of just under $6 million, including the two loans under consideration here.

The subject property which secures both loans consists of a single tenant medical office building. The local hospital has leased out the entire building with a lease which runs for 15 years. At the current level of indebtedness (i.e. before any principal curtailment) the property supports the debt on the 15 year amortization proposed for the term loan at a debt coverage ratio of 1.26x. The loan to value is also reasonable, coming in at 71%. You conclude that the property is self-supporting, but the issues of concern are the fact that a portion of the debt is floating, and the low level of personal liquidity for the guarantor.

Quickly, you jot down the following table in order to collect your thoughts:

	Current proposal	Counter proposal
Term loan	$ 3,821,435	$ 3,571,435
Line of credit	$ 800,000	$ 1,000,000
Total	$ 4,621,435	$ 4,571,435
Exposure reduction	$ 200,000	$ 250,000

As you ponder your decision, you consider the competing aims of this situation. The client would like to lower his monthly principal payment on the term loan, but apparently has no ability to decrease the line of credit over the next year. The bank would like for the line of credit to be paid out in the next year, and this represents the first opportunity to curtail some of that exposure. Given the level of personal liquidity, the bank's desire would be to curtail the total indebtedness to the borrower, something which the customer's counter-offer does provide. Lastly, you think about the fact that the lower payments on the term loan based on the customer's counter-offer will make it easier for the client to pay the loan, but will also make the client less motivated to find an alternative lender over the next year.

Having completed your research, it is now time to call back the client and let him know your intentions.

Questions for Discussion
1. Show whether the client's claim that he can save over $2,100 per month under the reduced term loan is accurate.
2. Which is the best alternative in this case: lowering the overall exposure to the client, or lowering the commitment on the line of credit?
3. What specific recommendations would you provide to this client other than what he has proposed (consider the various types of repayment structures covered in Chap. 2)?

4. Research what 1 month contract LIBOR is today to determine what the variable interest rate would be at the spread noted in the case.
5. Define the terms "guarantor" and "contingent liability".

References

Brueggeman, W. B., & Fisher, J. D. (2010). *Real estate finance and investments* (14th ed.). New York: McGraw-Hill/Irwin.

Economist. (2010). Repent at leisure: special report on debt, June 26, 2010.

Fitch, T. (1993). *Barron's business guides: Dictionary of banking terms* (2nd ed.). Hauppage, NY: Barron's Educational Series, Inc.

OECD. (2010). *Fisheries: While stocks last*. Paris: OECD Publishing. September 2010.

Vanguard. (2009). Annual Report of Vanguard Index 500 Trust.31 Dec 2009.

Vanguard. (2010). Vanguard REIT mid-year performance information, https://personal.vanguard.com/us/funds/snapshot?FundId=0123&FundIntExt=INT#hist=tab%3A1. Accessed 27 July 2010.

Wells Fargo Securities. (2011). Derivative Desk 25 year view of LIBOR and of US Prime. Accessed 12 Sept. 2011. Information accessed via the Bloomberg database.

Finance and Real Estate Valuation 3

The value of a principle is the number of things it will explain.
Ralph Waldo Emerson

Contents

3.1	Return to Fundamentals	49
3.2	Determining the Investor's Yield	50
	3.2.1 Basics of Investment Yield	50
3.3	Holding Period and Investment Strategy	52
3.4	The Concept of Compound Interest	53
3.5	Net Present Value (NPV)	55
3.6	Internal Rate of Return (IRR)	58
	3.6.1 Partitioning the Internal Rate of Return	59
	3.6.2 Weaknesses in the Internal Rate of Return Model	60
3.7	Valuing Real Estate Versus Homogenous Assets	61
References		65

3.1 Return to Fundamentals

In the aftermath of the "Yes Era", we believe that investors must return to fundamental analytical tools in order to help ensure that a property has the desired characteristics which allow an investor to achieve their required rate of return. Investing in non-speculative properties will help the investor in obtaining financing at a bank, as lenders are less willing to take risks than they were only a few years ago.

In this chapter we will introduce techniques utilized by commercial real estate investors to value investment property. Readers may be generally familiar with financial terminology such as the investor's yield, net present value, and the internal rate of return, but may not be as familiar with how these techniques are applied in a real estate context. As the next few chapters represent the quantitative component of the book, Chap. 3 will help to illuminate valuation of investment property in light of known financial metrics.

As we begin this chapter, it will be helpful to define a few terms that will be discussed in the coming pages. A discount rate is the required rate of return (or rate of interest) for the investor taking into account opportunity costs, inflation, and the certainty of payment (Phyrr and Cooper 1982). The internal rate of return (IRR) is the rate that equates the present value of the cash inflows with the present value of the cash outflows. The net present value is a technique that discounts the expected future cash flows at the minimum required rate of return. We will provide examples of both IRR and NPV in this chapter, and will discuss how the two techniques can be utilized together when choosing between competing investment projects.

The internal rate of return on equity (IRR_e) is the time adjusted yield that represents an expected rate of return to the equity investor. The IRR_e is the expected rate of return on an investment, which is then compared to the investor's required rate of return (discount rate) to determine the attractiveness of the investment. The IRR_e is also known as the investor's yield, and it is discussed below.

3.2 Determining the Investor's Yield

A benefit of this book is that we are describing valuation from both an investor perspective as well as from a lender's viewpoint. For an investor, even before the selection of an investment property has been made, the individual will determine generally the level of risk that can be tolerated, and they must also determine the minimum required rate of return that must be achieved in order to make the investment worthwhile. During the completion of a recent paper on cap rates for the RMA Journal, the authors were looking for an objective mechanism for determining the investor's yield (Goddard and Marcum 2011). From a lender perspective this question seldom arises, and in situations where the investor's yield is needed in order to obtain a value, third party market sources are utilized to obtain market averages which are found via survey. Many investor surveys are completed quarterly. The purpose of the survey is to gauge current market expectations regarding investor yields (also known as the internal rate of return), but the survey does not typically involve disclosure as to how these investment yields are determined by the investor.

3.2.1 Basics of Investment Yield

Real estate as an investment category has historically performed better than fixed income investments (bonds) but not as well as growth oriented equity investments (stocks). Thus the first step that an investor may do to determine their equity yield requirements is to locate a benchmark for what is considered a risk free security. Latent budget issues aside, the U.S. Treasury bills of the U.S. government can serve

3.2 Determining the Investor's Yield

Stocks	1 Year	3 Years	5 Years
Russell 1000 Index (Large Caps)	23.33%	0.45%	2.51%
Russell 2000 Index (Small Caps)	31.36%	4.57%	2.64%
Dow Jones U.S. Total Stock Market Index	24.31%	1.22%	2.86%
MSCI All Country World Index ex USA (International)	18.50%	-0.96%	4.08%
Bonds			
Barclays Aggregate Bond Index (Broad Taxable Market)	5.06%	5.36%	5.82%
Barclays Mutual Bond Index	1.10%	3.39%	3.88%
Citigroup 3-Month Treasury Bill Index	0.14%	0.59%	2.23%
CPI			
Consumer Price Index	1.63%	1.42%	2.12%
Real Estate Investment Trusts	**1 Year**	**5 Years**	**10 Years**
Vanguard REIT Index Fund (as of 01/31/2011)	40.02%	2.56%	10.89%
MSCI US REIT Index	40.23%	2.52%	10.92%
Real Estate Fund Average	38.56%	1.32%	9.96%

Fig. 3.1 Comparative investment returns (Vanguard 2011a)

as a useful benchmark for the risk free rate, as can corporate bond yields for AAA rated corporations. The risk free rate can serve as the bottom of the range of investor returns for real estate, as the interest rate associated with these investments are typically less than 4%. Since real estate will most typically have components of risk higher than these "risk free" assets, the equity yield rate should be superior to the risk free rate.

Over the last 100 years, the return of the U.S. New York Stock Exchange (NYSE) has averaged around 8%. Since these returns include all asset classes of securities listed on the exchange (i.e. from small cap stocks to the largest large cap stocks), this can be seen to represent an average return from an equity standpoint. Figure 3.1 presents the returns for various asset classes over various holding periods as of January 31, 2011.

As you can see from the table above, an investor has the opportunity to reference numerous other returns for competing investments when determining their investment yield hurdle rate. Figure 3.1 serves to update Fig. 2.1 and shows that asset returns over the 3–10 year periods are beginning to return to normalized levels. This information, when coupled with information obtained from investor real estate specific sources, allows the investor to select from a multitude of sources when determining their return preferences. Since risk and return are typically related, the higher the associated risk for a given investment option, the higher the required rate of return.

3.3 Holding Period and Investment Strategy

From a financial perspective, the various valuation techniques that we will discuss in this chapter allow an investor to determine whether the expected net operating income received over the holding period of the investment enhances the wealth of the investor. The holding period of the investor is similar to the "investment horizon" concept for stock investing. The holding period depends on the investment strategy of the investor. If an investor is hoping to time the market for a quick sale, a shorter holding period is to be expected. If the investor has a longer term investment strategy, the holding period would more than likely increase. In this section, we will discuss the various forms of investment strategy as related to the holding period of the investment.

Market timing strategy is often referred to as rotation strategy. Since the desire is to buy a property at a low price point, and to then sell at a higher price point after improvements and/or a market recovery has occurred, this particular strategy would typically involve shorter investment holding periods.

Another form of investment strategy which often leads to an abbreviated holding period for the investor is arbitrage investing. This form of investing is motivated by the desire to capitalize on differences in prices paid in different geographic areas for similar investment properties by purchasing a large property in the private market, and then taking it public via a real estate investment trust (REIT). Once an appropriate level of profit has materialized, the investment may then be sold. In any case, the holding periods under this form of investment are shorter than for other investment strategies.

A third form of investing leading to shorter holding periods would be opportunistic investing. This form of investing involves changing or modifying the use of a property for a profit opportunity unforeseen by other investors. Investors looking for properties in foreclosure might be representative of this form of investment strategy. Depending on the amount of improvements necessary in order to profit from an investment, the holding period here may be from very short (i.e. less than 2 years) to moderate (from 3 to 5 years), but the motivation for the investment is not a "buy and hold" consideration.

Other forms of investment strategies lead to longer holding periods. Investors that are utilizing a growth strategy may be inclined to a longer holding period given that their investment horizon is longer. Growth oriented investors are typically motivated to invest in the "next big thing" in a given market. For example, in many medium sized cities in the United States, urban renewal along with an increased desire to live and work in downtown areas has lead to a redevelopment of warehouse and office space for residential purposes. Investors that position themselves well may experience strong profits over a long period of time.

Another investment strategy which generates a longer holding period is value investing. As should be evident from the preceding chapters in this book, this is the primary investment strategy advocated by the authors. Value investing, which is similar to fundamental analysis in stocks, looks for properties which contain an income stream that has strong quantity, quality, and durability (QQD).

This methodology will be elaborated in the remainder of the book. For now, it should be noted that value investing tends to involve longer holding periods (from 7 to 10 years).

Other forms of investment strategy are not as easy to categorize by the length of the holding period. Contrarian investing, which involves investing in properties that are currently out of favor, could result in either shorter or longer holding periods, depending on the amount of time that it takes for the property type to return to favor (Rothchild 1998). In the aftermath of the "Yes era", many investors may find themselves in this category of investment strategy whether they like it or not. As a bear market is defined as a 20% drop in prices over a 2 month period, many markets met this classification in 2010–2011 (Vanguard 1997; 2011a). It remains to be seen how long recovery will take.

Property sector investment involves only investing in one type of property (i.e. multi-family units). Since many small investors get their experience first in rental housing, apartment investing is the logical next step. This form of investment strategy is akin to mutual funds that invest in one industry, such as healthcare or energy (Bogle 1999). Other investment strategies, such as investing in a particular price point of properties (size strategy), are also hard to classify based on investment strategy alone in relation to the expected holding period. Some REITs focus on secondary cities for their investment opportunities rather than on the more highly competitive primary cities. Thus, they might focus on properties in Heidelberg rather than in Berlin, with a noted difference in property value given the less competition in the smaller markets. The price differential between primary cities and secondary cities is also seen in emerging markets, and not just in the developed world (Suarez 2009; Lynn and Wang 2010). The holding period for each investment will thus depend on specific circumstances of the property and the location.

Tenant strategy involves purchasing investment properties that have specific tenants deemed desirable by the investor. Some investors may wish to concentrate on properties built for only investment grade credit quality tenants. The holding period for this type of investment strategy will depend on the durability of the income stream, which is defined as the length of the remaining lease term.

A similar strategy is blue chip property investing. Investors following this strategic paradigm invest in very visible, well located, unique properties. These properties have a timeless appeal with typically a long track record of success. For example, an Asian based REIT might focus on a few high quality properties that are prominent features of the Hong Kong sky line.

3.4 The Concept of Compound Interest

As you may recall from your finance courses in college, finance revolves around the concept of present value and future value. As Wimpy from the Popeye cartoon was wont to say, "I will gladly pay you Tuesday for a hamburger today". (Higgs 1955) It is implied here that a dollar today is worth more than a dollar in the future. The

n = Number of periods
i = interest rate
PV = present value
pmt= payment
FV = future value

Fig. 3.2 Basic requirements of financial problems

primary reasons that one dollar today is worth more than in the future are the opportunity costs of foregoing consumption today as well as the effects of inflation. Essentially, Wimpy was requesting a zero interest short term loan for the purchase of the hamburger. Whether you purchase the hamburger for Wimpy might depend on his capacity and willingness to repay the debt, but it might also depend on actually being able to profit from the situation in the form of accrued interest. This is where compound interest comes in.

Compound interest is the benefit which accrues to the owner of capital for allowing their capital to be borrowed by someone for a period of time. The more times that interest compounds per year, the more opportunity there is to earn interest on interest. The more opportunity to earn compound interest will allow for a higher effective annual yield (EAY). The annual yield is defined as follows:

$$\text{EAY} = \frac{\text{Future Value} - \text{Present Value}}{\text{Present Value}}$$

Compound interest is the foundation for concepts such as NPV, IRR, and discounted cash flow analysis. For residential mortgages in Canada, interest compounds semi-annually, compared to monthly compounding for similar mortgages in the United States. Semi-annual compounding leads to less interest accruing to the principal balance, and leads to a quicker amortization of principal. This basic tenet of mortgage structure could have contributed to the lack of residential mortgage defaults in Canada during the most recent financial crisis (Financial Times 2009).

The number of compounding periods is crucial in determining the answer to financial questions. Figure 3.2 itemizes the items which are required in order to solve financial problems. As shown in Figure 3.2, n, i, and pmt must correspond to the same period. Periods can be monthly, quarterly, semi-annually, or yearly in most financial calculators. For payments of an unusual structure, most financial calculators will be able to solve the problems correctly as long as "n" is set for some period. Many of our finance students prefer to use spreadsheet programs such as excel to solve financial problems. The ubiquitous "solver" feature makes for an easier time in calculating the break-even performance for investment property. At least one of the authors still prefers to teach with a Hewlett-Packard 10-B II calculator, given its relatively low cost and since we are often prisoners of those who taught us a given subject. The movement of financial calculators to the I-phone bodes well for their continuation in a fast moving world. It is still sometimes easier

3.5 Net Present Value (NPV)

	Investment A			Investment B	
Period	ATCF	ATER	Period	ATCF	ATER
0	$(100,000)		0	$(100,000)	
1	$ 5,000		1	$ 45,000	
2	$ 12,000		2	$ 35,000	
3	$ 20,000		3	$ 25,000	
4	$ 26,000		4	$ 25,000	
5	$ 30,000	$225,000	5	$ 15,000	$125,000

Fig. 3.3 Expected future benefits of competing investments

to use a financial calculator during a meeting with an investment real estate client than having to depend exclusively on spreadsheets. Thus, there will be some consideration given to financial calculators for the problem solutions in this book.

3.5 Net Present Value (NPV)

A primary method of assessment in investor real estate is the net present value (NPV) concept. A basic tenant of any investment decision is the concept of discounting income received at a future date at an appropriate discount rate. Thus if a project is expected to produce an income cash flow of $25,000 in year 3, and the investor's discount rate is 10%, the value today of the cash flow is

$$PV = \frac{\$25,000}{(1.10)^3} \text{ or } \$18,782.87$$

Income received in the future can be less valuable today due to opportunity costs, inflation, and the risk of non-payment. Opportunity costs in this context would consist of foregone earning today for agreeing to receive payment at a future date. Inflation involves dollars received in the future being worth less today due to intervening increases in the prices of goods and services. The risk of non-payment involves the uncertainty of receiving payment in the future. As we have discussed earlier in this chapter, the U.S. government has historically been considered to be the closest to a risk-free credit risk (jokes concerning budget deficits and possible inflation-inducing stimulus packages aside), while Wimpy would not exactly provide the same level of comfort. Opportunity costs, inflation, and the risk of non-payment must be considered when determining the discount rate being employed.

Suppose that an investor was contemplating two different investment property opportunities. After the investor has completed their analysis of the properties in question, the investor settles on a 5 year holding period for both investment alternatives. Figure 3.3 summarizes the expected cash flow receipts for both projects.

Period	Total CF	Discount Factor	Net CF
0	$(100,000)	-	$ (100,000)
1	$ 5,000	1.10000	$ 4,546
2	$ 12,000	1.21000	$ 9,917
3	$ 20,000	1.33100	$ 15,026
4	$ 26,000	1.46410	$ 17,758
5	$ 255,000	1.61051	$ 158,335
		Total Net CF	$ 105,582

Fig. 3.4 Investment A discounted at 10%

In Figure 3.3, ATCF is defined as "after tax cash flow", or the residual income left to the investor after operating expenses, taxes, and debt service obligations have been paid. ATER is defined as "after tax equity reversion", which means that this is the equity flow, after the mortgage has been extinguished, to an investor for selling the property at the end of the fifth year.

If the investor ignores entirely the concept of interest and discounting, the equity flows over the holding period would be $218,000 for investment A, and $170,000 for investment B. Since the reader should infer from our prior discussion that it is not advisable to simply take expected future benefits at face value, let's assume that the investor employs a 10% discount rate for each investment. What effect might this have on potential wealth enhancement for the investor?

Figure 3.4 illustrates the NPV concept for investment A discounted at 10%. For purposes of calculating the NPV, the period five after tax cash flow and period five after tax equity reversion can be combined as both are assumed to be received at the end of the fifth and final year of the holding period. Each successive year is discounted by a factor of 1.10 to correspond with the chosen discount rate. Thus the discount factor for the second year is $(1.10)^2$ or 1.21, and so on for each successive year in the holding period. By employing the 10% discount rate, the cash flows received by the investor drop from $218,000 in the absence of any discounting to $105,582. This is quite a large difference and this helps illustrate the problem associated with projecting future income receipts at today's dollar value. For those that need further proof that discounting makes sense in the real world, one only has to look back at the erosion of the purchasing power of $1.00 over time.

If the same 10% discount rate is employed for Investment B, the resultant net cash flow to the investor would drop from $170,000 to $92,622. To ensure that you understand how NPV works, attempt to duplicate the process sketched out for Investment A to arrive at the correct net cash flow amount for Investment B.

One of the key messages from our earlier discussion of determining the investor's required rate of return was that it can impact the selection decision between competing investments. Figure 3.5 shows how the net cash flow for both investment options change as the assumption regarding the discount rate (K_e) changes.

3.5 Net Present Value (NPV)

K$_e$	A	B
0%	$218,000	$170,000.00
5%	$154,112	$126,460.34
10%	$105,582	$ 92,621.90
15%	$ 68,218	$ 65,931.94
20%	$ 39,091	$ 44,592.34
25%	$ 16,128	$ 27,315.20
30%	$ (2,168)	$ 13,163.86
35%	$ (16,887)	$ 1,447.37
40%	$ (28,836)	$ (8,350.69)

Fig. 3.5 Investments compared at various discount rates

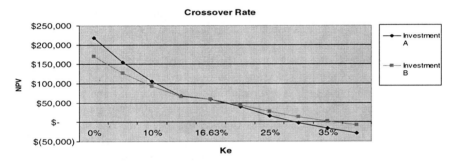

Fig. 3.6 Crossover rate for competing investments

As you can see from Figure 3.5, Investment B becomes the more favorable investment somewhere between 15% and 20%. Since the bulk of Investment A's operating cash flow comes toward the end of the holding period, having a higher discount rate affects this investment more than that of Investment B. As it turns out, the "crossover rate" is 16.63% as is shown in the graph below. This is the discount rate where Investment B becomes superior to Investment A (Fig. 3.6).

Another thing that bears comment concerning Fig. 3.5 is that at some level of Ke (30% for Investment A, 40% for Investment B), the cash flows accruing to the investor become negative. While the low interest rate environment of the current day might make the high discount rates appear remote, investors only have to look back to the 1970s for rates in the vicinity of 20%.

Once the investor decides upon the appropriate discount rate, the profitability index can be calculated. This is calculated by taking the present value of the cash flows and dividing by the investment cost.

Suppose that Investment A is chosen at the 10% discount rate. The profitability index would be calculated by dividing $105,582 by $100,000, for a result of 1.05582. The profitability index shows how much more additional equity could be invested to still achieve the desired 10% return (in this case 5.6%).

Fig. 3.7 IRR for competing projects

Period	A Total	B Total
0	$(100,000)	$ (100,000)
1	$ 5,000	$ 45,000
2	$ 12,000	$ 35,000
3	$ 20,000	$ 25,000
4	$ 26,000	$ 25,000
5	$ 255,000	$ 140,000
IRR	29.35%	35.69%

3.6 Internal Rate of Return (IRR)

Another primary method of assessment in investor real estate is the internal rate of return (IRR). This method is popular among real estate investors as the final result is fairly easily benchmarked with other properties and investment options. Similar to the NPV technique discussed earlier, the IRR involves utilizing the equity cash flows from operations and reversion to determine a time adjusted yield. Popular spreadsheet programs such as Microsoft Excel allow for easy computation of the IRR, while the CF_j button on the HP 10Bii financial calculator achieves the same result. As stated at the outset of this chapter, the IRR is the rate that equates the present value of the cash inflows with the present value of the cash outflows. Once a projected income stream has been determined, the resultant IRR can then be compared to the required rate of return for the investor to determine whether the project should be pursued. If the IRR is greater or equal to the required rate of return, then the NPV would be positive and the project would be accepted. If the IRR does not at least equal the required rate of return, then the NPV would be negative and the project would be rejected.

Let's return to our two investment alternatives previously discussed. Figure 3.7 shows the timing of the cash flows for both projects along with the resultant IRR for each investment alternative.

For period zero above, the cash flow sign is negative as the investor is required to pay this amount in initial equity in order to participate in the project. Additionally, the after tax cash flow and after tax equity reversion is combined in year 5, as both are assumed to be received at the end of the fifth and final year of the holding period.

The IRR for investment B is somewhat higher than that of investment A as is shown in Fig. 3.7. *To make sure that the reader understands how to calculate the IRR for both projects, use either the spreadsheet or financial calculator of your choice to verify the results.*

3.6 Internal Rate of Return (IRR)

Year	BTCF	PV		Year	BTCF	PV	
0	$ (100,000)			0	$(100,000)		
1	$ 5,000	$ 3,866		1	$ 45,000	$ 33,165	
2	$ 12,000	$ 7,173		2	$ 35,000	$ 19,011	
3	$ 20,000	$ 9,242		3	$ 25,000	$ 10,008	
4	$ 26,000	$ 9,289		4	$ 25,000	$ 7,376	
5	$ 30,000	$ 8,286		5	$ 15,000	$ 3,262	
5	$ 225,000	$ 62,145	62%	5	$ 125,000	$ 27,179	27%
		$100,000				$100,000	

Fig. 3.8 IRR partition for Investment A and Investment B

3.6.1 Partitioning the Internal Rate of Return

If the "50,000 foot view" of the cash flow patterns for both investments is taken into account, it will be clear that the period five cash flow is much larger for investment A than is the case for investment B. In Chap. 1, we reviewed the definition of speculative investments. Investments that rely more heavily on the reversion, or final sales price, rather than on the income from operations annually throughout the holding period, are by definition more speculative. An investor seeking to lower their investment risk should consider how much of the present value of the cash flows is expected based on the annual income produced by the property relative to the amount which is expected from the future sale of the investment. The higher the percentage of the cash flows from the sale relative to the total net cash flows over the holding period, the more inherent risk is in the projected income stream. By viewing the cash flows in this way, an investor is able to "partition" the net cash flows from operations from those that are from the expected sale. Figure 3.8 provides the partitioning data for both investment options.

By reviewing the table above, a few things should be apparent. As expected, the percentage of the net cash flows coming from the reversion is much higher for investment A than for investment B. Investment A has 62% of the net cash flows dependent on the future sale of the property, while investment B only is dependent on the reversion by 27%. Additionally, the present value of the cash flows for both projects is equal to the amount of the cash out flow. This links with our earlier definition of the internal rate of return. The IRR is simply the rate that equates the cash inflows with the cash outflows. Thus for each project, the discount factor is based on the calculated IRR. For investment A, with an IRR of 29.35%, the discount factor for the first year is 1.2935, while for investment B, with an IRR of 35.69%, the discount factor for the first year is 1.3569.

Partitioning the IRR allows the investor to quantify the risk inherent in an income stream. This process also highlights some of the weaknesses in the IRR model.

Period	Total CF	Discount Factor	Net CF	Period	Total CF	Discount Factor	Net CF
0	$(100,000)	-	$(100,000)	0	$(100,000)	-	$(100,000)
1	$ 5,000	1.46410	$ 7,321	1	$ 45,000	1.46410	$ 65,885
2	$ 12,000	1.33100	$ 15,972	2	$ 35,000	1.33100	$ 46,585
3	$ 20,000	1.21000	$ 24,200	3	$ 25,000	1.21000	$ 30,250
4	$ 26,000	1.10000	$ 28,600	4	$ 25,000	1.10000	$ 27,500
5	$ 255,000	1.00000	$ 255,000	5	$ 140,000	1.00000	$ 140,000
MIRR	18.24%	Total Net CF	$ 231,093	MIRR	16.02%	Total Net CF	$ 210,220

Fig. 3.9 MIRR calculations for Investment A and Investment B

3.6.2 Weaknesses in the Internal Rate of Return Model

One of the primary weaknesses in the IRR model is that the formula is an nth degree polynomial given the need to divide by $(1 + IRR)^n$. There is the possibility that numerous IRRs can result when the expected cash flows change from positive to negative and back to positive over the course of a holding period. For a real estate transaction, when might this be possible? Certainly the initial period zero number would be a negative in most cases, but when might a subsequent annual cash flow become negative after period zero? There might be a case where a new construction project is underway, and during the period of construction the property would not be generating income while at the same time expenses might be incurred. In most cases tenants do not start paying rent until they occupy the property, and during the "Yes era" some financial institutions approved loans for new construction without tenants in place. In any event, the "multiple IRR" problem has plagued this investment technique since its inception.

Another weakness in the IRR technique is more pervasive and fundamental. This has to do with the assumed reinvestment of the cash flows from operation during the holding period at the calculated IRR, rather than at the required return of the investor as per the NPV approach. The NPV method is considered a more conservative approach as the IRR is project specific and may not be able to be duplicated in other investments at a future date.

The proposed solution to these problems, in finance circles, is the Modified Internal Rate of Return (MIRR). The MIRR assumes a reinvestment rate equal to the required rate of return, or some other realistic rate. The MIRR is also known as the terminal value IRR as the annual cash flows, including the reversion, are equated with the original equity at the modified IRR (Jaffe and Sirmans, 1982). Thus the MIRR compounds forward to a terminal period (when the property is sold) all periodic cash flows at the required IRR, and then finds a compound interest rate that equates the terminal value and the initial investment cost.

Figure 3.9 shows how the MIRR would be calculated for our two investment options. As would be expected, modifying the IRR calculation to reinvest the cash flows at a required rate of return of 10% would decrease the resultant MIRR for both investments. Rather than solving for the rate that would equate the positive and negative cash flows, the MIRR provides an additional comparison of these two investments with a more conservative reinvestment rate assumption.

The MIRR has historically been viewed as mathematical overkill for investors (Phyrr and Cooper 1982). While the techniques reviewed in this chapter are useful when selecting among competing investments, the investor should spend more of their analytical energy evaluating the underlying economic and market assumptions rather than multiple reinvestment rate assumptions. The NPV and IRR are two of many pieces of information to consider when making an investment decision. Too exclusive a focus on one single investment criteria is generally unwise.

3.7 Valuing Real Estate Versus Homogenous Assets

A last topic for discussion in this chapter concerns the heterogeneity of real estate when compared with more homogeneous assets such as stocks and bonds. When assets are fairly homogeneous, it is much easier for investors to estimate the cost of capital. Since most stocks and bonds are frequently traded, there is naturally more data available in order to assess market expectations regarding those investment classes.

As was discussed in Chap. 2, the many forms of commercial mortgages lead to various debt service patterns being available in the marketplace. Additionally, prepayment penalties and maturity dates for existing mortgages vary considerably adding to the heterogeneity of the real estate investment class. When the various debt scenarios are coupled with diverse management, markets, submarkets, and tenant mix scenarios, the ability to find pure comparable properties is very difficult. When this is combined with the lower frequency with which real estate is bought and sold in comparison with stocks and bonds, the cost of capital becomes much harder to estimate than for more homogeneous asset classes.

While heterogeneity does present difficulty, if real estate was more easily identified as a homogeneous investment class, then much of the analysis in this book might not be necessary! As will be made apparent in the chapters to follow, even identical properties with identical debt and equity components can achieve different values should the investors purchasing the properties utilize differing assumptions when compiling their pro-forma operating statements. Should one investor utilize a 10% vacancy factor while another utilizes a 5% vacancy factor, differences in value will result. The beauty of a locally determined property value can sometimes be problematic when the frequency of comparable sales declines, or when no truly comparable properties exist. In the chapters that follow, we describe techniques and methodologies which can be utilized to partially compensate for the heterogeneity problem.

Questions for Discussion

1. Describe the various investment strategies outlined in this chapter. Make sure to include the QQD framework in your answer. Which of the investment strategies most aligns with your personal investment objectives?
2. Describe the internal rate of return (IRR), net present value (NPV), and the modified internal rate of return (MIRR). Which of these methods seems the most reasonable to use when making an investment decision?
3. How can the risk inherent in a stream of cash flows be judged by partitioning the internal rate of return?
4. Outline a process that an investor might use in order to determine the appropriate holding period for an investment property.
5. Elaborate on how inflation effects and opportunity costs are involved with time value of money considerations.

Problems

1. Babette makes a deposit of $50,000 in a bank account. The deposit is to earn interest annually at 3% for 5 years.
   (a) How much will she have on deposit at the end of 5 years?
   (b) Assuming the investment earned 7% compounded quarterly, how much would she have at the end of 5 years?
   (c) What are the respective effective annual yields for both investments?
2. Suppose that you have the opportunity to invest $70,000 today in an investment that will earn $17,000 at the end of year 1, $22,000 at the end of year 2, $13,500 at the end of year 3, $18,000 at the end of year 4, and $26,500 at the end of year 5.
   (a) What is the internal rate of return (IRR) for this investment?
   (b) What is the net present value (NPV) of this investment assuming an 11% discount rate?
   (c) What does the information in parts a and b above tell the investor about this investment opportunity?
3. Suppose that you have an opportunity to invest in a real estate venture that expects to pay investors $2,450 at the end of each month for the next 7 years. You believe that a reasonable return on your investment should be 10% compounded quarterly.
   (a) How much should you pay for this investment?
   (b) What will be the total sum of cash that you receive over the next 7 years?
   (c) Why is there such a difference between a and b?
4. Lois Lane has projected that an investment real estate opportunity could achieve an expected return of $220,000 after 7 years. Based on a careful study of competing investment alternatives, the investor believes that a return of 9% compounded quarterly is reasonable.
   (a) How much should she pay today for this investment?

3.7 Valuing Real Estate Versus Homogenous Assets

    (b) How much should she pay if the required rate of return was 10% compounded annually?

    (c) What considerations might lead Lois to change her expected rate of return assumption?

5. Bobby is considering an investment that will provide the following returns at the end of each of the following years: *1*: ($26,500), *2*: $13,000, *3*: $17,500, *4*: $21,000, *5*: $27,500.

    (a) How much should he pay for this assuming a discount rate of 12% annually?

    (b) What is the IRR assuming an investment today of $30,000?

    (c) What is the NPV assuming the period zero investment in part b and a required rate of return of 10%? What is the profitability index for this investment?

6. McAllister Development Company is considering the purchase of an apartment project for $750,000. The investors estimate that the apartments will produce net cash flow of $125,000 at the end of each year for the next 10 years. At the end of the tenth year, they will sell the apartments for a net profit of $200,000.

    (a) What is the IRR?

    (b) If the company considers 10% to be a reasonable return on investment, is this a good deal?

    (c) What would the IRR be if the company breaks even on the resale of the property at the end of the holding period?

Mini-Case #1: Rent Roll Analysis

You have received two rent rolls for two different investment properties. You are interested in assessing the strength of these two properties as an investment by starting with an analysis of the rent rolls.

Your first potential investment is a retail shopping center located in your hometown.

The listing of tenants is as follows:

Tenant	Unit #	Sq. ft.	Monthly rent	Lease maturity
Food Lion	#100	25,000	$31,250	11/15/2014
Family Dollar	#101	5,500	5,500	1/15/2015
Lucie's Nail Salon	#102	2,000	2,700	3/15/2015
Szechuan Palace	#103	3,000	4,000	7/15/2021
Vacant Space	#104, 105	4,000		

Your second potential investment is an office building also located in your hometown.

The tenant listing is as follows:

Tenant	Unit #	Sq. ft.	Monthly rent	Lease maturity
Dr. Ricardo, DDS	#100	3,000	$5,000	01/01/2014
Galbraith and Friedman, Attorneys at Law	#101	4,000	$6,000	03/15/2014
Novant Health	#102	12,000	$16,700	1/1/2021
Dr. Mudd and Dr. Sick, M.D.	#103	1,000	$1,850	01/01/2021
Dr. Sen's Acupuncture	#201	3,000	$5,000	04/15/2017
Veblen, Hirschman, and McNeely, PLLC	#202	4,000	$5,000	05/30/2016
We (really) care, not-for profit	#203	3,000	$3,800	MTM
Vacant Space	#204–206	10,000		

Questions for Discussion

1. Is there anything that seems troubling about the durability of the income stream for either property given an investment horizon of 3, 5, or 7 years?
2. What are your thoughts about the tenant mix for both properties?
3. What specific questions would you ask the seller concerning each property?

Mini-Case #2: Operating Expense Analysis

When evaluating an existing investment property, it is common that the last 3 years financial statements are analyzed in order to determine the most likely level of operating expenses on an on-going basis. The following exercise will serve to facilitate discussion on what expenses to include, and what assumptions to make when evaluating the historical operating performance of an investment property.

Consider the following investment property that has achieved the following level of operating expenses over the last 3 years:

Fiscal year end	Year 1	Year 2	Year 3	Pro-forma
General and administrative	76,819	65,323	68,134	
Repairs and maintenance	54,000	75,652	51,000	
Utilities	85,729	87,242	88,926	
Depreciation/amortization	75,000	75,000	75,000	
Salaries and commissions	153,000	157,500	161,000	
Management fee	50,000	50,000	50,000	
Property taxes	73,500	75,425	77,800	
Insurance	12,500	18,754	21,315	
Interest expense	115,700	121,100	128,762	
Advertising	4,500	7,200	8,995	
Total operating expenses	700,748	733,196	730,932	

Questions for Discussion
1. In the space under the pro-forma column, indicate your assumptions for the "going-forward" operating expenses.
2. Where might it not be prudent to take an average of the last 3 years in terms of expense calculations?
3. Which expenses should not be included in a pro-forma operating statement for an investment property?
4. What questions might you ask of the owner or property manager in order to aid you in this process?
5. How are good results determined here?

References

Bogle, J. (1999). *Common sense on mutual funds.* New York: John Wiley and Sons.

Freeland, C. Financial Times (1999). (Chrystia Freeland), View from the top with Julie Dickson, Canadian Bank Regulator. Accessed 18 Dec 2009.

Goddard, G. J., & Marcum, B. M. (2011). The crowd is untruth: The problem of cap rates in a declining market. *RMA Journal,* March 2011, Copyright 2011 by RMA pp. 26–32.

Jaffe, A. J., & Sirmans, C. F. (1982). *Real estate investment decision making*. Englewood Cliffs: Prentice Hall.

Lynn, D. J., & Wang, T. (2010). *Emerging market real estate investment: Investing in China, India, and Brazil*. Hoboken: John Wiley & Sons Inc.

Phyrr, S. S., & Cooper, J. R. (1982). *Real estate investment: Strategy, analysis, decisions*. New York: John Wiley & Sons.

Higgs, M. (1955). *Popeye* (M. Higgs, ed. and designed). Edison, NJ: Chartwell Books.

Rothchild, J. (1998). *The bear book: Survive and profit in ferocious markets*. New York: John Wiley and Sons.

Suarez, J. L. (2009). *European real estate markets*. New York/Hampshire: Palgrave Macmillan.

Vanguard Group. (1997). *Bear markets: A historical perspective on market downturns* [Plain Talk Library brochure].

Vanguard Group. (2011). *Annual report of Vanguard REIT index fund*. Accessed 31 Jan 2011.

Vanguard Group. (2011). *The active-passive debate: Bear market performance*, Philips, C.B., https://personal.vanguard.com/us/literature/search?searchInput=S601&search_mode=true. Accessed 3 Jan 2011.

Real Estate Valuation

4

To give up one's principles is to give up oneself.
Søren Kierkegaard

Contents

4.1	The Appraisal Process	67
	4.1.1 Beginning the Appraisal Process	68
4.2	The Three Approaches to Value	69
	4.2.1 The Sales Approach to Value	69
	4.2.2 The Cost Approach to Value	71
	4.2.3 The Income Approach to Value	74
4.3	Introduction to Cap Rates	80
	4.3.1 Market Extraction Method	81
	4.3.2 Market Survey Method	82
	4.3.3 Lender's Yield Method	83
	4.3.4 Band of Investments Technique	84
	4.3.5 Ellwood, Akerson, and Archeological Finance	86
	4.3.6 Conclusion	89
References		92

4.1 The Appraisal Process

A traditional step in the evaluation process for investment real estate is the completion of an appraisal by an independent appraiser. The investor may desire an unbiased valuation for purposes of agreeing on a fair sales price for the property. If the buyer of the property requires financing by a bank, the bank will utilize the appraisal in order to set their loan to value ratios. Federal regulations mandate that in order for a financial institution to utilize an appraisal for lending purposes, that the appraisal must have been engaged by a financial institution and not the customer (CFR 2011).

If the seller of the property has engaged the appraiser, the assumption is that there may be some level of influence to fulfill whatever final value is desired by the seller. In order to remove the potential of further influence on the appraisal process, most large commercial banks now have a separate appraisal department that bids out potential new appraisal requests to appraisers on the bank's approved list. The customer of the bank cannot influence the bank lender in the choosing of the appraiser, as the list of bid recipients is blind to all but the appraisal department staff at the bank. Gone are the days where one favored appraiser would receive the lion share of the business of any one given bank, at least for the medium to large financial institutions.

The goal of the appraisal process is to obtain a market value for the property that represents the most probable price in a competitive and open market. In order to achieve a market oriented value, a few things are required. The sale of the property must be an arm's length transaction, where the buyer and seller are not related. As is the case in any market-oriented transaction, both the buyer and the seller are assumed to be well-informed, rational market participants. In light of some of the deals transacted during the "Yes Era," some of the market participants may have been more rational than others. A third requirement for a "market value" for an investment property is that the appraisal be performed by someone with professional credentials. Many appraisers have sought the Member of Appraisal Institute (MAI) designation, while others have also sought the designation of Certified Commercial Investment Member (CCIM). In order to obtain these designations, individuals must pass rigorous examinations which culminate from both classroom and field experience (Appraisal Institute 2011; CCIM 2011).

4.1.1 Beginning the Appraisal Process

Once an appraiser secures the assignment from the bank or individual, the next step is to collect as much data as is possible and is relevant for the valuation at hand. The appraiser will obtain the physical and legal description of the property, will identify the property rights to be valued, and will specify the purpose of the appraisal. The appraiser will want to know if the purpose of the appraisal is a pending sale, estate settlement, or perhaps the beginning of a foreclosure process.

The appraiser will then begin to collect and analyze market data similar to what we discussed in Chap. 2. For investment real estate income producing properties, the appraiser will desire to determine the market vacancy, market rental, and market operating expense averages so the property specific information collected from the current owner of the property can be compared with *the collective* market averages.

The appraiser will then visit the property, which should include entering the property to obtain pictures and a description of the internal property improvements. The appraiser will specify the effective date of the appraisal, will typically apply three approaches to value, and then will reconcile the results for a final appraised value determination. This value represents the appraiser's opinion as to the value of the property on a specified date.

4.2 The Three Approaches to Value

An appraisal for commercial investment property will typically contain three approaches to value: sales, cost, and income. As the income approach is represented by the bulk of what follows in this book, we will only discuss this at a high level in this section. The sales (or market) approach is the primary approach utilized in residential valuations, and is a secondary approach in commercial appraisals. Most readers would be familiar with this approach if they have ever reviewed the appraisal performed on their primary residence. The cost approach is utilized in order to determine the remaining economic useful life for the subject property. Valuation theory suggests that the three forms of value should come to similar conclusions. During the height of the "Yes Era," the sales approach values often far exceeded the other two methods, which provided evidence that markets were imbalanced and that the prices that investors were paying during this time were not supported by the net operating income produced by the property. The following sections will discuss each of the three valuation methods.

4.2.1 The Sales Approach to Value

The sales approach has the goal of estimating the sales price per square foot of properties deemed comparable to the subject property being valued. As mentioned earlier, in order to be considered a comparable sale, properties must have been sold via arm's length transactions, with properly motivated and informed buyers and sellers, and the sale must not have been made in duress (such as in a foreclosure or forced liquidation). Pending sales should generally not be used when compiling a list of comparable sales.

The geographic fit for possible comparables will depend on the type of property in question. For residential properties, or indeed for any property that is subject to local market risk, the comparables selected should be proximate to the property being valued. Since "location" represents each of the top *three* most important factors for real estate valuation in the old joke, properties with local tenants should be compared with other properties in the same general area in order to be considered quality comparables.

It may be valid to consider market transactions outside the immediate market if the property has a single tenant with a regional, national, or international reputation. For example, if an appraiser is valuing a Walgreens retail drug store location, and the lease is to Walgreens Corporation, as opposed to being to a local operator as franchisee, then it is entirely reasonable to extend the geographic sampling of recent transactions when selecting comparable sales.

Sometimes properties located in close proximity to the subject property are not truly representative of the subject property. Fort Lee in Bergen County, New Jersey is the town which connects the George Washington Bridge to New York City. Properties located in Fort Lee, which is considered a vibrant market, are not truly comparable to properties located just across the Hudson River. Fort Lee sales may nevertheless come up when an appraiser conducts a ring study of property transactions within a 5 mile radius to the New York City property being valued. This example could be replicated in many markets where suburbs contain different underlying economic conditions than do the cities that the suburbs support.

Another important issue in the sales approach concerns how recent is the transaction. The timing of the transactions may depend on the quality of the performance of the local market rather than any hard and fast rules. If the subject property is located in a stable market, the age of the sales can be older than if the market was depressed or declining. In the aftermath of the "Yes Era," finding quality comparable sales proved difficult in many markets as the volume of transactions generally declined, and there was a large difference between sales prices per square foot made before and after the financial crisis which ensued in late 2008 (Goddard 2010).

Once a reasonable list of properties has been compiled, the next step is to apply an adjustment factor per square foot for any material differences between the comparable property and the subject. Adjustments might be made for differences in size, location, age, condition, quality of construction, and amenities. Since the comparable sales are historical, a negative adjustment to the final sales price per square foot for the comparable should be made when the comparable exhibits a more favorable component relative to the subject (Appraisal Institute 1996). If the subject is deemed superior to the comparable property, the sales price per square foot may be increased in compensation. The sales approach does have an element of subjectivity, and how effectively the appraiser documents their rationale impacts the effectiveness and quality of this approach to value.

As an example, assume that the subject property consists of a 52,000 square foot single-tenant industrial property located at 2501 Moya Lane, Norfolk, Va. The appraiser located four properties, deemed comparable, that will serve as the basis for the sales approach. The four properties are shown in Figure 4.1.

As you can see from Figure 4.1, the weighted average sales price per square foot is $23.58. If this sales price per square foot was used to value the 52,000 square foot subject property, the value would be $1,226,000 (rounded). A glance at the proposed comparables in Fig. 4.1 reveals that the second property, on Abernethy Street, is almost twice the size of the subject property. The first property in the comparables list sold for a higher price per square foot, which could be due to the prestigious "Ghent" location in Norfolk. The fourth property is located in Portsmouth, which is separated by a river from Norfolk, and is generally considered a lesser quality location as compared to Norfolk. In any event, the appraiser might decide to adjust the sales price per square foot based on the size and location differences between the subject property and the comparables. Figure 4.2 shows the adjustments and final sales prices per square foot which are relevant for the subject property.

The Ghent property was adjusted downward as location advantages were deemed superior for this property relative to the subject property. The Portsmouth property on Britt Avenue received a 50% upward adjustment as its location was deemed inferior to the subject property by the appraiser. The Abernethy Street

4.2 The Three Approaches to Value

Address	City	Size	Sales Price	Date	Price psf
1000 Ghent Avenue	Norfolk	50,000	$1,500,000.00	Nov-10	$ 30.00
1500 Abernethy Street	Norfolk	100,000	$2,250,000.00	Feb-11	$ 22.50
3200 Kivett Drive	Norfolk	75,000	$1,875,000.00	Jan-11	$ 25.00
2500 Britt Avenue	Portsmouth	88,000	$1,750,000.00	Oct-10	$ 19.89
		313,000			$ 23.56

Fig. 4.1 Example of the sales approach valuation method

Address	Reason	Adjust %	Adjusted Price psf
1000 Ghent Avenue	Area/Quality	-15%	$ 25.50
1500 Abernethy Street	Size	10%	$ 24.75
3200 Kivett Drive	None	0%	$ 25.00
2500 Britt Avenue	Area/Quality	15%	$ 22.87
			$ 24.40

Fig. 4.2 Adjusted price per square foot for sales approach valuation

property received a 10% positive adjustment to its market determined sales price per square foot as the appraiser deemed the larger location to be a disadvantage when compared with the subject property. Appraisers should document the reasons for their adjustments to the market based sales prices per square foot, and they should also cite the rationale for why a certain percentage adjustment was applied. Investors might question the logic of the sales price adjustments, and this represents an opportunity to seek further clarification from the appraiser and/or the bank lender.

Based on the adjusted price per square foot of $24.40, the subject property would be valued at $1,270,000. Appraisers typically round their estimates of value as sales prices are usually rounded in this fashion. This would represent the value of the property based on the sales approach to value.

4.2.2 The Cost Approach to Value

The second form of valuation is the cost approach. This form of valuation is primarily useful for new construction properties. The cost approach is also useful for existing properties, as the appraiser will offer comment on the effective age of the property via the cost valuation method. The effective age of a property is the remaining economic useful life of the property given the quality of construction and improvements made to the property since it was originally constructed. The effective age can differ from the actual age of the property depending on the level of maintenance and modern amenities of the property. Thus the old adage applies that you are only as old as you look and feel! For example, the Empire State Building in New York City was constructed in 1931, and at the time of opening was the tallest building in the world (New World Encyclopedia 2011). While this sounds like a property that might not serve as a good long term investment given its calendar age, the property clearly represents a "timeless" property that would prove to be a strong long term investment in any portfolio. While the property is certainly old in actual years, the property continues to thrive as an office property, which should continue for the foreseeable future.

The theory based on the cost approach is that a rational investor would not pay more for a property than the value of the land plus the replacement cost to build a similar property. Since the cost approach requires that land and improvements are valued separately, the approach is also useful for insurance purposes (Appraisal Institute 2008). Hazard and flood insurance policies are traditionally insured up to the replacement cost of the building, not including the land value. Keep in mind that replacement cost does not necessarily mean the cost of reproducing an exact replica, with the same construction components that were used to construct the original building. For an existing property, the idea behind the cost valuation method is to take the replacement cost of the building based on construction materials and methodologies available in the current day, and subtract out for wear and tear to the building. In the real estate investments course, we tell our students that unless they become real estate appraisers, they will most likely not have to ever construct the cost based valuation of a real property. In any event, an example of this technique is instructive as it will elucidate the practical usage of terminology associated with the cost approach, and will allow the reader to understand how this approach to value may be configured.

We will now illustrate how the cost approach might be utilized when valuing our 52,000 square foot industrial property in Norfolk. First, a series of assumptions heretofore not disclosed must be made concerning the subject property. These are summarized in Figure 4.3.

The building was constructed 5 years ago with an initial economic useful life of 50 years. The building to value ratio was 85%, which let's you know how much of the value of the property is from the building as opposed to the land value. The appraiser would assume that given the age of the property and given forecasts for competing space in the area, that the building over time would experience a functional obsolescence of $15,000 per year in rent loss. This is another way of saying that over time, as the property ages, there will be some level of deterioration in the value of the property which cannot be recouped via internal or external improvements to the property. There could be a situation where the height of the roof is no longer of the level desired by business tenants, or it could be due to the shape or size of the building generally in the future not being considered "modern" by business tenants. The appraiser would also estimate the cost for deferred outlays annually, so that the non-repairable rent loss can be projected over the remaining economic useful life of the building.

Once these initial assumptions are made, other assumptions must be made concerning the level of curable physical depreciation and the level of curable

4.2 The Three Approaches to Value

Building Age	5	Years
Original Econ Useful Life	50	Years
Building to Value Ratio	85%	
Functional Obsolescence (non-repairable) rent loss	$ 15,000.00	Per year
Cost for deferred outlays annually	10%	

Fig. 4.3 Initial assumptions for constructing a cost approach valuation

Calculation of Incurable Physical Depreciation	
Reproduction Cost of Building	$ 1,500,000
Less: Curable Physical Depreciation	$ 100,000
Less: Curable Functional Obsolescence	$ 50,000
Balance subject to Depreciation	$ 1,350,000
Incurable physical depreciation (5/50 or 10%)	$ 135,000
Calculation of Depreciated Cost	
Reproduction Cost of Building	$ 1,500,000
Physical Depreciation:	
Curable	$ 100,000
Incurable	$ 135,000
Functional Obsolescence:	
Curable	$ 50,000
Incurable	$ 147,942
Value of Building	$ 1,067,058
Land Value	$ 225,000
Total Value Estimate	**$ 1,292,058**

Fig. 4.4 Calculation of the cost approach valuation

functional obsolescence. These are things that can be improved via repairs and improvements to the building. Figure 4.4 summarizes this approach. The cost approach starts off by estimating the reproduction cost of the building. This is where the appraiser estimates the current cost associated with building the property today. Then curable and incurable physical depreciation and functional obsolescence should be calculated to arrive at the value of the building today. Physical depreciation includes any deficiencies in structural quality or issues associated with the aesthetic appeal, while functional obsolescence involves weaknesses in design owing to technology, style or current tastes and preferences. Obsolescence can also be external, which relates to externalities associated with pollution, noise, and any neighborhood issues. The final value then includes an estimate for land values. The land value is determined in much the same way as in the sales approach. Comparable land parcels are found in the area, which are zoned for a similar purpose as the subject property. Unlike a building structure, land does not depreciate. Based on the sales price per square foot, the value of the land is estimated.

Once the appraiser has determined the level of curable and incurable physical depreciation, as well as the curable level of functional obsolescence, the incurable functional obsolescence is calculated based on the remaining economic useful life of 45 years (N), with annual rent loss of $15,000 (PMT), and a future value of zero. The present value under these assumptions is $147,942. The cost approach typically begins with an assessment of reproduction cost, which is defined as the cost to construct, at current prices, a duplicate or replica of the property being appraised. The reproduction would use the same materials, construction standards, design, layout, and quality of workmanship that is present in the subject property. Given this assumption, there will be some level of depreciation and obsolescence which will be calculated in order to determine the replacement cost of the building.

Building costs may be estimated using the comparative-unit method, the unit-in-place method, or the quantity survey method. In the comparative-unit method, costs are expressed in dollar terms per unit of area or volume, and are based on known costs of similar structures that are adjusted for market conditions, location, and physical differences. The unit-in-place method, also known as segregated cost, estimates the total building cost by adding together the unit costs of the various building components as installed. The quantity survey method is the most comprehensive method of estimating total cost. The quantity and quality of all materials used and all categories of labor required are estimated and unit cost figures are applied to arrive at a total cost estimate for labor and materials (Appraisal Institute 2008).

The value of the property based on the cost approach shown above is calculated at just under $1.3 million. While calculations in the cost approach are not particularly intuitive, the approach has relevance for lenders as it helps them to determine how long a loan amortization should be for a given property.

4.2.3 The Income Approach to Value

The third approach to value is the income approach. The case studies and chapters that follow in this book will utilize this valuation methodology. Depending on the complexity of the property in question, there are various methods utilizing the income to evaluate an investment property. It is important to specify that the property must be an investment property in order to utilize the income approach to value. Should the property serve as the primary residence of the owner, or serve as the primary site of business operation for a company, the income approach should not apply. Similar to the requirement that both appraisers and transactions be independent or at an arm's length, an investment property should have a tenant or tenants that have separate ownership than does the real property. In Chap. 5, we will discuss leases at length, but for our purposes now it should be noted that an investment property is defined as a property that receives rental income from a third party tenant.

4.2.3.1 Gross Income Multiplier
The simplest method of performing the income approach valuation is the gross income multiplier (GIM). The GIM is a blend of the sales and income approaches, and is typically utilized in appraisals for single family residences or for generally

Address	Sales Price	Gross Income	Multiplier
1000 Ghent Avenue	$1,500,000.00	$ 150,000	10.00
1500 Abernethy Street	$2,250,000.00	$ 250,000	9.00
3200 Kivett Drive	$1,875,000.00	$ 175,000	10.71
2500 Britt Avenue	$1,750,000.00	$ 185,000	9.46
		Weighted Average	**9.70**

Fig. 4.5 Gross income multiplier calculations

smaller properties with stable income streams. In this approach to value, the sales price is divided by the gross income of the property, in an effort to obtain a market based multiple which can be used to set the current value of a property. The GIM differs slightly from the gross rent multiplier (GRM), as the GIM does not specify that the income must be from rents, as does the GRM. While the GIM is categorized as an income approach valuation, it only considers the gross income of the property, and not the effect of any expenses paid by the property owner. If an appraiser had a subject property with a single tenant with a strong national credit rating and a very long lease, the GIM might be employed. Given its simplicity, other income approaches as discussed in this section should also be utilized in order to obtain a quality valuation. Comparables similar to those in Fig. 4.1 would be used, but the appraiser might find that not all of the comparable properties were investment properties, further limiting the GIM approach.

For illustration purposes, we will revisit our Norfolk example utilizing the comparables used in Fig. 4.1. This time, we will also include the gross income for each property in an effort to obtain the gross income multiplier as shown in Figure 4.5.

Based on the comparable properties in the market, a gross income multiplier of 9.70 is obtained. Our subject property consists of a 52,000 square foot building on 2 acres of land and is (hypothetically) located at 2501 Moya Lane, Norfolk Va. The property has a single tenant paying annual rent of $3.00 per square foot with a lease that matures in 10 years. Based on the gross income multiplier approach to value, the property would be valued at $1,513,200 as is shown below:

$$52,000 \text{ square feet} \times 3.00 \text{ per square foot rent} \times 9.70 \text{ GIM}$$

The value produced by the GIM is higher than what was obtained for the sales and cost approaches to value. We will now review two more detailed income approaches that consider the effect of property expenses when determining value.

4.2.3.2 Direct Capitalization

Another common form of income approach valuation is the direct capitalization model. This model utilizes both the income and expenses from the property, along with market determined characteristics in order to obtain a property value. As was discussed in the mini-case on operating expenses at the end of Chap. 3, not all expenses that are reported on the tax returns for a property (whether held individually or in a corporate entity) would be included in the analysis. Expenses such as interest, depreciation, amortization, and depletion are commonly not a part of the property valuation. Interest expense is not included as a primary tenet of real estate valuation is that the value is irrespective of the amount of indebtedness on the property. In other words, the value of the property should not change if the owner has a mortgage on the property or not. As discussed in Chap. 3, an investor's return would change depending on whether a property is financed, but the value of the property should not be affected. Expenses such as depreciation and amortization are non-cash expenses, so they are not part of the valuation process for real estate valuation. In a later chapter, we will discuss the implications of depreciation and interest expense on taxes and the internal rate of return.

The direct capitalization model applies a capitalization rate to the estimated net operating income in order to obtain value. The net operating income is what is expected over the next year, and the cap rate should be reflective of market conditions. A capitalization rate is any rate that applies value to a stream of income, and is found by dividing the net operating income of the property by the sales price. The higher the cap rate, the more risk is inherent in the property, and the lower the property value. How an appraiser determines the cap rate will be discussed later in this chapter.

Returning to our 52,000 square foot industrial property in Norfolk, the property contains one tenant occupying the entire property paying annual rent per square foot of $3.00. Let's assume that the tenant is in good financial condition and is thus expected to remain as tenant for the duration of their remaining lease term. The lease has 10 years remaining, and the tenant is responsible for paying all operating expenses other than a management fee and a replacement reserve. The management fee is the cost to the owner for managing the property on a daily basis. Since the property has one tenant with a long lease term, the primary responsibility of the property manager would consist of keeping the tenant happy and free from general disturbance in their business operations. The fee is typically assessed as a percentage of effective gross income (EGI), which is the gross income from the property minus a vacancy factor to account for market vacancy rates for comparable properties, and a reduction for collection loss. Collection loss is defined as a reduction in expenses due to trouble collecting rent, or possibly due to lease rate concessions during the year. Since the property owner is responsible for repairs to the structure of the building over the term of the tenant's lease, the replacement reserve is similar to an escrow account where funds are accumulated over time for the purpose of ameliorating any structural damage which might occur. Replacement reserves will be further described in the mini-case at the end of the chapter.

Figure 4.6 illustrates the direct capitalization approach for our Norfolk industrial property. The property produces gross income of $156,000 annually, and the effective gross income is calculated at $152,880, after a reduction for frictional vacancy of 2%. Frictional vacancy is defined as the small amount of vacancy that exists in even the best markets given the movement in and out of tenants in investment properties. Should the subject property be a multi-tenant property, the

4.2 The Three Approaches to Value

Gross Potential Income	$	156,000	
Vacancy Factor		3,120	2% Frictional Vacancy
Effective Gross Income		**152,880**	
Management Fee		4,586	3% of EGI
Replacement Reserve		3,058	2% of EGI
Total Operating Expenses		**7,644**	
Net Operating Income		145,236	
Cap Rate		11.50%	
Value		1,260,000	

Fig. 4.6 Direct capitalization for Norfolk industrial property

vacancy factor employed would likely be higher to reflect the true market averages. Since we have a single tenant of assumed financial soundness and a long remaining lease term, the frictional vacancy rate was applied rather than the market vacancy rate. As an investor or a lender valuing the investment property, the vacancy factor accounts for the probability that even the best located and most modern facilities will eventually revert back to the mean market performance over time. Once the vacancy factor (and any collection loss) has been subtracted from the gross potential income of the property, the result is the effective gross income.

Variable expenses such as a management fee and replacement reserve are typically calculated as a percentage of EGI, or possibly on a per-unit or per square foot basis. Calculating management fees as a variant of EGI should be intuitive, as property managers should be incentivized to keep the property as fully leased as possible, thus having their fee as a variable expense helps to keep them property motivated. Typically, investment property operating expenses would also include real estate taxes, insurance, utilities, repairs and maintenance, and any other property related expenses. Based on the structure of the lease agreement with the single tenant, only a small management fee and a replacement reserve is subtracted from EGI to arrive at net operating income (NOI).

Once the total operating expenses have been determined, they are subtracted from EGI in order to arrive at NOI. The property value for the income approach is determined by dividing NOI by the market based cap rate. For the Norfolk property, a value of $1,260,000 is determined via the direct capitalization approach.

4.2.3.3 Discounted Cash Flow

The final variation on the income approach which we will introduce in this chapter is the discounted cash flow (DCF) model. The DCF is a multi-year view of the income and expenses of the property which are received during the defined holding period of the investment. The cash flows received by the investor are shown in a similar fashion to the IRR figures from Chap. 3, where the first year of the projected income and expenses are shown in period one, while the income received from the eventual sale of the property is shown in the final year of the holding period of the investment. The net cash flow shown in each year is discounted to the present time to determine value.

Similar to the prior reviewed income approach models, the DCF technique values the property irrespective of whether the property is indebted and regardless of any consideration of tax benefits from accrued depreciation. Thus the appraisal DCF does not include interest, depreciation, or amortization in the operating expenses. For multi-tenant retail, office, and industrial properties, the DCF does include additional expenses typically not calculated in the direct capitalization model: *leasing commissions* and *tenant improvements*. These items are best estimates for the costs incurred during the holding period in order to attract and obtain new tenants in a given property. In order to calculate the leasing commissions and tenant improvements, assumptions must be made for the probability of tenants renewing their leases, as well as the costs of improving the vacated space, and the costs paid to brokers in order to find replacement tenants.

For an overview of the DCF approach, let's return to our Norfolk industrial property. Figure 4.7 summarizes the results of the DCF valuation model. The DCF calculation is for the same single tenant property with the long lease that we have discussed in this chapter. Given the length of the lease for the single tenant, leasing commissions and tenant improvement expenses are assumed to be zero. The holding period of the investment was assumed to be 5 years. This considers the 10 years remaining on the lease, and the desire for the eventual buyer to have a certain cash flow for the first few years of their holding period. Utilizing the approach discussed in Chap. 3, a discount rate of 15% was selected. Given the relatively high discount rate, it can be assumed that there may have been a gap between the appraiser's knowledge of the single tenant's financial condition and the level of information that would have provided more comfort for their sustained profitability over the holding period.

The appraiser assumed an annual revenue increase of 2.50%, which may have been based on actual stated increases in the lease, or assumptions about the future of

	Year 1	Year 2	Year 3	Year 4	Year 5	Reversion	Year 6	Notes
GPI	$ 156,000	159,900	163,898	167,995	172,195		176,500	Rise by 2.5% per year
Vacancy	3,120	3,198	3,278	3,360	3,444		3,530	2% of EGI
EGI	152,880	156,702	160,620	164,635	168,751		172,970	
Mgmt Fee	4,586	4,701	4,819	4,939	5,063		5,189	3% of EGI
Rep Res	3,058	3,134	3,212	3,293	3,375		3,459	2% of EGI
Total Op Ex	7,644	7,835	8,031	8,232	8,438		8,648	
NOI	145,236	148,867	152,589	156,403	160,313	1,643,212	164,321	10% Terminal Cap
PV	$ 126,292	$ 112,565	$ 100,329	$ 89,424	$ 79,704	$ 816,967		
Value	$ 1,330,000							

Fig. 4.7 Discounted cash flow for Norfolk industrial property

4.2 The Three Approaches to Value

the consumer price index (CPI). The CPI is a typical benchmark for lease rate increases, with the goal of the lease payment received at least keeping up with expected price increases over the term of the lease. The frictional vacancy assumption of 2% was held constant over the holding period, as were the expense assumptions relative to EGI for management fees and replacement reserves.

The primary point of discussion in this simple DCF example is the reversionary calculation. Since the property is assumed to be sold at the end of the 5 year holding period, the appraiser assumed that the terminal cap rate, or the cap rate based on the sale of the investment, would be 10%. The reversion value of $1,643,212 was calculated by dividing the NOI for period six by 10%. Period six is here defined as the first projected year of operating performance for the eventual buyer of the property. Since the reversion is assumed to occur at the end of period five, both the year five NOI and the reversion are discounted back to the present time at $(1.15)^5$ or 2.011. The DCF model has estimated the value of the property to be $1,330,000. This is slightly higher than the estimated value under the direct capitalization. The difference appears to be due to the reversionary cash flow calculated at the end of the fifth year.

A final point on the income approach deserves comment. The appraiser will also provide commentary as to whether the lease rate being paid by the tenant is within the market averages or not. Additionally, the appraiser will make commentary on the overall market vacancy rates. Should the $3.00 per square foot lease rate be deemed to be above the market lease rate, the appraiser would provide an income approach analysis using market rental rates to value the subject property. This form of valuation is called a fee simple valuation, while the income approach based on the leases currently in place is called a leased fee valuation. For our example, it is assumed that the lease is within the market averages, so no variance exists between the leased fee and the fee simple valuation models.

4.2.3.4 Final Value Conclusion

Once all three forms of value have been completed, the appraiser will reconcile the values into a final conclusion of value. This final value represents the appraiser's opinion of value. During the "Yes era," many times the sales approach valuation was much higher than the other two forms of value. Based on the preceding discussion of the valuation methods, it is clear that the sales approach would produce higher valuations when markets are vibrant. For our Norfolk industrial property, the values and reconciled value conclusion are shown in Figure 4.8.

Our analysis actually produced four values, as we employed both the 1 year snapshot (direct cap) analysis as well as the multi-year (DCF) analysis for the income approach. The appraiser might take a rounded average of the four values to arrive at a final value conclusion of $1,290,000. In cases where the forms of value return widely disparaging results, the appraiser would use their judgment of the highest and best use of the property in order to determine which value is best representative of the subject property in question.

Fig. 4.8 Reconciliation of value

Sales Value	$1,270,000.00
Cost Value	$1,292,058.00
Direct Cap Value	$1,260,000.00
DCF Value	$1,330,000.00
Average	**$1,290,000.00**

From the preceding discussion, it should be obvious that the selection of a capitalization rate (either a "going in" cap rate at the beginning of an investment, or a terminal cap rate) is an important consideration of value for investment properties. In the following section, we will discuss various methods used in order to derive cap rates for investment real estate.

4.3 Introduction to Cap Rates

Cap rates are essential in valuing an investment property utilizing the income approach (RMA 2011). A cap rate is essentially any rate used to convert income into value. Cap rates are used in the income approach valuation for commercial investment properties. Once the net operating income (NOI) for a subject property is determined, this figure is divided by the cap rate in order to estimate a property's value. Lenders and investors spend a lot of time validating the numbers that comprise NOI, but often do not spend much time understanding how the cap rate was calculated. The cap rate is a very important component of the total value of a property. For example, if a given property produces net operating income of $250,000 based on an appraiser's estimates of what is *anticipated* for the next year, and a 9% cap rate is applied to that NOI, the resultant property value is $2.77 million. If cap rates increase to 11%, the property value based on the same $250,000 NOI would drop to $2.27 million. Thus a 200 basis point (bp) movement upward in cap rate erodes the property's current value by $500,000. Anticipation is a very important element in the income approach, as the selling price of an investment property is based on what a reasonable investor would be willing to pay, which is based on the future earnings potential of the property. Similar to the requirement of realistic projections of revenues and expenses being needed in order to obtain a reasonable estimate of NOI, having a realistic basis for cap rate construction is necessary to obtain a reasonable opinion of property value.

During the run-up in bank lending from 2001 to 2007, the appraisal industry moved closer toward the singular reliance on market averages for cap rate determination than seen in prior years. While most appraisers today use more than one method to derive a cap rate, each of these methods is based on utilizing the recent past as means to project the current and future environment. Common methods include market extraction, lender's yield, and citing commonly referenced market surveys. Each of these methods utilizes market conditions seen in the recent past to

4.3.1 Market Extraction Method

This approach, also known as market abstraction, attempts to find comparable properties relative to the subject property, which were sold in the recent past, to validate the cap rate assumption today. The following table provides an example of this approach:

Address	City	Sale price	Sale date	Size	Year built	NOI	CR (%)
100 North main street	High point	$ 950,000	Apr-08	10,000	2005	$ 72,000	7.58
200 Spring garden street	Greensboro	$ 2,000,000	Mar-08	22,000	2001	$ 150,000	7.50
300 Reynolda road	Winston-Salem	$ 4,400,000	Oct-08	50,000	1990	$ 345,000	7.84
400 Tabor view lane	Greenville	$ 1,950,000	Nov-08	15,000	2006	$ 155,000	7.95
						Min	7.50
						Max	7.95
						Mean	7.72

This is typically one of the primary methods of obtaining cap rates for commercial investment properties. During stable or rising markets, this method appears adequate, but what about its relevance in a declining market?

As discussed previously, the volume of commercial real estate transactions since the end of the "Yes era" have slowed down considerably. There may not be recent sales in some markets in order to utilize market extraction. Unfortunately, a lack of recent sales has not stopped appraisers from continuing to dip into the well. Assume that an appraiser is valuing a property that consists of a 20,000 square foot office building built in 2008 located in Winston-Salem, North Carolina. As you can see from the table above a few items come to the surface:
- All of the sales are old and are pre-financial crisis
- Only one property is located in the same city as the subject property
- Only one property is approximately the same size as the subject property
- All properties are older than the subject property

This is an example of the weakness of market extraction. Nonetheless, in a market without current sales, many appraisers are still utilizing this approach based on pre-financial crisis sales. Before accepting an appraisal, lenders and investors should determine if old comparables, or perhaps even pending sales, have been used in order to validate the cap rate. Sometimes appraisers attempt to update the market extraction numbers with pending sales, but these are not finalized and should not be used as the basis for cap rate construction.

Regarding the comparables used above, it is sometimes reasonable to use out of market comparables if the NOI being produced is from tenants with a regional or national reputation. For example, if the subject property was a 10,000 square foot Walgreens retail store location, it very well could be reasonable to use comparables from another city, although the date of the sales in this case are still highly questionable.

Because the volume of transactions has fallen, and since many of the novice investors, or investors unable to comply with stricter bank lending parameters requiring more equity in deals than in the past, have fallen out of the market, the few transactions that are occurring today are of better quality than those seen pre-financial crisis. These deals often represent a *flight to quality* among investors, so even the few sales that have occurred in your home market may be of a better quality than those transactions in the recent past.

4.3.2 Market Survey Method

Given the issues with the market extraction method, appraisers often will follow up this approach by citing national publications such as the PwC Real Estate Investor Survey or some other publication. These publications provide quarterly updates for metro market vacancy rates, rental rates, cap rates, and discount rates by property type. If your home market is located outside of a major metro market, these surveys do not provide market specific information. For example, if you are reviewing a retail property located in Boone, NC or Asheville NC, two thriving markets in western North Carolina, some well known third party market research providers have historically not surveyed those specific locations. Limited information is available for other non-metro markets. Market participants in these smaller, but often high transaction volume markets, such as Savannah, GA or Greenville, NC, are left to extrapolate the current cap rates based on either utilizing the national average, or the survey results for the closest major metro market.

A recent appraisal employing market extraction used Korpacz surveys to show that the national average for warehouse cap rates increased between 2008 and 2010. The older comparable sales were added to the average national increase in cap rates since the time period in between the current day and the dates of the comparables, and a "current" cap rate was extrapolated. For example, the national average for warehouse space per the Korpacz market surveys increased by 170 basis points between the second quarter 2008 and second quarter 2010 (Korpacz 2010). If your dated comparables produced cap rates of 7%, the national average could be used to increase the cap rate in the current appraisal to 8.7%.

4.3 Introduction to Cap Rates

While using market surveys to increase cap rates in this fashion helps to improve the validity of the cap rate being constructed, it still is relying on prior sales data which is often simply not pertinent to the market being surveyed. National averages in terms of cap rate movements are helpful in order to assess high level trends in the market, but using these national averages to update local markets where sales have not occurred seems less than desirable. It should be noted that the method of adding to the market extraction cap rates based on national averages typically has been seen by the authors when appraisers are valuing a property in a foreclosure or workout situation.

4.3.3 Lender's Yield Method

A third method used to derive cap rates in commercial real estate appraisals is the lender's yield method. This approach is based on the assumption that current lending requirements can be used in order to estimate overall cap rates (CR) by multiplying the debt coverage ratio (DCR) by the mortgage constant (MC) and the loan to value ratio (LTV). Appraisers often query lending officers as to the current LTV and DCR that they are using for particular types of commercial property. The mortgage constant is the ratio of the annual debt service to the initial loan amount.

For example, the mortgage constant on a loan charging 7%, amortized over 20 years is 0.0930. If typical bank lending requirements were providing a DCR of 1.25 and an LTV of 75%, the lender's yield cap rate would be:

$$0.0930 * 1.25 * 0.75 = 8.72\%$$

As many readers will attest, the *credit crunch* was appropriately named, because the lending parameters seen in many markets today differ substantially with the pre-financial crisis levels: amortizations are shorter for most property types, and loan to value ratios are lower. This "crunch" has led to more equity being required in deals, and ceteris paribus lower loan to value ratios.

Given the "credit crunch," your bank may now be lending on retail properties with debt coverage ratio requirements of 1.40 and LTV of no more than 65%, and these properties might be financed with loans charging 6% with amortizations of 15 years. What effect would this have on the lender's yield cap rate method?

$$0.1013 * 1.40 * 0.65 = 9.22\%$$

Consequently, the cap rate under the lender's yield method has increased, but unfortunately, many appraisers are still using a DCR of 1.25 and LTV of 75% as estimates for this approach. It seems that during the heady days of 2001–2007, appraisers followed banks out on the risk curve as shown in this approach, but the appraisal industry has been slow to adjust during the down market.

When applying lender's yield, it is important to query your appraiser if the DCR and LTV assumptions do not appear to represent the reality of the current environment.

In summary, the three most popular methods of cap rate determination in commercial real estate appraisals are all applying the recent past, and have not proven as reliable in a market environment of sustained reduction in property values.

4.3.4 Band of Investments Technique

It would seem that the sustained period of increasing property values (2001–2007) has led to a reliance on market based cap rates, which have been lower than should be expected going forward. These deflated cap rates have, in turn, generated inflated property values in many appraisals. As the appraisal industry moved with the times during the boom years, more academic cap rate models were abandoned as they were considered outmoded or too cumbersome to be of use in a market that was expected to continue to appreciate into the foreseeable future.

One approach used more frequently in the past is the band of investments technique. This is a variation on the lender's equity method, and it can be used in place of the method discussed above. In applying the band of investment, both a debt component and an equity component are used in order to construct the cap rate (CR).

$$\text{The formula is as follows: } CR = d * MC + (1 - d) * R_E$$

Where d is the debt ratio, or loan to value (LTV) ratio, MC is the mortgage constant, $(1-d)$ is the equity ratio, and R_E is the investor's required rate of return on the equity investment, or the equity dividend rate. Appraisers today are using benchmarks to estimate the components of this technique. There are various internet sites that provide aggregate information concerning interest rates, loan amortizations, and equity dividend rates that are seen in the market. These various sites obtain their information from survey data, which is similar to the weaknesses identified in other approaches. As mentioned previously, national averages simply are not reflective of many local market conditions.

We propose amending the band of investment technique so that transaction specific factors can be used to derive the appropriate cap rate. For example, assume that you are reviewing a property with an assumed LTV of 65%. The purchase price of $1,400,000 includes a $910,000 loan paying 7% amortized over 15 years. The property is expected to produce pre-tax annual cash flow (NOIe) of $50,000 during the first year.

One of the difficulties of applying the band of investment technique, or any method that explicitly employs the required return on equity, is estimating R_E. In the following example, allow the estimated NOIe from the project to determine the rate of return expected by the investor, so that $R_E = $ NOIe/equity investment.

4.3 Introduction to Cap Rates

The rationale for substituting for R_E is summarized as follows: the bank will know the investor's projected NOI, and will also know how much equity the investor wishes to invest when purchasing the property. Knowing the equity requirements along with the purchase price allow the appraiser to estimate NOIe, which is the pre-tax annual cash flow; or NOI after debt service. In situations where the bank determined NOI differs significantly from the investor determined NOI, the loan would not be approved and an appraisal would not be engaged. Thus the appraiser's NOI, in conjunction with the borrower's equity investment and the projected annual debt service on the loan, can be used to set the required return on equity percentage. Rather than relying on third party sources for nationally determined return on equity percentages, this alteration allows the local market to help set the equity component of the calculation.

Debt component	Equity component
LTV% * (annual debt service/loan amount)	1-LTV% * (NOIe/$ equity invested)
0.65 * ($98,152/$910,000)	0.35 * ($50,000/$490,000)
0.65 * 0.1079	0.35 * 0.1020
= 0.0701	= 0.0357
Thus the total cap rate is 10.58% as per this approach.	

This variation on the band of investments eliminates the need for survey data in order to estimate the required return on equity, because this is now defined in terms of the specific transaction being valued. An appraiser will have a knowledgeable estimate of the projected NOI for the property, and will certainly know the purchase price. All that is needed when applying the approach is the bank's lending parameters, which is something that can be highly localized, in order to determine NOIe. Furthermore, the appraiser can use local market conditions to set the leverage ratios rather than relying on aggregate data from third party sources.

How would this approach fair in the aftermath of the financial crisis? This, of course, depends on how the equity component is constructed. If the equity component is determined via the specific transaction (i.e., NOIe/$ invested), then a decreased LTV might lower the equity dividend rate far below the initial preferences of the investor. If this is the case, the investor may not decide to proceed with the transaction. If the investor's equity dividend rate remains 10.2%, applying an LTV of 50% and lowering the interest rate to 6.5% results in a lower cap rate because less relatively expensive equity is used to finance the project. The calculation is shown in the following table.

Debt component	Equity component
LTV% * (annual debt service/loan amount)	1-LTV% * R_E
0.50 * ($73,172/$700,000)	0.50 * .1020
0.50 * 0.1045	0.50 * 0.1020
= 0.05225	= 0.05100
Thus the total cap rate is 10.4% as per this approach.	

Although the band of investment technique allows an appraiser to adjust for local market conditions and incorporates the investor's expected return, it lacks flexibility in that it fails to incorporate expected future outcomes such as declining income and/or property depreciation.

4.3.5 Ellwood, Akerson, and Archeological Finance

The mortgage equity technique is an older, equation based approach developed by L.W. "Pete" Ellwood in the late 1950s (Appraisal Institute 1996). The benefit of the mortgage equity technique (as well as a follow-up method by Charles B. Akerson) is that it can accommodate declining property values and/or falling income over the projection period of investment. Furthermore, Ellwood's equation explicitly incorporates the impact of financing choices made by the investor or required by the lender. Most current appraiser education does not include in depth discussion of the Ellwood or Akerson techniques as both are considered too cumbersome or too academic to be of practical use.

The Ellwood formula is written as:

$$CR = \frac{R_e - \text{LTV}\{R_e + \text{AP}(1/S_n) - \text{MC}\} - \Delta \text{PV}(1/S_n)}{1 + \Delta_I J}$$

Where CR is the capitalization rate, R_e is the investor's required return on equity, or the equity yield rate, LTV is the loan to value ratio, AP is the amount of the loan that has been amortized (paid off), S_n is the balance of a sinking fund set up to pay off expenses, MC is the mortgage constant, ΔPV is the expected change in the property's value during the holding period and the denominator, $(1 + \Delta_i J)$ adjusts for anticipated changes in the income generated by the property.

The Ellwood formula allows the cap rate to be adjusted for expected changes in future NOIe as well as changes to the property's value. In fact, if the NOIe and property value are expected to remain constant, the Ellwood formula produces the same result as the band of investment.

The term $R_e - \text{LTV}\{R_e + (AP * 1/S_n) - MC\}$ represents the traditional cap rate, which is the required return on equity with adjustments for debt financing. A higher LTV means that more relatively inexpensive debt is employed, which thereby lowers the overall rate. The next term has the same effect as AP captures the risk mitigating effects of an amortizing loan, while the sinking fund factor, $1/S_n$, serves as a discount factor for the investor's growing equity. However, as the interest rate on the debt increases or the amortization period shortens (i.e., to 15 years from 20 or 30 years), the mortgage constant (MC) grows, causing the cap rate to increase.

Let's now apply the Ellwood formula to the previous example, in which the loan is newly originated, so there is no pay down of the loan. As in the previous example, the interest rate is 7.00% on a 15-year amortizing loan, the required LTV is 65%, and the investor's required return on equity remains at 10.2%.

4.3 Introduction to Cap Rates

This generates the following cap rate:

$$50{,}000/490{,}000 - 0.65 * \{(50{,}000/490{,}000) - 0.1079\} = 10.58\%,$$

which is exactly the cap rate produced by the band of investments technique.

The remaining term in the numerator of the Ellwood formula, $\Delta PV(1/S_n)$, allows the appraiser to adjust the cap rate for anticipated changes in the property's value, where ΔPV is the expected percentage price change and S_n is once again the sinking fund factor. Notice that the anticipation of a decrease in the property's value increases the cap rate.

The denominator, $1 + \Delta_I J$, incorporates expected changes to the income generated by the property over the life of the project (Δ_I) as well as the potential for nonlinearity in its growth (J). For example, Δ_I would equal 0.30 if the appraiser expects the NOIe of the project to increase from \$100,000 in the first year to \$130,000 in the last. The J-factor is used for curvilinear accumulating income – the J-factor serves to discount accumulating NOIe in accordance with a sinking fund. While the formula for the J-factor is hardly intuitive, it is entirely identified by the required return on the equity investment (R_e) and the number of periods of the project's life. Consequently, J-factor tables and even websites with Ellwood calculators are available, so it can be easily applied if the appraiser believes a curvilinear income stream will be generated by the property. Because the J-factor is always positive, the impact of the denominator is entirely determined by the expected growth rate of the income (Δ_I). For example, if NOIe is expected to remain constant, $\Delta_I = 0$ and the denominator becomes $1 + 0(J)$, and regardless of J's value, the denominator is 1, which has no impact on the cap rate. Conversely, if $\Delta_I < 0$, the denominator is less than 1 and the cap rate will be driven upward to reflect the falling value of the project income.

Akerson also defined a "K-factor" that accommodates a linear increase in the project's NOIe. In this case, $1 + \Delta_I J$ is replaced by K. K is a function of the appraiser's assumption regarding the annual growth rate of the income, the required return on equity (R_e) and the number of periods the project will generate the income. Once again, tables are available to determine K.

So, what happens to the cap rate as explicit assumptions regarding the property's changing value and NOIe are incorporated? The following table provides comparison of Ellwood cap rates under various assumptions regarding expected growth rates of NOIe and property value. The calculated cap rates are based on a property priced at \$1.4 million that is partially financed with a fully amortizing 15-year loan, a required return on equity of 10.2%, and a projection period of 5 years. The base case NOIe is \$50,000.

Panel A of the table presents Ellwood cap rates given a loan rate of 7% with an LTV of 65%. The column labeled "Expected Change" provides the various changes in either the expected NOIe generated by the project, or the property's value during the projection period. The "NOIe Growth" column shows cap rates given the various changes in NOIe while holding the changes in the property's price to zero; "Property Change" holds NOIe changes to zero. Panel B is constructed the same way, but assumes an LTV of 50% and an interest rate on the proportionally smaller loan of 6.5%.

Ellwood cap rates based on various changes to NOIe and property value

	Panel A		Panel B	
	LTV = 65%, Loan rate = 7%		LTV = 50%, Loan rate = 6.5%	
	NOIe growth	Property change	NOIe growth	Property change
Expected change	No change in property value	No change in NOIe	No change in property value	No change in NOIe
20%	0.0741	0.0492	0.0763	0.0517
15%	0.0759	0.0574	0.0782	0.0598
10%	0.0778	0.0655	0.0801	0.0680
5%	0.0798	0.0737	0.0822	0.0745
0	0.0819	0.0819	0.0843	0.0843
−5%	0.0841	0.0900	0.0866	0.0925
−10%	0.0864	0.0982	0.0889	0.1006
−15%	0.0888	0.1063	0.0914	0.1088
−20%	0.0914	0.1145	0.0941	0.1169

Both Panel A and Panel B show that the cap rate is more sensitive to changes in the property's appreciation (or depreciation) than to changes in income. That is, an expected increase (decrease) of 20% in property value produces a much lower (higher) cap rate (i.e., 249 bp lower; 231 bp higher) than an equivalent expected change in the NOIe. This effect is more pronounced over shorter projection periods, because the change in property value arrives much earlier – a 20% appreciation in the property's value has much more impact on the viability of a project if it occurs in 5 years rather than 15 years. Consequently, over longer projection periods, a steady increase in income can have as much importance as an increase in the property's reversion value.

Comparing Panels A and B shows that shifting the LTV to 50% and lowering the rate by 50 bp does not create dramatically different cap rates. While the impact of the property's changing value remains quite high over short projection periods, the increase in the cap rates is relatively small. Interestingly, an LTV of 50% and a loan rate of 6% produces almost identical cap rates to those in Panel A.

The Ellwood and Akerson techniques have fallen out of favor ever since the creation of the discounted cash flow income approach technique that was described earlier in this chapter. We contend that these equations do still have some relevance for cap rate construction when markets experience significant property value declines or in an absence of market transactions. In the process of writing the paper for the RMA Journal where the information in this section initially appeared, it felt like we were trying to resuscitate a mummy, as the use of the Ellwood and Akerson techniques was relegated to the past. In a way, the attempt to go backwards to revisit these equations represents a key thesis of this book: in the aftermath of the "Yes era," going back to fundamentals is required so that the mistakes of the past are not repeated. This section on cap rates thus concludes with the creation of a new term: archaeological finance.

4.3.6 Conclusion

When you receive appraisals as part of the due diligence requirements of a loan, secured with an investment real estate property, the first step in assessing the quality of the cap rate assumption is to review the types of methods utilized in light of the *current economic environment*. Cap rate analysis today, which typically does not include a discussion of the probability of declining property values or changing future income, may be missing the mark. Cap rate valuation that includes the revised band of investments or possibly the Ellwood equation may lead to more meaningful results in a declining market or in a market without many current comparable sales. Relying on the recent past in a sustained down market provides an example of where, to quote the Danish philosopher Søren Kierkegaard, "the crowd is untruth" (Kierkegaard 1847).

Chapter 4 has served as an introduction to the three forms of value in an appraisal. As we have seen, there is much subjectivity in the determination of the current value of an investment property. Readers are reminded that the quality of any valuation is entirely driven by the quality of the inputs, and that the components of value should be based on the current market conditions.

Questions for Discussion
1. Describe the importance of an arm's length transaction when assessing the value of a property by the sales approach.
2. In your own words, describe the three approaches to value for investment real estate.
3. Elaborate on the various methods of cap rate construction. Make sure to discuss the strengths and weaknesses of each approach, especially in uncertain or declining markets.
4. Describe situations when the cost approach is helpful for investors, lenders, and insurance companies.
5. Discuss the strengths and weaknesses of the various income valuation techniques. When might the various techniques be appropriate from the perspectives of the investor, lender, and appraiser?

Problems
1. A retail strip center is expected to generate NOI of $175,000 for the first year. Rents and expenses are expected to grow at 2% per year. The property is expected to be sold at the end of 5 years.
   (a) What is the value of this property if the investors require a rate of return of 12%?
   (b) What is the value of this property if rents and expenses grow at 4% annually?
2. The net income for a single tenant office building is expected to remain constant at $200,000 per year for the next 5 years. Starting in year 6, the

lease calls for gross rent to increase which will increase NOI to $220,000. Thereafter, net income is expected to increase by 2% annually.
   (a) If investors require a 10% return, what is the estimated value of this property?
   (b) What would the value of the property have been if net income began at $200,000 and grew by 2% annually?
3. Clarke Investments is considering the purchase of an apartment building. After an extensive market study, the investment manager has estimated the annual NOI as follows: (year 1): $800,000 (year 2): $815,000 (year 3): $830,000 (year 4): $840,000 (year 5): $845,000 (year 6): $855,000. It is expected that beginning in year six and thereafter, NOI will reflect a balanced, stable market and should grow at 3% per year. The investment manager believes that the firm should earn a 12% return on an investment of this kind.
   (a) Assuming that NOI is produced for 6 years and that the investment is held for 5 years, what is the value of the property today?
   (b) What is the value of the property at the end of year 5?
   (c) What is the terminal cap rate at the end of year 5?
   (d) What is the going in cap rate based on year one NOI?
   (e) What explains the difference between the going in and terminal cap rate?
4. Mandelbaum Appraisal Company has been asked to provide a cost approach valuation for a suburban office building that is 5 years old. When constructed, the building was estimated to have an economic life of 50 years, and the building to value ratio was 85%. Based on current cost estimates, the structure would cost $3 million to reproduce today. The building is expected to wear out evenly over the next 50 year period of its economic life. Estimates of other economic costs associated with the improvements are as follows:

Repairable physical depreciation:	$100,000 to repair
Functional obsolescence (repairable):	$50,000 to repair
Functional obsolescence (non-repairable):	$15,000 per year

   (a) What is the value of the land?
   (b) What is the value of the property if the appraiser believes that an appropriate cost for any deferred outlays is 10% annually?
5. The NOI for a small warehouse is expected to be $135,000 in year one. Financing will be based on a 1.25 debt service coverage ratio applied to year one NOI, and will have an interest rate of 7%, monthly payments, and a 20 year loan amortization. The NOI is expected to increase by 3% annually, and the holding period is expected to be 5 years. The resale price is estimated by applying a 10% terminal cap rate to year six NOI. Based on a study of the market, investors require a 12% rate of return (equity yield) on property of this type.
   (a) What is the present value of the equity interest in the property?
   (b) What is the total present value of the property (mortgage and equity)?
   (c) Based on your answer in part B, what is the implied overall cap rate?

4.3 Introduction to Cap Rates

6. Trend Lines Investment Company has had much success in valuing properties for investment via the sales approach. The investment manager is looking to expand the existing portfolio of residential rental houses, and has located a subdivision which has seen much development over the last 10 years. The investment company has traditionally stayed within the residential rental housing sector, and has preferred home prices in the $175,000–$250,000 range. Rental houses in the area are from 2,000 to 2,500 square feet in living space, and are similar in lot sizes, quality of construction, and all other aspects. The investment manager is considering the purchase of a 2,200 square foot rental home which has a current sales price of $110 per square foot. While this is within the target price point, the investment manager wants to make sure that the company does not overpay for the rental house. The following list itemizes the sales per square foot in the submarket over the last 10 years:

Price per square foot timeline						
1	$ 93.79	10 years ago	10	$ 112.45	4 years ago
2	$ 96.19	10 years ago	11	$ 156.45	4 years ago
3	$ 96.23	9 years ago	12	$ 121.56	3 years ago
4	$ 92.25	8 years ago	13	$ 96.28	2 years ago
5	$ 101.81	7 years ago	14	$ 104.22	2 years ago
6	$ 90.03	6 years ago	15	$ 109.85	1.5 years ago
7	$ 91.85	5 years ago	16	$ 110.63	1 year ago
8	$ 127.36	5 years ago	17	$ 104.05	1 year ago
9	$ 97.28	5 years ago	18	$ 118.03	6 months ago

(a) What is the average sales price per square foot over the last 10 years?
(b) What is the average sales price per square foot over the last 2 years?
(c) What other considerations should be made in order to determine an appropriate sales price per square foot for the new property?
(d) What should the investor use for sales price per square foot in this situation?

Mini-Case: A Primer on Replacement Reserves

A replacement reserve provides for the periodic replacement of building components that wear out more rapidly than the building itself and must be replaced periodically during the building's economic useful life. These components may include:

> Roof covering, carpeting, kitchen, bath or laundry equipment, compressors, elevators, boilers, specific structural items and equipment that have limited economic life expectancies, interior improvements to tenant space made primarily at lease renewal by landlord, sidewalks, driveways, parking areas, and exterior painting.

The extent of replacement reserves is based on the annual repair and maintenance expenses of the property. The replacement reserve would be above what would be necessary as shown in the actual repairs and maintenance expenses in the pro-forma operating statement of a subject property. Appraisers often include replacement reserves as either a percentage of the Effective Gross Income (EGI) of the property, or on a per square foot basis. The amount that is included as replacement reserves depends on how the leases are structured. If the tenant pays for all repairs and maintenance, then there is not much that the landlord will be responsible for paying.

Since the calculation of replacement reserves is an estimate of the annual replacement cost of various items in the subject property, real estate investors should have adequate cash reserves to cover any unforeseen expense items.

Here is an example of a replacement reserve calculation for a 55-unit apartment complex in average condition, constructed in 1988:

Replacement item	Annual cost	Calculation
Kitchen and bath equipment	7,150	($1,300*55)/10
Carpeting	8,250	($900*55)/6
Roof	900	$18,000/20
Total replacement reserve	16,300	% of EGI
Replacement reserve/unit	296	Per unit

As you can see from the table above, the consideration is the average cost of a given replacement ($1,300 for a kitchen appliance) multiplied by the total number of units (55) and then divided by the economic useful life of the item (10 years for kitchen appliances).

Questions for Discussion

1. How might these calculations be performed for an office, retail, or warehouse property?
2. Should the replacement reserve include items that already appear in the repairs and maintenance expense numbers or only additional items?
3. What other (borrower specific) information should be considered when making this calculation?

References

Appraisal Institute website. (2011). http://www.appraisalinstitute.org/designations/MAI_Designations.aspx. Accessed 6 Jan 2011.
Certified Commercial Investment Member (CCIM) website. (2011). http://www.ccim.com/. Accessed 6 Jan 2011.
Code of Federal Regulations (CFR). (2011). Title 12 (Banks and Banking), Chapter 3 (Federal Deposit Insurance Corporation), Subchapter B (Regulations and Statements of General

References

Policy), Part 323 (Appraisals), 323.5 Appraiser Independence, CFR website, http://cfr.vlex.com/vid/323-5-appraiser-independence-19623506. Accessed 6 Jan 2011.

Goddard, G. J. (2010). The global housing boom. In G. Raab, G. J. Goddard, R. Ajami, & A. Unger (Eds.), *The psychology of marketing: Cross-cultural perspectives*. Aldershot: Gower Publishing.

Institute, A. (1996). *The appraisal of real estate* (11th ed.). Chicago: Appraisal Institute.

Institute, A. (2008). *The appraisal of real estate* (13th ed.). Chicago: Appraisal Institute.

Kierkegaard, S. (1847). Upbuilding discourses in various spirits. In *Kierkegaard's writings* (Vol. 15). Copenhagen. Princeton NJ (2009): Princeton University Press.

Korpacz. (2010). Commercial Real Estate Market Survey, Second Quarter 2010.

New World Encyclopedia. (2011). Website. http://www.newworldencyclopedia.org/entry/Empire_State_Building. Accessed 11 Jan 2011.

Risk Management Association (RMA) Journal. (2011). The crowd is untruth: the problem of cap rates in a declining market, Mar 2011 (pp. 26–32). Excerpts from section 4.3 were taken from the RMA Journal article authored by Goddard and Marcum.

The Anatomy of a Lease 5

The human race is divided into two classes: landlords and tenants.
Gustave Flaubert

Contents

5.1	The Lease and Its Impact on Cash Flow	95
	5.1.1 How the Price of Rent is Determined	96
	5.1.2 How Rent is Calculated	97
	5.1.3 Expense Sharing	100
	5.1.4 Impact of the Lease on Cash Flow	101
5.2	Various Leases Available in the Market	102
	5.2.1 Ground Leases	103
	5.2.2 Various Forms of Traditional Leases	104
5.3	Components of a Lease	105
	5.3.1 Setting the Boundaries	106
	5.3.2 Other Lease Contents	106
5.4	Lease Rollover Risk	108
References		117

5.1 The Lease and Its Impact on Cash Flow

The review of commercial leases is a key component of an investor's (or a lender's) due diligence prior to purchase (or financing) of an investment real estate property. The length of the lease income, in conjunction with the amount of competing, available space in the market, represents the durability of the income stream in our aforementioned QQD framework. The lease is a contractual obligation between the lessor (owner) and the lessee (tenant). The lease documents how much rent will be paid, for how long it should be paid, and how expenses may be shared. In this section, we will discuss how the price of rent is determined, how rent is calculated, and how expenses can be shared between the lessor and the lessee.

5.1.1 How the Price of Rent is Determined

Assume that you have purchased a 10,000 square foot retail property which consists of four retail units of unequal size. One unit consists of 5,000 square feet, while the remaining space is allocated among three tenants. You are lucky enough to have the other three units currently leased, and have just received a solid lead for a tenant wishing to occupy the largest space in the property. How might an investor go about determining a reasonable lease rate for this space?

As should be apparent from the discussion in Chap. 2, one starting place is to determine what the market rental rate is for comparable space. This could start with what annual rent per square foot is being paid by the other three tenants in the subject property, but it is preferable to determine what annual rent per square foot is being received in other comparable properties in the market. As should be apparent from the discussion in Chap. 3, another important consideration is the profit expectation of the investor. Based on the discussion in Chap. 1, another alternative is to outsource the process to a commercial broker, who will negotiate a market based lease rate with a prospective tenant in exchange for a leasing commission.

When negotiating an annual lease rate with a tenant, the assumptions concerning which expenses, if any, will be reimbursed by the tenant is crucial. Operating expenses are typically reimbursed by the tenant rather than paid directly, so the property owner does not assume the risk that the tenant fails to pay the expenses in a timely manner. If the leases deviate from the financial projections of the investor, the profitability of the investment is at stake. As an example, let's review the direct cap projections for our retail shopping center where it is assumed that the tenants do not reimburse the lessor for any operating expenses.

The investor has projected expenses for the first year in relation to the current leases in place for the three smaller tenants, and has projected a lease rate of $14.75 per square foot for the prospective tenant, Marvelous Marvin's. At this rent level, the weighted average rent per square foot for the subject property is $15.75. Prior to adding the tenant in the final 5,000 square feet of space, the weighted average rents were $16.75 per square foot for the other three existing tenants. These lease rates could have been signed in better economic times, or the lower lease rate chosen for the final tenant could represent the investor's projection of their worst case scenario. In any event, the lease rates per square foot should be checked with market averages for comparable space during the lease rate determination process. Based on the analysis in Fig. 5.1, the net operating income after projected expenses is right at $50,000, which could serve as a trigger point for the investor. Based on a cap rate of 9%, the property value under the direct cap is $560,000.

As discussed in Chap. 4, the direct cap numbers can serve as the first projected year in the discounted cash flow model. The DCF model in Fig. 5.2 carries forward the direct cap assumptions assuming a 3% annual increase in most operating expenses, and a two and one-half percent annual increase in gross revenue. The investor's assumed holding period is 10 years, and a 25% equity injection relative to the property value is assumed for the period zero equity investment.

5.1 The Lease and Its Impact on Cash Flow

Tenant	Sq. Ft.	Rent psf	Gross Rent
Uncle Bobby's Toy & Hobby	2,500	$ 18.00	$ 45,000
Marvelous Marvin's	5,000	$ 14.75	$ 73,750
Stop and Rob	1,000	$ 14.00	$ 14,000
The Eatery	1,500	$ 16.50	$ 24,750
Total	10,000		$ 157,500
Vacancy Factor		10%	$ 15,750
Effective Gross Income			$ 141,750
Taxes	$ 25,000		
Insurance	$ 12,500		
Repairs & Maintenance	$ 24,000		
Utilities	$ 18,500		
Management Fee	$ 7,088	5%	
Replacement Reserves	$ 4,253	3%	
Total Op Ex	$ 91,340		
Net Operating Income	$ 50,410		
Cap Rate	9.00%		
Estimated Value	$ 560,000		

Fig. 5.1 Direct cap calculation for tenant lease rate

A loan equal to 75% of property value is also assumed, at an annual interest rate of 7% with an amortization period of 20 years.

Based on these and other assumptions concerning tenant rollover risk during the holding period, a property value of $580,000 and an internal rate of return of 13.06% are estimated. If the investor's target return was less than or equal to this IRR, the lease rate for Marvelous Marvin's would be acceptable to the investor, assuming that the tenants do not reimburse for any operating expenses. As discussed in Chap. 4, in order to calculate the investor's internal rate of return, the cash flows for the entire holding period must be considered, including the termination of the investment at the end of the holding period.

How a lease is structured often depends on the party that is drafting the lease document. If the property owner, or their agents, is drafting the lease, the initial document will naturally reflect items favorable to the tenant. The opposite is true when the tenants, or their agents, craft the lease document (Zankel 2001).

5.1.2 How Rent is Calculated

The calculation of rent is the product of a negotiation between the lessor and lessee. The rent is typically a reflection of the market rent for comparable space, but this is not always the case. The tenant may agree to reimburse the landlord for tenant improvements, made at the landlord's expense, over a specified period of time. The rent paid may be fixed over the lease term, or it may increase on a predetermined

	Growth%	Year 0	Year 1	Year 2	Year 3	Year 4	Year 5	Year 6	Year 7	Year 8	Year 9	Year 10	Reversion	Year 11	Comments
Gross Potential Income	2.5%		157,500	161,438	165,473	169,610	173,851	178,197	182,652	187,218	191,898	196,696		201,613	2.5% Annual Increase
Vacancy Factor	10%		15,750	16,144	16,547	16,961	17,385	17,820	18,265	18,722	19,190	19,670		20,161	10% Annual
Other Income	0%		-	-	-	-	-	-	-	-	-	-		-	
Effective Gross Income			141,750	145,294	148,926	152,649	156,465	160,377	164,387	168,496	172,709	177,026		181,452	
Taxes	3%		25,000	25,750	26,523	27,318	28,138	28,982	29,851	30,747	31,669	32,619		33,598	
Insurance	3%		12,500	12,875	13,261	13,659	14,069	14,491	14,926	15,373	15,835	16,310		16,799	
Repairs & Maintenance	3%		24,000	24,720	25,462	26,225	27,012	27,823	28,657	29,517	30,402	31,315		32,254	
Management	5%		7,088	7,265	7,446	7,632	7,823	8,019	8,219	8,425	8,635	8,851		9,073	5% of EGI
Utilities	3%		18,500	19,055	19,627	20,215	20,822	21,447	22,090	22,753	23,435	24,138		24,862	
Other	3%		-	-	-	-	-	-	-	-	-	-		-	
Replacement Reserves	3%		4,253	4,359	4,468	4,579	4,694	4,811	4,932	5,055	5,181	5,311		5,444	3% of EGI
Total Expenses			91,340	94,024	96,786	99,630	102,558	105,572	108,675	111,870	115,158	118,544		122,029	
Net Operating Income			50,410	51,270	52,140	53,019	53,908	54,805	55,711	56,627	57,550	58,482	660,000	59,423	
Annual Debt Service			40,122	40,122	40,122	40,122	40,122	40,122	40,122	40,122	40,122	40,122			
Tenant Improvements			-	-	3,750	2,500	12,500	3,750	6,250	2,500	3,750	12,500			
Leasing Commissions			-	-	2,250	1,500	7,500	2,250	3,750	1,500	2,250	7,500			
Debt Coverage Ratio			1.26	1.28	1.30	1.32	1.34	1.37	1.39	1.41	1.43	1.46			
Net Cash Flows		(143,750)	10,288	11,149	6,018	8,897	(6,214)	8,683	5,590	12,505	11,429	357,365			
BTIRR On Equity	13.06%														
Property Value		$580,000.00													

Fig. 5.2 Discounted cash flow to determine if IRR meets investor hurdle rate

basis. For example, the prevailing market lease rate may be $15.00 per square foot, but in markets with high vacancy rates, a landlord may agree to a lease rate of $13.50 per square foot, with the provision that the lease rate should increase by $1.00 per square foot annually over the term of the lease.

The term of a lease varies based on the property type. Some properties, such as apartments or self-storage facilities, have monthly or annual leases, while office, industrial, and retail properties can have lease terms varying from monthly to over 20 years in length depending on the strategy and tolerance for risk of the tenant and the investor. Some investors desire longer lease terms as these longer leases provide some level of comfort that shorter leases do not provide. While longer lease terms add a level of certainty to the cash flow projections for the investor (and the lender!), longer lease terms do not necessarily guarantee that the tenant will remain in occupancy over the term of the lease. In the aftermath of the "Yes Era", financial markets were witness to two economic truths: what goes up will eventually come down, and signing a lease is not a guarantee of financial success.

An additional issue with longer lease terms concerns the effects of inflation. If a lease is set at a fixed amount per square foot for 5 years, inflation could somewhat erode the profitability of the income stream. This is especially true if the operating expenses are not fixed over the lease term. Ceteris paribus, an investment with a fixed revenue stream and increasing expenses will erode investment profitability

5.1 The Lease and Its Impact on Cash Flow

over time. To counter this effect, landlord's will either designate specific rental increases over the term, or will utilize the first year's rental rate as a base, and then allow for annual increases relative to some benchmark index. Benchmark indexes include the consumer price index (CPI), or other indexes that represent the change in general price levels. It is important for the tenant that the index not be something controlled by the landlord. In a financing situation, the landlord will further desire that the index referenced in the lease, and in the loan documents relative to the loan interest rate, not be something controlled by the lender.

Some investors prefer shorter lease terms. If the strategy of the investor is to renovate an existing property in order to increase rents, or if the investment is made during a period of rising rents (with correspondingly low market vacancy rates), it may make sense to the investor to keep their lease rates shorter for the retail, office, or industrial properties. As mentioned previously, apartments and self-storage facilities will tend to have shorter lease terms regardless of investor strategy. Lenders typically prefer longer lease terms where possible, as lenders seek certainty in cash flows as repayment for their loans.

Another form of rent calculation exists for retail properties. Percentage (or overage) rent allows the parties to set a base lease rate, and then to allow the landlord to recoup additional rent once a tenant's sales hit a certain benchmark. This sort of arrangement is typically structured for anchor tenants, defined as those tenants that serve as a primary draw for a property and occupy a large percentage of the total leasable area in a given property.

An example of a percentage rent agreement is as follows. A tenant signs a 5 year lease at $20.00 per square foot. The lease specifies that the tenant will pay 1% annually once store sales achieve $400,000 at the specific location. The percentage rent target sales level is the result of negotiation between the landlord and the tenant. Typically, tenants with a national or regional presence will have numerous locations from which to draw sales per square foot data for a given store location. The break point should be set at a reasonable level, where the break point represents a successful, but not overly unrealistic, performance level for the tenant.

If the percentage rent break point is set at a reasonable level, the landlord can participate in the success of a tenant, and is property motivated to act in the interests of tenants when considering renewal lease terms (Wheaton 2000).

Since percentage rent is only for retail tenants, lease structures for office, industrial, and apartment occupants do not have this particular feature of variable income. If tenant leases are for a fixed rental amount over the term of the lease, the landlord would bear the risk of income foregone, should market lease rates for comparable space rise during the term of the lease. As earlier alluded to, this risk also involves property operating expenses rising during the lease term, while rent remains constant. This is similar to investing in a multifamily property in a market with rent controls. In a rent controlled market, rental rate increases are capped at some desired level. The same cannot be said for the operating expenses; they are market determined. While there can be compelling reasons for property investment in rent controlled markets, ceteris paribus, an investment that caps the revenue growth but not the operating expenses is inferior to an investment with market determined revenues and expenses during the investment holding period.

Thus most investors prefer rate increases, which are either specified on a per square foot basis over the lease term, or increase relative to a fixed spread over an index which is outside of the control of the property owner, tenant, and lender.

5.1.3 Expense Sharing

As was exhibited in Chap. 4, the net operating income is an important consideration in the determination of property value using the income approach methodology. Thus, having tenants share a portion of the expense burden will improve the value of an investment property. As long as a tenant is paying their rent, any expense sharing agreement, as specified in the lease, will accrue benefit to the owner. If a portion of the tenants occupying space in a given property vacate the space prematurely, the landlord is responsible for that portion of the tenants operating expenses.

When considering expense sharing, the first question to answer is which operating expenses should be shared, while the second question is how they should be shared. For commercial tenants, landlords typically require the tenants to reimburse any expenses which can be considered a primary cost driver of the tenant's business. For example, office tenants specializing in the medical industry will typically pay for their own water and electric usage (so called utilities). This would also be the case for retail tenants, such as restaurants. Since the use of utilities is higher per square foot for certain industries than others, lease arrangements for expense sharing of these costs items are both prudent for the investor, and expected by the tenants. Since business operators who are leasing space also seek certainty in their cash flows, prudent managers are budgeting for expenses that are required when running their businesses. Typically, expenses eligible for reimbursement by the tenant include property taxes, property insurance, utilities, repairs, and maintenance. These expenses are typically not shared with tenants on multi-family or self-storage properties.

How expenses are shared between the property investor (landlord) and tenant can take several forms. It is in how expenses are shared that helps to classify the lease types that we will discuss in Sect. 5.2. For now, an important consideration is that a tenant is typically expected to pay their pro-rata share of whichever expense is being reimbursed.

Returning to our four tenant retail property, the allocation of pro-rata expenses are shown in Fig. 5.3. This expense sharing arrangement is known as common area maintenance (CAM). CAM is typically an agreement for tenants to reimburse the landlord for their pro-rata share of repairs, maintenance, and sometimes utilities expenses. The CAM calculation can include any expense item as defined in the lease. While it is easier to classify leases into a few broad categories, the lease represents the conclusion of a negotiation between the landlord and the tenant. Thus, the lease can take many forms to suit individual situations and preferences.

5.1 The Lease and Its Impact on Cash Flow

Sq. Ft.	Tenant	Taxes	Insur	Repair/Main	Utilities
2,500	Uncle Bobby's Toy & Hobby	6,250	3,125	6,000	4,625
5,000	Marvelous Marvin's	12,500	6,250	12,000	9,250
1,000	Stop and Rob	2,500	1,250	2,400	1,850
1,500	The Eatery	3,750	1,875	3,600	2,775
10,000	Totals	25,000	12,500	24,000	18,500

Fig. 5.3 Pro-rata expense reimbursement for multi-tenant retail property

Since both the tenant and the landlord are seeking certainty in their revenues and expenses by entering into the lease agreement in the first place, the reimbursement of an expense, or of expenses, can also be set to begin or end at a certain amount per square foot. This is known as an expense stop. In Fig. 5.3 above, the total operating expenses were $8.00 per square foot on an annual basis. This is calculated by adding up the totals for each expense item and dividing by the 10,000 square feet of building space. The local market may not support a complete pass-through of expenses as shown in Fig. 5.3. Instead, the convention might be for the landlord to pay for the expenses up to a certain amount, and then for the additional expenses to be paid by the tenants. The expense stop might kick in at $5.00 per square foot, where any additional expenses are passed through to the tenants. This provides certainty in the expenses for the landlord, and motivates the tenants to keep their controllable expenses within reasonable levels.

5.1.4 Impact of the Lease on Cash Flow

Depending on how operating expenses are shared, the impact on the investor's net operating income could be significant. It should be evident that the lease rates charged when the tenant is responsible for reimbursing the owner for the majority of operating expenses is significantly less than what the tenant would pay when the landlord is responsible for the operating expenses of the property.

If we return to our four tenant retail property, the average rents received were $15.75 per square foot, as shown in Fig. 5.1, when all expenses were borne by the property owner. Also shown in Fig. 5.1 was that the estimated property value was $560,000. In Fig. 5.4, the expenses are assumed to be paid by the tenants. Rents are reduced to arrive at the same estimated property value as shown in Fig. 5.1.

As is implied in Figure 5.4, the weighted average lease rate with CAM reimbursements can drop to $7.75 per square foot annually and still achieve the same value as was shown in Fig. 5.1. The difference is that the $8.00 in CAM was originally born by the investor. In this example, the tenants are responsible for this portion of the operating expenses, in the same arrangement as was itemized in Fig. 5.3. Thus any increase in average rents above $7.75 would accrue positively to the investor, assuming CAM remains at $8.00 per square foot. It should be noted that if any of the tenants vacate their space prior to the end of their lease term, the

Tenant	Sq. Ft.	Rent psf	CAM	Gross Rent
Uncle Bobby's Toy & Hobby	2,500	$ 9.00	$ 8.00	$ 42,500
Marvelous Marvin's	5,000	$ 7.00	$ 8.00	$ 75,000
Stop and Rob	1,000	$ 8.00	$ 8.00	$ 16,000
The Eatery	1,500	$ 8.00	$ 8.00	$ 24,000
Total	10,000			$ 157,500
Vacancy Factor			10%	$ 15,750
Effective Gross Income				$ 141,750
Taxes	$ 25,000			
Insurance	$ 12,500			
Repairs & Maintenance	$ 24,000			
Utilities	$ 18,500			
Management Fee	$ 7,088	5%		
Replacement Reserves	$ 4,253	3%		
Total Op Ex	$ 91,340			
Net Operating Income	$ 50,410			
Cap Rate	9.00%			
Estimated Value	$ 560,000			

Fig. 5.4 Direct cap calculation with CAM reimbursements

owner is responsible for that portion of the building's operating expenses. Thus, how a lease is structured does have a material impact on property value, and the investor's return. How much value can be generated based on the lease structure depends on the negotiating strength of the landlord and tenant, the current vacancy level in the local market, and the market averages for rental rates in consideration of which party bears the brunt of the operating expenses.

5.2 Various Leases Available in the Market

There are many forms of commercial leases. Sometimes a lease may only pertain to the right of occupying a specific parcel of land for a specified period of time. In most cases, the landowner (the lessor) will also own the vertical improvements that will be incorporated into the leased premises occupied by the tenant (the lessee). Leases can vary in terms of duration. An estate for years is a lease with a specified duration of tenancy. Lease terms under this scenario can be up to 99 years for land contracts. Most lease terms will be from 3 to 25 years in situations where the leased property consists of both land and building. An estate from year to year involves periodic tenancy, which continues until either the lessor or the lessee provides a notice of termination. These basic leasehold estates were introduced in Chap. 1.

5.2.1 Ground Leases

The first form of lease that we will discuss is called a ground lease. These leases are involved in situations where the landowner allows the use of their land for a fee. Some examples of this include a parcel of land located adjacent to an existing retail complex, where a nationally known tenant wishes to construct a building, but is only interested in being committed to the location for a certain period of time. Ground leases of this fashion allow the owner of the building to not have to purchase the land outright, thus saving money in terms of constructing the new location.

There are some important questions to ask concerning ground leases. The first question relates to which form of ownership the investor is considering purchasing, or to whom the bank is considering for financing. Will the investor become the owner of the land or the owner of the improvements? Ceteris paribus, being the owner of the land is the safer approach, due to the fact that at the end of the ground lease (and any renewal options), all improvements revert back to the owner of the land. If the investor is to become the owner of the improvements, it is of paramount importance to understand a few key facts concerning the ground lease.

One important question deals with the length of time remaining on the ground lease. If the investor is the owner of the improvements, at some time in the future those improvements will revert back to the owner of the land (Hagen et al. 1994). Thus, savvy investors should make sure that there is a reasonable period of time between when their desired holding period ends (and when the bank loan expires, which is defined by the amortization period offered on the loan as opposed to just the loan term), and when the ground lease expires. Prudent bank lending policy would be to verify if the ground lease contains renewal options at the discretion of the borrower, as then the effective length of the ground lease should be calculated to include the renewal options. If renewal options are at the discretion of the landowner, i.e. not the bank's borrower, then the lender should consider the effective length of the lease to end at the expiration of the original term of the lease. This same approach would be prudent from the investor's perspective as well.

As mentioned previously, there should be a reasonable period of time left on the ground lease. A general rule of thumb is that the length of time on the ground lease should exceed the amortization of the loan, and the investor's holding period, by 20 years. There are two reasons for this. The first reason is that while the bank may provide only a 5 or a 7-year term, situations could arise in the future whereby the customer asks the bank to extend the amortization of the loan (for example at renewal or if they wish to lower payments in the future via an extension strategy), so simply having the ground lease equal to the amortization of the loan is not acceptable from the lender's viewpoint. It may impede the bank's ability to be taken out by another lender in the future. The second reason is the investor's desire to sell the property before the end of the ground lease. If there is not a reasonable length of time left on the ground lease, the investor will have difficulty selling the building to another investor owing to the risk that the ground lease could expire during the investment holding-period of the new purchaser.

Ground leases are deemed unsubordinated if the landowner does not subordinate the ownership to the leasehold interest. This means that if the ground rent is not paid, the landowner can foreclose and terminate the leasehold rights. In this situation, the improvements could be claimed by the landowner, thus defeating any claim by the lender holding only a pledge of the leasehold interest. If the lease is subordinated, the landowner grants the lease and then encumbers the fee title with a subordination agreement. This means that the landowner subordinates the ownership in the land in favor of the mortgage holder (Wiedemer 1995).

A final important aspect of ground leases is to verify that the right to cure is acceptable to the bank. The bank will want to make sure that the owner of the land provides the right for the bank as mortgagor to step in and pay the ground lease payments (which should be accounted for in the underwriting analysis) in case the owner of the improvements defaults on the loan.

5.2.2 Various Forms of Traditional Leases

Now that we have covered ground leases, we will discuss leases that involve both the land and improvements. Since the bank considers the net operating income as the primary source of repayment for an investor real estate loan, the lease income represents the duration of the income stream for an investment property. The duration of the income stream for leases can vary according to the property type. For apartments and self-storage facilities, average lease terms vary from monthly to annually, with lease durations seldom exceeding 1 year. For office and retail properties, lease terms can typically vary from monthly to up to 5 years. For warehouse properties or for single tenant properties in general, the lease terms can be much longer than 5 years. These are only general guidelines, but in most cases having a longer lease term commitment is better than having a shorter time commitment. This is due to the desire for certainty in terms of the future cash flows of a given project

The best-known method of classifying leases is via the arrangement between the owner of the land and improvements (lessor) and the tenant (lessee) that occupies the property. Most leases are classified by how property expenses will be paid. What is important to remember is that while we can categorize leases in broad categories, the lease is a negotiation between the lessor and the lessee, and thus may not be the same structure even within the same property. Figure 5.5 illustrates the most common forms of expense sharing arrangements: Gross, modified gross, triple net, and absolute net.

In returning to our four tenant retail property example, the analysis outlined in Fig. 5.1 represents a situation where the tenants are paying gross leases, while the situation outlined in Fig. 5.4 represents a situation where the tenants are paying triple net leases (NNN). The absolute net lease would require the tenant to pay all property operating expenses. In viewing the analysis in Fig. 5.4, this would include the tenant paying the management fee as well as the replacement reserve for structural repairs. Should the tenants negotiate an expense sharing arrangement

Gross (or Full Service) Lease	Owner pays all expenses
Modified Gross Lease	Tenant pays for some expenses
Absolute Net Lease	Tenant pays for all expenses
Triple Net Lease (NNN)	Tenant pays for taxes, insurance, and maintenance

Fig. 5.5 Primary categories of commercial leases

somewhere in between the triple net lease and gross lease structures, this is typically referred to as a modified gross lease.

Throughout the discussion in Chap. 5, we have illustrated other lease types. The index lease is an arrangement where annual rent and expense reimbursement changes are scheduled in reference to a public record of cost changes, such as the consumer price index (Friedman et al. 1993). Another form of lease is a graduated rental lease. This lease form allows for the increasing or decreasing of rent or reimbursed expenses over the term of the lease. An escalator lease is typically structured where the lessor pays expenses for the first year, and the lessee would pay any increase in expenses over the established level as additional rent over the subsequent years of the lease.

Another form of lease is the revaluation lease. This lease structure allows for periodic rent adjustments based on the revaluation of the real estate, at prevailing market conditions, during specific intervals during the lease term (Appraisal Institute 1993). The revaluation lease structure is preferred when the condition of the property at lease origination is in flux. If a property is being redeveloped for an alternative purpose, or if the property is undergoing significant cosmetic improvement, or if a property's occupancy is expected to change significantly over the lease term, the revaluation lease structure may be agreeable by both the landlord and the tenant.

All of these lease structures are based on negotiation between the landlord (property investor) and the tenant, and represent various methods of dealing with the uncertainty of rising rents and expenses over the agreed lease term, plus any renewal options.

5.3 Components of a Lease

Thus far in Chap. 5, we have primarily reviewed how rent and expenses are computed in the various forms of leases. While these are important elements in leases, the *anatomy of a lease* would not be complete without itemizing other lease provisions. Without lease provisions beyond revenue and expenses, tenants have a natural tendency to under-maintain or to over-use a rental property (Benjamin et al. 1995). As discussed earlier, the lease must elucidate how revenues and expenses will be shared or paid over the lease term. What follows below is a broadened discussion of other elements of importance in commercial leases.

5.3.1 Setting the Boundaries

As the lease is a legally binding agreement between the property owner and the tenant, the first requirement is that the lease discloses the commencement date, or when the lease takes effect, as well as the maturity date. This could be different than the occupancy date, when the tenant officially begins to utilize the premises. The parties to the lease should also be indentified, and both parties to the lease, or their assigns, must execute the lease document in order for it to be a legally binding contract. The lease must also clearly state the term of the lease, and whether any renewal options exist, and how many and for how long of a term. Renewal options are typically at the discretion of the tenant.

Lenders often inquire whether a tenant will likely renew their lease upon maturity. Some leases provide escape clauses whereby a tenant can end their lease prior to the stated lease expiration, as long as they notify the landlord by a specified date. From an investor's perspective, as well as that of the lender, should any escape clauses exist, this date should be considered the actual lease termination date. Observant readers might question how an investor or a lender can estimate the probability that a tenant may exercise their renewal option in a lease. While not immutable, the psychology of the renewal decision has some primary considerations. If a tenant has spent a large amount of money in leasehold improvements, if the rent that they are paying is under or within the market averages, and if the tenant is experiencing success at the specific location, the chances are good that they will renew their lease. Obviously this is not true in every case, as sometimes tenants are so successful that they need to expand their business operation into a larger space. In any event, keeping abreast of the market rental rates and market success of the commercial tenants certainly has its advantages.

Other important considerations are the approved use and legal description of the property (Brueggeman and Fisher 2010). The approved use allows the landlord to restrict the use of the tenant during the lease term. Landlords with social responsibility concerns will mandate that ethically objectionable activities not occur on site, and landlords can also protect their property from over-use, or harmful use, by mandating that environmental contaminants are not utilized by the tenant on-site. The legal description of the property ensures that both parties are clear on the amount of space being leased by the tenants. As stated earlier, this is necessary in order to compute any pro-rata expense reimbursements for the tenant. This load factor should be stated in the lease, and is defined as the rentable area per floor in a subject property divided by the useable area per floor.

5.3.2 Other Lease Contents

Leases should generally contain provisions itemizing the landlord and tenant's responsibilities concerning the upkeep and repair of the property. Both parties should maintain adequate insurance coverage in case of an unforeseen circumstance

5.3 Components of a Lease

affecting the property. Property insurance would consist of hazard insurance of typically up to at least the replacement value of the property. In cases where the property is located in a designated flood zone, flood insurance will be required by the lender in order for a mortgage to be placed on the property. Subrogation represents assuming the legal rights of another party and has relevance in how insurance claims are filed in a tenancy situation. The lessor and lessee may release each other from any and all liability from loss due to fire or other damage, and essentially agree to file a claim with the insurance company.

The lease should also disclose any concessions. This is defined as free rent and other discounts that typically arise when the tenant has made improvements to the property, or if the landlord is competing with other properties and had to make concessions to obtain the tenant. In the aftermath of the "Yes Era", lease concessions are more common than in years past. Given the increased vacancy rates in many locations, tenants are more likely to see a rental concession in order to renew their leases. This is a risk that some property investors did not foresee: that rental rates could actually drop below the original levels at lease execution.

Some leases will contain a non-compete clause. In this section, it will be specified that the landlord cannot lease adjacent space to a competitor. Related to this idea is the non-dilution/radius clause. This is common in retail leases, and states that the tenant cannot lease another location within a certain radius. These last two clauses provide further evidence that the lease is a negotiation between the landlord and the tenant, as these provisions help protect both parties from the competition.

A provision that is common in retail leases is referred to as the domino clause. Also known as the go-dark clause, this provision allows for in-line (or supporting) tenants in a retail property to break their leases should the anchor tenant vacate the property. It is very important to obtain all leases for a retail property, so as to make sure that this clause is not contained in the leases. Since the in-line tenants require the anchor tenant as a draw for customers, if the anchor tenant should vacate the property, the in-line tenants may follow.

Another important provision in a lease is the assignment and sublease section. This agreement forbids the tenant from subleasing the space to another tenant, typically without the prior consent of the bank. In similar fashion, leases will also typically require that the lender be consulted prior to changes to the existing tenant mix, which is something also referenced in the bank's closing documents. Another important clause deals with subordination, non-disturbance, and attornment (SNDA). Subordination is an agreement by the tenant to subordinate its leasehold interest to the lender's mortgage including renewals, modifications, and extensions. This allows the lender to be in primary position relative to the tenant in a foreclosure situation. Non-disturbance requires the lender to acknowledge the tenant's right of possession after foreclosure. Thus the tenant can remain in occupancy in the unfortunate event that the lender was required to foreclose on the property owner. Attornment is an agreement whereby the tenant must recognize the lender as the rightful landlord following a foreclosure. The SNDA, when combined with an estoppel certificate, ensures that all of the parties to the lease (lender, borrower,

Tenant Rollover Risk							
Property One		**Property Two**		**Property Three**		**Property Four**	
Sq Ft	**Year**	**Sq Ft**	**Year**	**Sq Ft**	**Year**	**Sq Ft**	**Year**
25,000	2013	42,206	2019	29,284	2014	32,040	2018
7,680	2013	9,600	2014	6,400	2013	5,500	2012
7,200	2013	1,200	2013	1,600	2012	1,300	2012
6,400	2013	**1,320**	**Vacant**	1,600	2012	**1,300**	**MTM**
1,600	2013	**1,680**	**MTM**	1,600	2012	1,300	2015
1,600	2013	1,200	2012	**800**	**MTM**	2,600	2013
1,600	2012			800	2012	1,300	2012
51,080 Total		**57,206 Total**		1,600	2012	2,250	2012
				43,684 Total		1,300	2014
Property Five		**Property Six**				**48,890 Total**	
Sq Ft	**Year**	**Sq Ft**	**Year**				
33,000	2021	30,280	2015	**Rollover Risk**			
9,295	2014	**7,020**	**Vacant**	Vacant	4.91%		
2,800	2012	**1,468**	**MTM**	MTM	2.84%		
1,500	2012	**1,298**	**MTM**	2012	9.22%		
1,500	2014	1,334	2012	2013	20.61%		
1,500	2013	2,050	2012	2014	16.76%		
1,500	**Vacant**	**1,500**	**Vacant**	2015	10.38%		
1,500	2013	**1,500**	**Vacant**	2016	0.00%		
2,100	**Vacant**	**2,100**	**MTM**	2017	0.00%		
54,695 Total		**48,550 Total**		2018	10.54%		
				2019	13.88%		
				2020	0.00%		
				2021	10.85%		
				Total	**304,105**		
				Total %	**100.00%**		

Fig. 5.6 Sample tenant lease rollover risk calculation

and tenant) understand and accept the bank's position as mortgagor and subsequent landlord in the event of foreclosure. Lenders may also require that the lease require that any major changes to the property's tenant mix be pre-approved by the lender prior to executing the lease document.

5.4 Lease Rollover Risk

To conclude our chapter on commercial leases, we will discuss the importance of calculating lease rollover risk for both the property investor and the lender. Imagine that you own a portfolio of retail properties. After you have successfully negotiated leases with your tenants, there is still the risk that the properties may experience higher vacancy owing to the maturity of the leases of the tenants occupying the properties in your portfolio. The lender will care about this as lease rollover risk can affect the overall cash flows that a property investor has on hand available to pay mortgage debt. The property investor will care about lease rollover risk as well, as based on how a portfolio of properties leases are "arrayed", a given year may bring very different expectations for leasing commissions and tenant improvement allowances than another year. While Fig. 5.6 analyzes a hypothetical portfolio of six retail properties based on the percentage of the leases that mature in a given year, what is not considered based on this analysis is the probability that the tenants

5.4 Lease Rollover Risk

will renew their leases, or what might happen should tenants vacate the properties prior to the expiration of the lease.

As can be seen from Fig. 5.6, each property appears to have at least one anchor tenant that occupies the largest space in each retail property. The year 2013 has the highest percentage of tenant leases maturing within one calendar year, followed by 2014. This appears to be due to the expiring anchor tenant leases in property one and three. Hopefully, based on our earlier discussion on domino clauses, none of the tenants' leases contain these provisions!

Regardless of the types of properties held in the portfolio, viewing the collective risk posed by the possibility of tenants not renewing their leases is a pivotal component to risk management concerning commercial leases.

Questions for Discussion

1. Differentiate between an absolute net, triple net, gross, and modified gross lease.
2. All other things being equal, in what inflationary, rent and expense growth environment would a landlord prefer each of the lease alternatives described in question one?
3. Elaborate on the various components of an SNDA.
4. Describe the process of quantifying tenant rollover risk. How might this risk affect both the investor as well as the lender? How might the lender be affected even if the property that they have financed does not experience tenant rollover?
5. Define a domino clause and how its presence in retail leases could present problems for the property owner as well as the lender.

Problems

1. Jimmy's Burger Shack leases a 2,000 square foot free standing building located in an urban setting with a high level of pedestrian traffic. Jimmy recently renewed his lease at a rate of $25 per square foot for the first year, absolute net. The annual lease rate will increase annually as per movements in the consumer price index (CPI). The CPI is expected to increase by 3% annually over the term of this lease which is five years long.
   (a) Determine the present value of the lease assuming a 10% discount rate.
   (b) What is the average lease rate annually over the term of this lease?
   (c) Determine the effective rent for this lease.
2. A property owner is evaluating the following alternatives for leasing space in an office building for the next 3 years:
   Net lease with CPI adjustment: $18 for the first year, with 3% assumed CPI increases annually
   Net lease with steps: $17 for the first year, increasing $1 in rent per square foot annually

Gross lease: $34 for each year for rent per square foot; operating expenses are expected to be $13 per square foot for the first year, increasing $1 per square foot annually

Gross lease with expense stop and CPI adjustment: $24 for the first year, assumed 3% annual CPI adjustments thereafter; there is an expense stop in place for the entire lease term at $7 per square foot

  (a) Calculate the present value of the lease, the average net income, and the effective rent to the owner after expenses for each leasing alternative. Assume a discount rate of 10%.
  (b) How would you rank the alternatives in terms of risk to the property owner?
  (c) How would you rank the alternatives in order of preference?
3. Kings View Retail Center consists of six retail bays, five of which are currently leased. The existing leases call for operating expenses to be reimbursed to the owner by the tenants on a pro-rata basis. The current rent roll and most recent annual operating expenses are shown below.

Sq. ft.	Tenant	Annual operating expenses	
11,500	Charming Charlie's	Tax	$32,000
10,000	Pro Sport Shoppe	Insurance	$18,000
12,000	Finnigan's Wake	Utilities	$45,500
10,000	CVS	Repairs and maintenance	$61,500
10,000	Hush Puppies		
10,000	Vacant		

  (a) How much CAM should be charged per unit for the year on a per square foot basis?
  (b) Itemize how much each tenant is required to reimburse the owner for each expense item.
  (c) How will reimbursement for the currently vacant space be handled?
4. Highland Management Company must compute the CAM charges for retail leases in a shopping mall that they manage. The retail mall consists of 2.1 million square feet, of which 750,000 square feet has been leased to anchor tenants. The anchor tenants have agreed to pay $2 per square foot in CAM charges. In-line tenants occupy 900,000 square feet with the remainder of the space considered common area. The landlord believes that $6 per square foot annually is required to maintain the common area.

5.4 Lease Rollover Risk

(a) How much will the anchor tenants pay in CAM?

(b) How much will in-line tenants have to pay per square foot to maintain the common area?

5. Cutler Investors is considering the purchase of an undeveloped track of land in Greenville, NC. The property is currently zoned for agricultural use, but the investors are interested in determining what the highest and best use for the land could be, so that they can decide how the land should be rezoned for commercial development. The company has made the following estimates concerning possible rents, expenses, and costs for construction as shown below.

	Office	Retail
Rentable sq. ft.	50,000	40,000
Rent psf	$20	$25
Operating exp ratio	50%	55%
Avg NOI growth	2%	2%
Required return	10%	10%
Construction cost psf	$100	$100

(a) Estimate the property value of each possible development alternative.

(b) Estimate the cost for each development alternative

(c) What is the highest best use of the land based on the proposed development alternatives?

6. Paradorn Consulting has been asked to help an investor in the preparation of their discounted cash flow analysis for a property that they own. The investor is having trouble calculating the tenant improvements and leasing commissions associated with a specific bay in the subject retail property. The investor assumes that there is a 65% probability that the tenant will renew their lease in year four of the holding period. If the tenant renews their lease, tenant improvements in the market cost $2 per square foot, while if a new tenant is required, market tenant improvements cost $8 per square foot. The tenant space in question is 10,000 square feet, which represents 13% of the total rentable area in the subject property. The investor is projecting annual inflation of 3% over the holding period. Regarding leasing commissions, the year four effective gross income is projected at $200,000, while the costs for leasing commissions if a tenant renews their lease is estimated at 2% of EGI. A new lease is estimated to cost 6% of EGI, with the average length of the leases in the property being 3 years.

(a) Calculate the cost of tenant improvements for this space

(b) Calculate the cost of leasing commissions for this space

Mini-Case: Interpreting a Sample Lease

Commercial Lease Agreement

This LEASE AGREEMENT is made and entered into June 30, 2012, by and between John Landlord, whose address is 100 Main Street (hereinafter referred to as "LESSOR"), and Jane Tenant, Inc., whose address is 150 Deacon Boulevard (hereinafter referred to as "LESSEE").

Article I: Grant of Lease

LESSOR, in consideration of the rents to be paid and the covenants and agreements to be performed and observed by the tenant, does hereby lease to the LESSEE, and the LESSEE does hereby lease and take from the LESSOR, the property described in Exhibit "A" attached hereto, and by reference made a part hereof (the "Leased Premises"), together with, as part of the parcel, all improvements located thereon.

Article II: Lease Term

Section 1. Total Term of Lease. The term of this lease shall begin on the commencement date, as defined in Section 2 of this Article II, and shall terminate 5 years from this same said date (the "Lease Term").

Section 2. Commencement Date. The "Commencement Date" shall mean the date on which the tenant signs this lease.

Article III: Extensions

The parties hereto may elect to extend this Agreement for two periods after the termination date, each period being 5 years in duration. Either party must give the other written notice 120 in advance of lease termination date or the previous extension termination date stating its desire to execute the said extension.

Article IV: Determination of Rent

The LESSOR agrees to pay the LESSEE and the LESSEE agrees to accept, during the term hereof, at such place as the LESSOR shall from time to time direct by notice to the LESSEE, rent at the following rates and times:

Section 1. Annual Rent. Annual Rent for the first year of this lease shall be Twenty-Five Thousand dollars ($25,000.00), plus applicable sales tax. For each successive year of the lease term, the rent shall increase by the same percentage increase in the annual CPI index.

Section 2. Payment of Yearly Rent. The annual rent shall be payable in advance in equal monthly installments of one-twelfth (1/12th) of the total yearly rent, which shall be 2,083 dollars and 33 cents ($2,083.33) during the first 12 months on the first day of each and every calendar month during the term hereof, and pro-rata for the fractional portion of any month, except that on the first day of the calendar month immediately following the Commencement Data, the LESSEE shall also pay to the LESSOR rent at the said rate for any portion of the preceding calendar month included in the term of this lease. Reference to the yearly rent hereunder shall not be implied or construed to the effect that this Lease or the obligation to pay rent hereunder is from year to year, or for any term shorter than the existing Lease term, plus any extensions as may be agreed upon. It also does not imply or construe that the payments of annual rents are fixed at the amount stated above, as each year the rental payments due will adjust with the CPI index as is noted in Section 1 above. A late fee in the amount of 200 dollars ($200.00) shall be assessed if payment is not postmarked or received by the LESSOR on or before the tenth day of each month.

Article V: Taxes

Section 1. Real Estate Taxes. During the continuance of this lease, the LESSOR shall deliver to the LESSEE a copy of any real estate taxes and assessments against the Leased property. From and after the Commencement Date, the LESSEE shall pay to LESSOR not later than twenty-one (21) days after the day on which the same may become initially due, all real estate taxes and assessments applicable to the Leased premises, together with any interest and penalties lawfully imposed thereon as a result of the LESSEE'S late payment thereof, which shall be levied upon the Leased premises during the term of this lease.

Article VI: Utilities

Section 1. Utilities. LESSEE shall pay for all water, sanitation, sewer, electricity, light, heat, gas, power, fuel, and other services incident to the LESSEE'S use of the Leased Premises, whether or not the cost thereof be a charge or imposition against the Leased Premises.

Article VII: Obligations for Repairs

Section 1. LESSOR'S Repairs. Subject to any provisions herein to the contrary, and except for maintenance or replacement necessitated as the result of the act or omission of sub-lessees, licenses or contractors, the LESSOR shall be required to repair only defects, deficiencies, deviations, or failures of materials or workmanship in the building. The LESSOR shall keep the Leased Premises free of such defects, deficiencies, deviations, or failures during the first twelve (12) months of the term hereof.

Section 2. LESSEE'S Repairs. The LESSEE shall repair and maintain the Leased Premises in good order and condition, except for reasonable wear and tear, the repairs required of LESSOR pursuant hereto, and maintenance or replacement necessitated as the result of the act of omission or negligence of the LESSOR, its employees, agents, or contractors.

Section 3. Requirements of the Law. The LESSEE agrees that if any federal, state, or municipal government or any department or division thereof shall condemn the Leased Premises or any part thereof as not in conformity with the laws and regulations relating to the construction thereof as of the commencement date with respect to conditions latent or otherwise which existed on the Commencement Date, or, with respect to items which are the LESSOR'S duty to repair pursuant to Section 1 and 3 of this Article; and such federal, state, or municipal government or any other department or division thereof, has ordered or required, or shall hereafter order or require, any alterations or repairs thereof or installations and repairs as may be necessary to comply with such laws, orders or requirements (the validity of which the LESSEE shall be entitled to contest); and if by reason of such laws, orders, or the work done by the LESSOR in connection therewith, the LESSEE is deprived of the use of the Leased Premises, the rent shall be abated or adjusted, as the case may be, in proportion to that time during which, and to that portion of the Leased Premises of which, the LESSEE shall be deprived as a result thereof, and the LESSOR shall be obligated to make such repairs, alterations, or modifications at LESSOR'S expense. All such rebuilding, altering, installing, and repairing shall be done in accordance with Plans and Specifications approved by the LESSEE, which approval shall not be unreasonably withheld. If, however, such condemnation, law, order, or requirement, as in this Article set forth, shall be with respect to an item which shall be the LESSEE'S own costs and expenses, no abatement or adjustment of rent shall be granted; provided, however, that LESSEE shall also be entitled to contest the validity thereof.

Article VIII: Use of Property by Lessee

Section 1. Use. The Leased Premises may be occupied and used by the LESSEE exclusively as a Hair Salon, to be known as Marvelous Marvin's. Nothing herein shall give the LESSEE the right to use the property for any other purpose or to sublease, assign, or license the use of the property to any sublessee, assignee, or licensee, which or who shall use the property for any other use.

Article IX: Signage

Section 1. Exterior Signs. Tenant shall have the right, at its sole risk and expense and in conformity with applicable laws and ordinances, to erect and thereafter, to repair or replace, if it shall so elect signs on any portion of the Leased Premises, providing that LESSEE shall remove any such signs upon termination of this lease, and repair all damage occasioned thereby to the Leased Premises.

Section 2. Interior Signs. LESSEE shall have the right, at its sole risk and expense and in conformity with applicable laws and ordinances, to erect, maintain, place and install it usual and customary signs and fixtures in the interior of the Leased Premises.

Article X: Insurance

Section 1. Subrogation. LESSOR and LESSEE hereby release each other, to the extent of the insurance coverage provided hereunder, from any and all liability or responsibility (to the other or anyone claiming through or under the other way by way of subrogation or otherwise) for any loss to or damage of property covered by the fire and extended coverage insurance policies insuring the Leased Premises and any of the LESSEE'S property, even if such loss or damage shall have been caused by the fault or negligence of the other party.

Section 2. Contribution. The LESSEE shall reimburse the LESSOR for all insurance premiums connected with or applicable to the Leased Premises for whatever insurance policy the LESSOR, at its sole and exclusive option, shall select.

Article XI: Damage to Demised Premises

Section 1. Abatement or Adjustment or Rent. If the whole or any part of the Leased Premises shall be damaged or destroyed by fire or other casualty after the execution of this Lease and before the termination hereof, then in every case the rent reserved in Article IV herein and other charges, if any, shall be abated or adjusted, as the case may be, in proportion to that portion of the Leased Premises of which the LESSEE shall be deprived on account of such damage or destruction and the work of repair, restoration, rebuilding, or replacement or any combination thereof, of the improvements so damaged or destroyed, shall in no way be construed by any person to effect any reduction of sums or proceeds payable under any rent insurance policy.

Section 2. Repairs and Restoration. LESSOR agrees that in the event of the damage or destruction of the Leased Premises, the LESSOR forthwith shall proceed to repair, restore, replace, or rebuild the Leased Premises (excluding LESSEE'S leasehold improvements), to substantially the condition in which the same were immediately prior to such damage or destruction. The LESSOR thereafter shall diligently prosecute said work to completion without delay or interruption except for events beyond the reasonable control of the LESSOR. Notwithstanding the foregoing, if LESSOR does not either obtain a building permit within ninety (90) days of the date of such damage or destruction, or complete such repairs, rebuilding, or restoration within nine (9) months of such damage or destruction, then LESSEE may at any time thereafter cancel and terminate this Lease by sending ninety (90) days written notice thereof to the LESSOR, or, in the alternative, LESSEE may, during said ninety (90) day period, apply for the same and LESSOR shall cooperate with LESSEE in LESSE'S application. Notwithstanding the foregoing, is such damage or destruction shall occur during the last year of the term of this Lease, or during any renewal term, and shall amount to twenty-five (25%) percent or more of the replacement cost (exclusive of the land and foundations), this Lease, may be terminated at the election of either LESSOR or LESSEE, provided that notice of such election shall be sent by the party so electing to the other within thirty (30) days after the occurrence of such damage or destruction. Upon termination, as aforesaid, by either party hereto, this Lease and the term thereof shall cease and come to an end, any unearned rent or other charges paid in advanced by the LESSEE shall be refunded to the LESSEE, and the parties shall be released hereunder, each to the other, from all liability and obligations hereunder thereafter arising.

IN WITNESS WHEREOF, the parties hereto have executed this Lease the day and year first above written or have caused this Lease to be executed by their respective officers thereunto duly authorized. Signed, sealed and delivered in the presence of:

_____   _____
"LESSOR"   "LESSEE"
Date: June 30, 2012

Questions for Discussion

1. What type of lease structure is this proposed lease?
2. What type of property is this? What business is conducted on site? Who pays the following expenses:

Expense type	Landlord pays	Tenant pays
Taxes		
Insurance		
Utilities		
Repairs (structural)		
Repairs (maintenance)		
Insurance		

3. What would a prudent lender require in terms of the amount of insurance coverage on the subject property?
4. Are their any risks to the landlord in structuring a lease in this manner?
5. What other lease provisions might you consider adding to this lease agreement prior to execution?

References

Appraisal Institute. (1993). *The dictionary of real estate appraisal* (3rd ed.). Chicago: Appraisal Institute.

Benjamin, J. D., De La Torre, C., & Musumeci, J. (1995). Controlling the incentive problems in real estate leasing, 1995. *Journal of Real Estate Finance and Economics, 10*, 177–191.

Brueggman, W. B., & Fisher, J. (2010). *Real estate finance & investments* (14th ed.). London: McGraw-Hill.

Friedman, J. P., Harris, J. C., & Lindeman, J. B. (1993). *Barron's dictionary of real estate terms* (3rd ed.). Hauppage: Barron's Educational Series.

Hagen, D. A., Palmer, R. A., & Keck, N. F. (1994). *North Carolina real estate principles and practices* (2nd ed.). Scottsdale: Gorsuch-Scarisbrick Publishers.

Wheaton, W. C. (2000). Percentage rent in retail leasing: The alignment of landlord-tenant interests, 2000. *Real Estate Economics, 8*, 185–204.

Wiedemer, J. P. (1995). *Real estate finance* (7th ed.). Englewood Cliffs: Prentice Hall.

Zankel, M. I. (2001). *Negotiating commercial real estate leases*. Fort Worth: Mesa House Publishing.

Risk Analysis 6

> *Nothing profits more than self-esteem founded on just and right.*
> *Aldous Huxley*

Contents

6.1	Risk in Real Estate Investment	120
	6.1.1 Business Risk	120
	6.1.2 Management Risk	120
	6.1.3 Liquidity Risk	122
	6.1.4 Legislative Risk	122
	6.1.5 Inflation Risk	123
	6.1.6 Interest Rate Risk	124
	6.1.7 Environmental Risk	128
	6.1.8 Financial Risk	130
6.2	Leverage Effects	131
6.3	Statistics and Risk	133
6.4	Partitioning the IRR	135
6.5	Due Diligence Analysis	135
6.6	Conclusion	136
References		140

In this chapter, the various forms of risk in real estate investment will be described. These forms of risk will be outlined in light of the recent financial crisis. The chapter discussion will elucidate various statistical analyses that the individual investor can perform in quantifying the level of risk in a particular property. As a component of risk analysis, the concept of leverage will be discussed. The internal rate of return (IRR) will be revisited in the context of risk determination. The chapter will conclude with discussion of risk measurement from the perspective of the lender.

There are numerous definitions of the term risk. Risk can be defined as the probability of loss, the probability of not receiving what is expected, a difference (or potential variance) between expectations and realizations, the possible variance of returns relative to expected (or most likely) returns, and a chance or probability that an investor will not receive the expected or required rate of return that is desired on an investment (Phyrr and Cooper 1982).

6.1 Risk in Real Estate Investment

6.1.1 Business Risk

As with any class of investment, real estate investment is exposed to certain possible risks. Some of these risks can be diversified away to a certain extent, while others cannot. The underlying goal is not the complete removal of risk from an investment, as risk and return are often related. If the various risks can be segmented into categories that can be controlled by the investor, and those that cannot, there exists the potential to increase investor return. Business risk is typically seen as arising from fluctuations in the economy (Brueggeman and Fisher 2010). Business risk will appear in the financial statements of the investment property in changes from original projections in capital expenditures, gross potential income, vacancy factors and credit losses, operating expenses, and in the final property value. Business risk is inherent in each form of investment. As discussed at the end of Chap. 5, real estate investors can assess this risk, at least partially, via tenant lease rollover analysis. We say partially, as tenant lease rollover analysis only considers when leases are scheduled to mature, versus when tenants are forced to leave early due to poor management or a general decline in economic conditions. Empty retail and office properties were certainly on the increase in the aftermath of the "Yes era." While prognosticating about the possible deleterious effects of mass vacancies in properties that comprise an investor's portfolio of for-lease properties is not for the light-hearted, the analysis of business specific risks is certainly a worthwhile endeavor. Beyond conducting an adequate screening of the prior operating performance of any potential tenants, investors should also pay close attention to the overall changes in market demand, population changes, and changes in an area's base employment. Business risk can be partially alleviated by investing in multiple geographic areas or countries, and by diversifying the tenant base from an industrial perspective.

6.1.2 Management Risk

Concomitant with the discussion of business risk is management risk. As should be apparent from the preceding discussion in the book, having quality management for rental property is of paramount importance. While the United States Government considers real estate to be a passive investment from an income classification standpoint, the best real estate investors know that real estate property management is hardly a passive activity. Competent management on a daily basis is less likely for investment options such as stocks or bonds, which are actually classified as non-passive activities by the IRS (IRS 2011). While successful investors in stocks and bonds require expertise during the investment acquisition and divestiture process, and during times of periodic asset realignment, there is typically not a need for daily management of stock and bond investments, as is often the case in real estate investment.

Some property investors have the ability to keep a performing property leased and possibly possess the ability to lease-up an under-performing property. Many investors simply choose to outsource the management of the investment property to a qualified property management firm. As was discussed in Chap. 5, the complexity of the management responsibility of the property depends on the number of tenants, as well as how the leases are structured. The management of a single tenant property under a triple net lease agreement will typically involve curing any structural deficiencies at the location, and not much more than collecting the monthly rent check. If the tenant vacates the property, the property manager function could be employed to find a suitable replacement tenant. Barring any property vacancy, the day to day management of this type of leasing scenario is akin to the passive income classification for investment property. For multi-tenant investment properties, utilizing the property manager function is more compelling. For some investment properties, such as multi-family complexes or self-storage units, an on-site property manager may be required in order to handle daily issues that arise from the existing pool of tenants, and that arise from the desire to maintain a sufficient level of occupancy. The leasing activity for multi-tenant properties is more arduous than for properties with only a few spaces for lease. Larger multi-tenant facilities will either require the full attention of the investor, or a qualified property management firm.

Sometimes an owner misjudges the level of daily management that is required for keeping existing tenants satisfied with their leased space. Absentee management is often the result of such a misjudgment. The tenants' desires may go unfulfilled or in the most severe cases, basic property maintenance may not occur at a frequency desired to maintain the aesthetic and functional appeal of the property. The appearance of poor management can influence the tenant's success at the location, and also negatively affect the investment return for the property manager.

The appraiser and lending officer will include a fee for property management in the projected expenses regardless of whether property management is a paid expense. Even if the investor is planning to manage the property themselves without a fee, the property value will include an estimate of the typical management fee for an investment property. As is mentioned in Chap. 4, the valuation of property should include expenses paid by the average investor. One benchmark for this expense is from 5% to 8% of effective gross income (Jaffe and Sirmans 1982). The property manager is typically paid in a variable fashion so that there is an incentive to maximize occupancy and income for the property owner.

6.1.3 Liquidity Risk

Another important risk consideration in investment real estate is liquidity risk. This risk is associated with a lack of a deep market with many buyers and sellers. During the "Yes era," the frequency of investment real estate transactions were high, with investors and appraisers alike having multiple comparable properties from which to gauge their financial projections. In the aftermath of the financial crisis, liquidity risk surfaced on numerous fronts. On the micro front, investors were unable or unwilling to close as many new deals as in years past. On the macro front, investment banks such as Lehman Brothers experienced insolvency issues from banks not willing to lend to one another in the interbank lending market. Commercial banks of various sizes also experienced financial trouble as the quality of their loan portfolios quickly and sometimes violently declined. The initial response to the liquidity crisis was to create a $70 billion pool of funds to lend to troubled banks (Story and White 2008). This preceded the US Government's involvement in the crisis. For our purposes, this serves as a recent and clear example of liquidity risk.

The less liquid a market is, the more likely it is that price concessions will be offered by the seller. Real estate is inherently less liquid than other forms of investment given the time needed to consummate a sale. From the beginning of the search for an investment property to purchase, until the property ownership changes legally, six months could expire, on average. For more special purpose properties, such as nursing home facilities, restaurants, and marinas, the cycle time could further increase as there are less alternative uses for these properties, and thus less potential buyers. A lack of a deep potential sales market could also affect more traditional property types. Owning the nicest home in town may help inflate the owner's sense of self-worth, but the value of the home will serve to limit the pool of potential buyers. This knowledge helped create the "jumbo mortgage" market in the United States, with the sole classification for this category of loan being the amount of the indebtedness on the property.

6.1.4 Legislative Risk

Government intervention in financial markets also is represented in legislative risk. This particular risk is defined as a change in the regulatory environment which increases costs or makes certain types of lending unattractive. The aftermath of each economic recession typically brings a wrath of new legislative requirements for lenders and investors. The Great depression helped usher in policies such as the Glass-Steagall Act, which separated commercial and investment banking activities, while the savings and loan meltdown of the 1980s led to FIRREA and FDICIA. The Financial Institutions Reform, Recovery and Enforcement Act (FIRREA) dealt with ensuring that real estate appraisals met certain standards. The Federal Deposit Insurance Corporation Improvement Act (FDICIA) was passed in 1991 and

6.1 Risk in Real Estate Investment

increased the surveillance powers of the FDIC in an effort to help consumers. The Basel Accord from the Bank for International Settlements (BIS) went into effect after the Asian Financial Crisis in the mid-1990s, and the wake of the most recent financial crisis introduced a host of regulatory reform. While the opinion as to whether such reform will improve or prove detrimental to the economy is an open question, what is of primary concern for investors and lenders is that the real estate sector is highly-regulated. Government mandates could stem from the municipal level, such as with rent controls on apartments in New York City, or the legislation could be of federal origins. Based on the prior trial and error of government intervention in financial markets, which has typically been conceived as a response to poor bank lending and speculative fervor, the only constant on the legislative front is change.

Prudent investors will follow possible changes in tax laws, rent controls, zoning allowances, and other restrictions that can seriously impact the return that an investor desires. Based on our earlier discussion of risks that can be controlled, savvy investors will attempt to eliminate the risks that they can indeed control, and will ensure that the necessary permits and zoning are in place when considering both new construction projects as well as existing investment properties.

6.1.5 Inflation Risk

Another area of concern in real estate investment is inflation risk. This risk surfaces when the income increase experienced during the investment holding period does not keep pace with overall price level or operating expense increases. In Chap. 3, we discussed the importance of present value in investment decisions. In Chap. 4, we illustrated how the concepts of present value and discounting could be utilized when valuing investment property. In Chap. 5, we discussed the importance of structuring leases so that the investor is not solely at the mercy of inflation. Prudent lenders will typically consider the impact of expected inflation when they are pricing loans. Considering the discussion from prior chapters with relevance for inflation risk, real estate performs well, as compared with other investments, in tempering the impact of inflation risk. Additionally, up until the recent financial crisis, real estate property appreciation has compared favorably with inflation over time. As mentioned in our Chap. 1 discussion concerning speculative investment, there are times when investors purchase an investment property not for its current income, but for the potential of increased property appreciation. In order for this to be a successful strategy, the property value appreciation will need to exceed the average increase in prices (i.e. inflation). While over the long run this has been a positive endeavor for many investors, the success of this strategy will largely depend on the strength of the market and location of the property within the market.

6.1.6 Interest Rate Risk

Interest rate risk is something that will effect most investments, with real estate not being an exception. Typically, real estate is highly leveraged as many investors seek partnerships with financial institutions in order to make their investment dream a reality. The lender and the investor desire as much certainty as is possible when it comes to the prevailing rate of interest. As interest rates rise, investors have less of a motivation to borrow money and may have trouble in continuing to remain current on their existing obligations if the rate is variable. The problem with variable rate mortgages has received an inordinate amount of press coverage in recent years at both the domestic and international level. While adjustable rate mortgages (ARM), are primarily a residential mortgage structure, commercial investment property mortgages can be structured on a variable rate basis. At its core, both fixed and variable rates are determined based on the lender's cost of funds.

Students of finance and economics will recall that there are three primary methods where the Federal Reserve affects the level of interest rates via the manipulation of the money supply (Mishkin 2009). One method is via open market operations, where the Fed buys and sells government bonds in order to increase or decrease the supply of money in the economy. If the policy objective is to increase the supply of money, the Fed will buy government bonds, in effect trading those bonds for new money in the economy. As the supply of money in the economy rises, the price of money, the interest rate, declines. If the Fed wants to contract the money supply (and effectively increase interest rates), the Fed will sell the same bonds to the banks.

The discount rate is a second mechanism for affecting the supply of money in the economy. The discount rate is the lending rate for commercial banks at the Fed's discount window. The discount rate is distinct from an investor's discount rate that we have discussed previously. It is also different from the federal funds rate, which is the bank to bank lending rate. In recent years the discount rate has been approximately 50 basis points lower for primary credit banks than for secondary credit banks (Federal Reserve 2011). Typically, banks will maintain the spread between the discount rate and the prime rate for lending purposes to their clients, as the discount rate changes. Thus, policy decisions for changing the discount rate are passed to customers via the bank's prime rate lending benchmark. During the height of the "Yes era," the Federal Reserve increased the discount rate 17 consecutive times from 2004 to 2006. The policy objective of increasing the discount rate was to quell what former Fed chairman Alan Greenspan called "irrational exuberance" in the markets. This example makes clear a primary point: moving the interest rates upward will only stem borrowing if the lenders experience credit losses in their loan portfolios. The amount of loan losses and defaults will temper the lending appetite quicker than any increase in borrowing costs. Another factor affecting the supply of credit is the cost of loan originations. If the cost of loan origination rises relative to other investment opportunities, the volume of credit on offer should decline. The desire to reduce costs in the face of a rising cost of funds led many commercial

6.1 Risk in Real Estate Investment

banks to outsource and off-shore backroom banking support positions during the "Yes era" (Goddard and Ajami 2008).

A final primary method of controlling the money supply and interest rates is via reserve requirements. The Fed can mandate that banks increase the level of reserves, which will then reduce the amount of funds that the banks can lend out. Increased reserve requirements leads to less of a money multiplier effect as less money is being injected into the economy for economic expansion. The Basel Accord noted earlier in this chapter was an internationally sanctioned effort to require banks to hold reserves in relation to the overall risk levels in their loan portfolios.

A final determinant of interest rates is on the demand side of the equation. The demand for credit, and thus money, will be affected based on the overall economic vibrancy of the general economy, and the stage of the economic cycle for a given market. This point links the discussion in Chap. 2 with interest rate risk.

Interest rate risk is present for both the investor and the lender. The lender, when pricing a loan, should consider the effects of inflation in the interest rate offered to the customer. The real rate of interest considers both the nominal, or stated interest rate, and a premium for inflation expectations. The real rate of interest was theorized by Yale economist Irving Fisher (Fisher 1930). The real rate of interest is the minimum rate where savers will agree to forego current consumption and to save. This increased savings adds to the supply of funds with which banks can lend.

Now that we have discussed the factors that influence the cost of borrowing, the next question for discussion is who might bear the risk when interest rates rise and fall. Investment real estate loans are priced on either a variable or fixed rate basis. Fixed interest rate loans are more common, but some investors prefer variable rate loans. During the "Yes era," there was a prolonged period of low interest rates. Thus investors paying variable interest rates were saving on interest expense that they would have had to bear had they locked into fixed rates at a higher level than the prevailing interest rate of the current time. Variable rate loans are typically priced relative to a market index, such as the 10-year U.S. Treasury, LIBOR, or the bank's prime rate. Assume that an investor has a loan on a variable rate. The loan is priced at 200 basis points higher than the 1-month contract LIBOR rate. Depending on the underlying value of the index, the variable rate of interest could vary considerably during the loan term. Table 6.1 shows actual movements of LIBOR for various dates over the last few years.

As you can see from the table below, the rate of interest paid by the investor ranges from a low of 2.25% to a high interest rate of 7.4375%. As a means of comparison, if the investor had a variable rate loan during the early 1980s, the rate paid on the loan would have approached 20%!

Obviously, as the variable rate index increases, so does the investor's interest rate risk. The lender is not exposed to interest rate risk under a variable rate pricing scenario. The lender is exposed to default risk, or the risk that the loan will not be repaid, should rates rise to precarious levels. The lender may also be exposed to prepayment risk, where the desired rate of return is impacted by the early repayment of loan principal. Variable rate loans do not have prepayment penalties which

Table 6.1 One-month contract LIBOR and loan rates over time

Date	LIBOR (%)	Loan rate (%)
4/4/2003	1.2500	3.2500
7/6/2004	1.3750	3.3750
4/25/2005	3.0625	5.0625
6/5/2006	5.1250	7.1250
12/5/2007	5.4375	7.4375
5/9/2008	2.6250	4.6250
6/6/2009	0.3200	2.3200
11/11/2010	0.2500	2.2500
2/8/2011	0.2600	2.2600

are found on many fixed rate loans. Prepayment penalties can be calculated in various ways. The penalty may be a set percentage of the outstanding loan balance, or a set percentage of the amount that the loan is prepaid. The percentage of penalty may change over time, like for example in a 3-2-1 step down over the first 3 years of the loan term. Another common form of prepayment penalty is "make whole." If the customer attempts to payout or prepay a loan where interest rates are higher than the current market rate of interest, the bank is entitled to obtain a fee which allows the bank to recoup the costs associated with the early payout. How prepayment penalties are calculated will be specified in the loan documents.

A traditional fixed rate loan eliminates the upside risk of interest rate movements during the term of the loan for an investor. The lender is partially insulated from default risk as the interest rate alone will not rise to a point where loan repayment is threatened. The fixed rate does not protect the lender from other reasons where a customer could not meet their payment obligations.

Products such as interest rate swaps allow the bank to earn a fee for providing the opportunity for an investor to obtain a fixed rate. In the traditional fixed rate (rate exposure remains on-balance sheet for the bank) scenario, an investor benefits from locking into a fixed rate prior to rates increasing. The lender would lose out on additional interest above the contract rate of interest. Thus the fixed rate poses interest rate risk for the lender, if the contract rate of interest is less than the current prevailing market rate. An interest rate swap is a derivative product as the swap is *derived*, or based on, the underlying loan in place.

The swap product allows the bank to move the risk associated with rates increasing off of their balance sheet, as the swap represents a contract between the borrower and a counter party. The counter party to the swap is an investor looking to profit should rates drop below the contract rate of interest. Each month, the borrower's swap is settled by the bank, whereby the borrower will have to pay additional interest to the counterparty (via the bank's derivatives area) should the prevailing rate of interest in the current market be lower than the contract rate of interest. For example, assume that a borrower locks in a fixed rate via a swap at 6%. The term of the rate commitment is 5 years. At the time that the swap was executed, the 6% rate represented the prevailing rate of interest for investment real estate properties of average risk.

6.1 Risk in Real Estate Investment

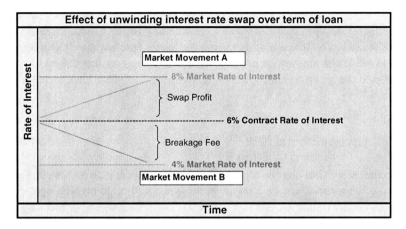

Fig. 6.1 Effect of unwinding an interest rate swap over the term of a loan

As shown in Fig. 6.1, over the term of the loan, market interest rates could follow two trajectories, as labeled market movements A and B.

Under market movement A, during the term of the loan, market rates rise above the contract rate of interest of 6%. Said another way, during the loan term, new swaps are locked in at interest rates above the 6% contract rate obtained by the borrower. Thus the original decision to lock in at a rate of 6% was prescient, and the borrower now has something of value. If the borrower decides to "unwind the swap," they will receive some level of profit given the upward movement in market interest rates since the swap was executed.

Under market movement B, interest rates take a negative trajectory during the term of the swap. Under this scenario, the borrower has locked in at a rate higher than what borrowers of similar risk profile might obtain in the current environment. Most borrowers would rue their original decision of locking in on the swap rate, and would prefer that the swap just go away! Unfortunately, since the rate that the borrower is paying is higher than the prevailing market rate of interest, the borrower would have to pay a breakage fee if they chose to unwind the swap.

Swaps have presented lenders with risk exposure in the aftermath of the financial crisis. As rates have fallen considerably in recent years, many borrowers who locked in swaps during the height of the "Yes era" (2006–2007), have found themselves in a situation exhibited in market movement B. Since many borrowers were likely to swap the entire loan amount, versus only swapping a portion of the loan exposure, lenders desiring for customers to find alternate lenders now have the added problem of the breakage fee for customers already in precarious situations. Swaps can have a portability feature, where if the underlying collateral is sold, the investor could utilize the swap for a similar property. How portable the swap is will depend on the sophistication level of bank considering the swap, and whether the swap is "in the money" or not.

The term "swap" was preceded in banking parlance with the term "hedge." The name change might have been necessitated by swaps being seen as an investment vehicle rather than simply as a hedging tool for interest rate exposure, at least for the borrower desiring movement from variable to fixed exposure. It remains to be seen if swap risk will continue as financial markets recover. It will also be left for time to tell if the concept of hedging returns to managing interest rate exposure, rather than the all or nothing speculative investment as sometimes seen during the "Yes era."

6.1.7 Environmental Risk

Environmental risk is one classification of risk that is specific to real estate. Environmental contamination can be a very expensive risk and could involve legal disputes between aggrieved parties. Environmental risk concerns the specific property and its effect on land, water, air, sewage, and aesthetics of the surrounding area and of the community at large (Pyhrr and Cooper 1982). Property investors must concern themselves with how their property and how the operation of business on site effects air, water, and noise pollution. This concern stemmed from the National Environment Policy Act of 1969 (NEPA) (Jaffe and Sirmans 1982).

Air rights affect the land owner via government mandated legally permissible maximum levels of pollution in the air. In countries subject to the Kyoto Protocol, aggressive government intervention is prevalent in order to help ensure that pollution levels do not rise above agreed upon levels (Ajami et al. 2006). The administration of air rights is not only for punitive purposes. High rise properties and properties located in urban areas allow the possibility of landlords leasing a portion of the air immediately above a property, and sufficiently below air space denoted for air travel, in the form of billboards. This allows for increased revenue opportunity.

Mineral rights, or subsurface rights, pertain to the landlord's rights regarding mineral, oil, and gas deposits on the property. Since deep excavation is typically reserved for governmental entities, the mineral rights issue will typically impact a property owner if deep below ground excavation impacts the property or land at the surface to a point where the land owner experiences a loss of their right to quiet enjoyment of their property or experiences a deleterious effect on property value. A famous example of this was when foundational damage occurred to several older buildings on Peachtree Street in Atlanta during the tunneling for construction of the MARTA subway system (Buchanan and Johnson 1988).

Riparian rights refer to the right to the use of water whereby all land owners should have equal access, as long as their use does not deprive others of equal use. The source of contamination could prove very costly to a land owner if contamination on their property enters into the water underground and is the cause of spreading contamination to adjacent properties.

Another water related risk is the threat of flood. The Federal Emergency Management Agency (FEMA) administers the National Flood Insurance Program in the United States which classifies some lower lying and coastal areas as being in designated flood zones (FEMA 2011). If a property is located in a flood zone,

Key manufacturing industries	Hazardous wastes produced
Metal finishing, Electro-plating, etc.	Heavy metals, fluorides, cyanides, acid and alkaline cleaners, abrasives, plating salts, oils, phenols
Leather tanning	Heavy metals, organic solvents
Textiles	Heavy metals, toxic organic dyes, organic chlorine compounds, salts, acids, caustics
Pesticides	Organic chlorine compounds, organic phosphate compounds, heavy metals
Pharmaceuticals	Organic solvents and residues, heavy metals (esp. mercury)
Plastics	Organic solvents and residues, organic pigments, heavy metals (esp. lead, zinc)

Fig. 6.2 Industries producing hazardous wastes

prudent investors and lenders will require that flood insurance be in place up to allowable limits. In general, property insurance should be in place up to the replacement value of the improved property in order to help alleviate the concerns of loss associated with insurable risks such as fire and other property hazards. If a property is located in a flood way, property owners may have trouble selling or encumbering the property, as if a flood comes to the area, the property has been estimated by FEMA to be in the direct line of where the flood waters will most likely flow.

The remaining environmental risks involve factors which are specific to past or current uses of the subject property. The environmental risk could emanate from how a property is constructed. Buildings with asbestos, lead based paint, or other contaminants may prove costly to remediate. The presence of storage tanks or heating oil tanks, whether above or below ground, can also be a source of environmental concern. Lastly, current or historical tenant uses of property can prove environmentally troublesome. Uses such as dry cleaning, the manufacture or disposal of hazardous materials, and any other use that involved the application of chemicals which may have contaminated the land, are problematic from an environmental risk standpoint. Business operations such as gas stations and auto repair shops can prove damaging just from the normal course of business. Figure 6.2 summarizes some key manufacturing industries that are known to produce hazardous wastes (Leonard 1986).

Management of environmental risk is an excellent example of the partnership between lenders and investors. Most commercial lenders have set requirements for which types of environmental risk are acceptable. Once all possible environmental risks are known, the investor will then evaluate the total risk of the property in comparison with alternative investments.

A process considered fairly standard will help illustrate environmental risk management. The investor must first answer a series of questions posed by the lender in order to risk rate a given property from an environmental perspective. If any probable issues are identified, the lender will require an environmental database report. This will identify any prior issues from an environmental standpoint reported in the past for a given property and the surrounding area. If an issue is identified, the lender will most probably require a phase one environmental report. This report is prepared by a certified expert in environmental risk assessment. The report will highlight prior usage of the property and whether any contamination is present on site. If no problems are noted, the process may stop at this point. This step will also involve disclosure of any underground storage tank leak tests that may have been performed, and whether the state authority has issued a "no further action letter" to confirm that a prior contamination issue has been remediated. If current issues remain unresolved, the next step is a phase two environmental report. This assessment includes soil testing, as well as the testing of groundwater and air, if necessary. The report will offer conclusions for estimated costs of any remediation that is required.

Any unresolved environment problem will reduce the appraised value of the subject property typically by the cost of the remediation. Such issues may also impact a lender's willingness to finance the property. It is recommended that environmentally suspect properties be avoided under the "life is too short" principle.

6.1.8 Financial Risk

The last risk that we will highlight in this section is financial risk. This is the risk associated with debt financing. During our discussion of interest rate risk, we itemized the problems posed for investors and lenders from a repayment standpoint when interest rates rise, or conversely when rates fall and a borrower is subject to breakage fees for interest rate swaps. Financial risk is concerned with the amount of leverage for a real estate property.

Since risk and return are positively related, then it should follow that there should be a similar relationship between risk and leverage. This relationship actually depends on the cost of debt relative to the expected return that the investor hopes to receive. Typically, the more leverage, the greater the risk for the lender. Psychologically speaking, a borrower with less of their own money at risk, via an equity injection, will have less of a motivation to keep their payments current as the property experiences occupancy or other performance related issues. Additionally, the higher the loan to value, the less room the lender has when negotiating a sale price for the property during a foreclosure situation.

Sometimes the comparison of expected costs and expected benefits for additional leverage for the investor is fairly simple. For example, if a certificate of deposit is currently paying 3% interest, and the commercial lender is charging an interest rate of 4% on the investment real estate loan, ceteris paribus it is better to increase the amount of equity into a given purchase rather than paying the higher rate of interest for a higher loan amount. Borrowers can justify saving in this situation owing to factors outside of the simple interest rate comparison, such as having available funds in the case of emergency. Sometimes the comparison point between increased leverage and expected return can be harder to discern. This decision path will be clarified in the following section on leverage effects.

6.2 Leverage Effects

Suppose that an investor is considering financing options at a commercial bank. The investor informed their lender that they would like to finance as much as possible on a purchase of an investment property for $1.6 million. The lender is willing to quote two loan amounts. The first represents a loan to value of 75%, for a loan offering of $1.2 million. This loan is priced at 6% with payments based on a 240 month amortization. The lender is also willing to quote the terms for a loan of $1.360 million, which equates to a loan to value of 85%. This loan is priced at a fixed rate of 7% with payments also based on a 240 month amortization. Both loan offerings have terms of 5 years, thus the rate is good for that amount of time.

Assuming that the investor has the equity requirements for each loan scenario, the question then becomes how the incremental cost of debt between 75% and 85% loan to value compares with either the investor's expected internal rate of return on the investment, or with any competing uses for the funds in terms of alternate investments. Rather than viewing the two loan offerings in isolation relative to the projected investment return, the effective interest rate over the 5 year rate commitment period should be calculated. This is shown in Fig. 6.3.

The first step in calculating the incremental cost of debt is to determine the differences in monthly payments under both loan options. The second step is to determine the difference between outstanding loan balances for both loans at the end of the 5 year rate commitment period. Once this is accomplished, we have identified the difference in starting loan amounts, the difference in monthly loan payments, and the difference in ending loan balances after 5 years. If the starting and ending loan amounts are viewed as present value and future value respectively, the determination of the effect of increased leverage is determined by solving for the interest rate. As shown in Fig. 6.3, the incremental cost of debt under this scenario is 14.10%.

Positive leverage is achieved when the investor is borrowing at an interest rate lower than the expected rate of return on the total funds invested in a property. Said another way, positive leverage is achieved when the before tax internal rate of return (BTIRR) for equity and debt exceeds the BTIRR of debt alone. Formally, positive leverage is defined as follows (Brueggeman and Fisher 2010):

$$BTIRR_e = BTIRR_p + (BTIRR_p - BTIRR_d)(D/E)$$

LTV	Loan	Interest	Payment	Future Value
85%	$ 1,360,000	7.00%	$ 10,544	$ 1,173,090
75%	$ 1,200,000	6.00%	$ 8,597	$ 1,018,795
Difference	$ 160,000		$ 1,947	$ 154,295
	PV	14.10%	PMT	FV

Fig. 6.3 Calculation of the incremental cost of debt for competing loan offerings

IRR	No Loan	75%	85%	
Before Tax	8.00%	10.50%	12.50%	
After Tax	6.50%	8.00%	9.00%	
Loan Cost	No Loan	75%	85%	Incremental
Before Tax	NA	6.00%	7.00%	14.10%
After Tax	NA	4.32%	5.04%	10.15%

Fig. 6.4 Decision point for incremental cost of debt

Where:
- $BTIRR_e$ is the BTIRR on equity invested
- $BTIRR_p$ is the BTIRR on the total property investment
- $BTIRR_d$ is the BTIRR on debt
- D/E is debt divided by equity

The formula is the same for the after tax IRR, as BTIRR is substituted with ATIRR in the formula.

Negative leverage is experienced when the cost of debt is higher than the expected rate of return on the total funds invested. In this situation, the more an investor borrows the worse their returns become. In a corollary to this, it would be similar to a company whose cost of goods sold is higher than their gross revenues attempting to make it up on volume!

In either case, the effect of increased leverage is increased financial risk since there is more indebtedness and increased variability of returns of before and after tax cash flow and equity reversion. Risk, as measured by standard deviation, will always increase with leverage. Whether the expected return increases depends on whether the leverage is positive or negative.

The maximum interest rate paid on debt before leverage becomes negative is known as the break even interest rate. The BTIRR of debt is equal to the ATIRR of the project divided by one minus the investor's tax rate. This is shown in the formula below:

$$BTIRR_d = \frac{ATIRR_p}{(1-T)}$$

In order to illustrate the decision point for whether increased leverage is favorable or not, Fig. 6.4 is presented.

The 75% return has positive financial leverage as the return increases on both a before and after tax basis. This is due to the "no loan" return being higher than the cost of debt. The "no loan" return after tax is 6.50%, and the after tax cost of debt is 4.32%. This is calculated by taking the before tax cost of debt of 6%, and multiplying by one minus the assumed tax rate of 28%.

The 85% return has negative financial leverage as the incremental cost of 14.10% is higher than the "75% return" of 10.50%. In this example, the investor is better off choosing the 75% loan to value proposal.

6.3 Statistics and Risk

Sensitivity analysis is an important component of financial risk management. Once an investor has selected a property for prospective investment and has completed their initial pro-forma operating projections as discussed in Chap. 4, the next step is to vary key assumptions in the possible operating performance of an investment property. The vacancy factor can be utilized to project the impact that vacancy increases have on the performance of the subject property. One method of initial assessment is break even analysis. This is performed for the first projected year of revenue and expenses. By taking the sum of annual operating expenses and annual debt service and dividing by the projected gross potential income, the break even occupancy can be determined. Additionally, property operating expenses, reversion cap rates, and the interest rate on the debt can be manipulated in order to arrive at three operating performance possibilities by using the multiple period projection view. These three alternate projections are known as the most likely case, the pessimistic (or worst) case, and the optimistic (or best) case. The investor would compute what the expected NPV and IRR would be under the three scenarios, while the lender would focus on the resulting debt coverage ratio and loan to value under the three case scenarios.

Since risk is defined as the variability of the return from expectations, properties with a higher variability in possible returns are of higher risk than properties with a narrower band of possibilities. Our "QQD" framework should be revisited during this discussion. If a property has tenants of a higher financial quality with leases of a longer length, this property should on average have less variability than properties not meeting the QQD criterion.

Once the investor's expected returns are computed based on the three iterations, the expected return is calculated. The investor should assign probabilities to each of the three scenarios so that a weighted average expected return can be calculated. As discussed in Chap. 3, the expected return is defined as the internal rate of return for the investment over the projected holding period. Included in this analysis is the projected resale of the property at the end of the holding period. The expected return for a property is analyzed in Fig. 6.5.

	IRR	Weight	Product
Worst Case	8.37%	30%	2.51%
Most Likely	10.79%	40%	4.32%
Best Case	13.64%	30%	4.09%
Expected Return			10.92%

Fig. 6.5 Calculation of expected return

Estimated BTIRR	Expected Return	Deviation	Squared Deviation	Weight	Product
8.37%	10.92%	-2.55	6.5025	30%	1.9508
10.79%	10.92%	-0.13	0.0169	40%	0.0068
13.64%	10.92%	2.72	7.3984	30%	2.2195
				Variance	4.1770
Range	8.88%	10.92	12.96	Std. Dev	2.044

Fig. 6.6 Calculation of variance, standard deviation, and range of returns

As you can see from Fig. 6.5, three scenarios have been calculated, with weights assigned to each. The resultant weighted average is 10.92%. Once the weighted average expected return is known, the investor should then compare the variability of returns between the three scenarios. This analysis is shown in Fig. 6.6. As mentioned previously, the higher the variability the higher the risk. The variance is estimated based on taking the squared deviation from the expected return for each of the three scenarios, and then multiplying by the probability of each scenario becoming a reality. The variance is calculated in Fig. 6.6 as 4.1770.

The next step would be to convert the variance into a standard deviation. The standard deviation is a measure of dispersion about the mean. If we assume that we have a normal distribution, the standard deviation can then be utilized to provide a specific range where the actual investment return will fall. In a recent book which has received much notoriety, Nassim Nicholas Taleb theorized that the normal distribution was too prevalent in financial metrics (Taleb 2010). This was partly to blame for modeling errors seen during the "Yes era." For our purposes, the normal distribution will be assumed, as at the individual investor level, if the investor was not able to gain confidence at some level that the financial risk modeling was not normally predictive, they would in all probability forgo investing in the property.

As shown in Fig. 6.6, the standard deviation is the square root of the variance. This produces a result of 2.044. If this amount is added to and subtracted from the expected return, a range of returns is revealed. This range is, statistically speaking, the range of return percentages that the investment revealed could produce with 68% probability. If this process is repeated for several other potential investment properties, the investor can determine which properties are more risky relative to competing investment properties.

6.4 Partitioning the IRR

As was discussed in Sect. 3.6.1, partitioning the internal rate of return is another method of assessing risk. Ceteris paribus, if a higher percentage of the projected cash flows from an investment property are from the prospective sale of the property at the end of the holding period, as opposed to cash flows received from operations during the holding period, then there is more risk inherent in the investment opportunity from a total return perspective. The most appropriate method for assessing this risk is to evaluate the present value of the cash flows from an investment property, and to then separate (or partition) the reversionary cash flows from the cash flows from operation. While this process was discussed in more detail earlier in the book, we would be remiss if we did not mention this technique in light of the present discussion on risk assessment. A variant to this technique is the risk absorption ratio. This is the annualized net present value divided by the initial equity investment, where annualized NPV is the maximum amount by which cash flow each year could be reduced without reducing the NPV to below zero. The risk absorption ratio measures the risk-absorbing ability per dollar of the investment. Competing projects can be ranked via this technique, as well as via the partitioning of the IRR.

6.5 Due Diligence Analysis

Due diligence is the process of discovering information needed to assess whether an investment risk is suitable given the objectives of the investor and the lender. This is an example of where the partnership between the investor and the lender can serve both parties well. The due diligence process attempts to evaluate all of the various risks that have been discussed in this chapter which are pertinent to the property being considered for investment.

Business and financial risk can be evaluated by reviewing the property rent roll and other financial documents. The rent roll lists the tenants occupying the subject property and should also itemize the square footage for each tenant, the beginning and ending occupancy dates, the annual rent paid, and any common area reimbursements. The investor and lender will review this document along with other financial documents in order to determine the QQD for the property. Copies of all commercial leases should be obtained, along with copies of service and maintenance contracts for the subject property. The historical profit and loss performance for at least the last 3 years (if applicable) should be obtained in order to assess trends regarding revenues and expenses. As discussed in Chap. 1, title and deed documents should be reviewed in order to assess if any issues exist regarding the ability to convey the property from the seller to the buyer.

Regarding physical property risk assessment, a survey of the subject property should be reviewed in order to assess whether any encroachments exist, or if the property may be near a flood zone or other hazardous areas from an environmental standpoint. On the legislative front, the investor and lender should ensure that zoning and other government mandated requirements are satisfactory, and both should verify that property taxes and insurance payments are up to date. Additionally, both the buyer and the lender should physically inspect the property in order to determine if an engineering study or other expert inspection is required. For new construction, a market study could be requested from an appraiser. This report, along with a feasibility study, is concerned more with the supply and demand dynamics in the given market than with simply valuing the subject property. These reports will allow the lender and investor to assess the current and future trends in supply and demand in the area, and to determine if new construction will be absorbed successfully by the market, at least in the near term.

When approaching a bank for financing of an investment property, the investor should expect to provide certain items to facilitate the lender's evaluation of the applicant and the subject property.

Items traditionally required by commercial banks include:
- Current/prospective rent roll
- Copies of leases
- Purchase contract
- 3 years business and personal tax returns
- Current personal financial statement for loan applicants or guarantors
- 3 years profit and loss for subject property (if not disclosed on tax returns)
- Prior appraisal, phase I, and survey (if applicable).

The investor should also expect to be asked to comment on their management ability and their relevant experience, and possibly to provide unbiased references.

6.6 Conclusion

At this point, the reader may question why, given all of the various risks associated with real estate investment, anyone would want to purchase an investment property. Expected returns and diversification aside, prudent real estate investors should perform due diligence *prior* to purchasing an investment property. Adequate risk analysis prior to purchase should limit the amount of surprises for the investor during the holding period. If the risks are seen as too high relative to the expected rewards, then the investment opportunity should not be pursued. Good due diligence and sound property management, along with risk transfer strategies such as insurance coverage, limited partnerships, and perhaps non-recourse financing, should serve the investor well as they contemplate the purchase of an investment property that has adequate returns relative to the risks known at the time of purchase. Other benefits from a taxation perspective will be discussed in Chap. 7.

Questions for Discussion
1. Elaborate on how leverage may lead to lower investment returns and how investors can verify when this might or might not be the case.
2. Differentiate between the many forms of risk in investment real estate.
3. Explain how partitioning the internal rate of return can help an investor differentiate between competing investments.

6.6 Conclusion

4. Itemize factors that should be considered when analyzing the environmental risk of an investment property. Make sure to include the business conducted by the tenants on site as part of your answer.
5. Explain how statistical measures such as standard deviation and variance can aid an investor in assessing the risk of a particular investment.

Problems

1. Von Julio Investments is considering the purchase of an investment office building located in Paramus, New Jersey for $2 million. NOI is expected to be $215,000 in the first year, increasing 3% annually. The property is expected to appreciate by 3% annually and is expected to be sold in 5 years. The building and improvements represent 80% of value and will be depreciated over 39 years. Assume a tax bracket of 32% for all income and 20% for capital gains.
   (a) Calculate the BTIRR and ATIRR assuming 70% financing at 7% for 25 years.
   (b) Calculate the BTIRR and ATIRR assuming 80% financing at 8% for 25 years.
   (c) Calculate the break-even interest rate for this project.
   (d) What is the marginal cost of the 80% loan? What does this mean?
   (e) Does each loan option represent favorable leverage? What would you recommend?
2. An investor is considering between three competing projects all requiring an equity investment of $125,000 at the beginning of the investment period. The first investment is projected to return NOI of $15,000 the first year, $18,000 the second year, $21,000 the third year, $27,000 the fourth year, and $30,000 the fifth year. Upon sale at the end of the fifth year, the investment is projected to return net income of $150,000. The second investment is projected to return NOI of $16,000 the first year, $12,000 the second year, $20,000 the third year, $28,000 the fourth year, and $25,000 the fifth year. Upon sale at the end of the fifth year, the investment is projected to return net income of $155,000. The third investment is projected to return NOI of $14,000 the first year, $16,000 the second year, $18,000 the third year, $20,000 the fourth year, and $24,000 the fifth year. Upon sale at the end of the fifth year, the investment is projected to return net income of $150,000.
   (a) What is the BTIRR for each investment option?
   (b) Partition the BTIRR for each investment option using a discount rate of 10%. What percentage of total income receipts is represented by operations versus sale at the end of the period?
   (c) Based on this information, which investment option contains the least risk?

3. An investor is considering two projects. He has estimated the BTIRR for each under three possible scenarios and assigned probabilities of occurrence for each scenario. The chart below itemizes the two investments.

	Investment A Prob (%)	BTIRR (%)	Investment B Prob (%)	BTIRR (%)
Optimistic	30	20	25	20
Most likely	50	14	50	15
Pessimistic	20	10	25	8

(a) Compute the range of expected returns for both investments
(b) Which is the best choice and why?

4. Market research has shown that the town where an investment property is located has a total of 500,000 square feet of office space, of which 375,000 square feet is currently occupied by a total of 3,500 employees in professional services such as real estate, insurance, and banking. Given the recent global recession, employment growth is expected to be somewhat slow over the next 4 years. It is projected that only 50 new employees will be needed in each of the next 4 years. The amount of space per employee is expected to remain constant. A new 50,000 square foot office building was started before the recession and its space is expected to become available at the end of the current year (1 year from now). No more space is expected to become available after that for quite some time.
   (a) What is the current occupancy rate for office space in this market?
   (b) How much office space will be absorbed each year for the next 4 years?
   (c) What will the occupancy rate be at the end of each of the next 4 years?
   (d) Based on the above analysis, do you think it is more likely that office rental rates will rise or fall over the next 4 years?

5. An investor has projected three possible scenarios for a project as follows:

Optimistic: NOI will be $150,000 for the first year and will then increase by 3% per year over a 5 year holding period. The property will sell for $1.7 million after 5 years.

Most Likely: NOI will be $150,000 for the next 5 years. The property will sell for $1.5 million after 5 years.

Pessimistic: NOI will be $150,000 for the first year and will then decrease by 2% per year over a 5 year holding period. The property will sell for $1.3 million after 5 years.

The asking price for the property is $1.5 million. The investor believes that there is a 30% probability for the optimistic scenario, a 40% probability for the most likely scenario, and a 30% probability for the pessimistic scenario.
   (a) Compute the IRR for each scenario.
   (b) Compute the expected IRR.
   (c) Compute the variance and standard deviation of the IRRs.
   (d) How might the investor determine if this project is worth pursuing?

Mini-Case: Sensitivity Analysis

When evaluating an existing investment property, a sound due diligence on the part of the investor and the lender includes sensitivity analysis. The sensitivity analysis is essentially a what-if scenario that seeks to find how the risk inherent in an investment property. The risk is defined by increases in vacancy rates and interest rates primarily.

Assume you have a property exhibiting the following financial characteristics:

Tenant	Sq. Ft.	Annual rent
Winn Dixie	22,000	$ 462,000
Dollar general	6,000	$ 123,000
Franco's Pizzeria	2,500	$ 60,000
Uncle Jimmy's Toy and Hobby	4,000	$ 92,000
Total gross revenue	**34,500**	**$ 737,000**
Market vacancy factor	10%	$ (73,700)
Effective gross income		**$ 663,300**
Property taxes		$ 75,000
Insurance		$ 27,000
Repairs and maintenance		$ 55,000
Utilities		$ 18,000
Management fee	5%	$ 33,165
Other expenses		$ 5,000
Replacement reserve	1%	$ 6,633
Total operating expenses		**$ 219,798**
Net operating income		**$ 443,502**
Annual debt service		**$ 305,000**
Debt coverage ratio		**1.45**

Questions for Discussion

1. What is the break-even occupancy percentage for the subject property?
2. Assume that annual debt service is calculated over a 20 year amortization for a loan amount of $3 million, with an interest rate of 7.97%. What would the interest rate have to rise to (under the current situation) in order for the net operating income to equal the annual debt service?
3. How many of the current tenants does the 10% vacancy factor cover in the current underwriting? How would the DCR look if both of the two local tenants vacated the property?
4. Research the financial health of Winn Dixie and Dollar General on the internet. How strong are they from a financial standpoint? Any concerns about the bulk of your NOI coming from these two tenants?

References

Ajami, R., Khambata, D., Cool, K., & Goddard, G. J. (2006). *International business theory and practice* (2nd ed.). Armonk: M.E. Sharpe.

Brueggeman, W. B., & Fisher, J. (2010). *Real estate finance and investments* (14th ed.). New York: McGraw-Hill.

Buchanan, M. R., & Johnson, R. D. (1988). *Real estate finance* (2nd ed.). Washington, DC: American Banker's Association.

Federal Reserve Website (2011). http://www.frbdiscountwindow.org/historicalrates.cfm?hdrID=20&dtlID. Accessed February 2, 2011.

FEMA Website. (2011). http://www.fema.gov/about/programs/nfip/index.shtm. Accessed February 10, 2011.

Fisher, I. J. (1930). *The theory of interest*. New York: Kelley.

Goddard, G. J., & Ajami, R. A. (2008). Outsourcing: Which way forward? An essay. *Journal of Asia Pacific Business, 9*(2), 105–120.

Internal Revenue Service. (2011). Passive activities, IRS website. http://www.irs.gov/businesses/small/industries/article/0%2C%2Cid=98881%2C00.html. Accessed February 1, 2011

Jaffe, A. J., & Sirmans, C. F. (1982). *Real estate investment decision making*. Englewood Cliffs, NJ: Prentice Hall.

Leonard, H. J. (1986). Hazardous waste: The crisis spreads. *National Development* (April): 44.

Mishkin, F. S. (2009). *The economics of money, banking, and financial markets* (2nd ed.). Boston: Addison-Wesley.

Pyhrr, S. S., & Cooper, J. R. (1982). *Real estate investment: Strategy, analysis, decisions*. New York: Wiley.

Story, L., & White, B. (2008). The Road to Lehman's Failure was littered with lost chances, *New York Times*, October 5, 2008.

Taleb, N. N. (2010). *The black swan: The impact of the highly improbable* (2nd ed.). New York: Random House Trade Paperbacks.

7 Taxation in Investment Real Estate

Every form of government tends to perish by excess of its own basic principles.
Will Durant

Contents

7.1	Calculation of Property Taxes	141
7.2	Effects of Interest Expense and Non-Cash Expenses on Taxable Income	142
7.3	Introduction of After Tax Internal Rate of Return (ATIRR)	145
7.4	Various Forms of Property Ownership	148
	7.4.1 The Global View	148
	7.4.2 Choose or Lose	149
7.5	The Good, the Bad and 1031 Exchange	152
	7.5.1 Types of 1031 Exchanges	152
	7.5.2 Exchange Economics	154
	7.5.3 Exchanges at High Noon	155
	7.5.4 The Good: A Fistful of Dollars	156
	7.5.5 The Bad: Unintended Consequences (For a Few Dollars More)	157
Appendix: Complete ATIRR Example Spreadsheet		161
References		162

In this chapter, we will explore the benefits and possible pitfalls of taxation reduction and deferment strategies in investment real estate. We will discuss the effect that interest expense and non-cash expenses can have on investment return, and we will introduce the after tax internal rate of return. We will then discuss the various forms of real estate property ownership, and its effect on taxation at the corporate and personal level. In the final section of the chapter, we will outline like kind exchange in a real estate context. The chapter concludes by returning to our sensitivity analysis mini-case from the last chapter to view the after tax effects.

7.1 Calculation of Property Taxes

Before we introduce the after tax internal rate of return, we will briefly discuss how property taxes are calculated for investment property. Property tax is an expense item that is included in operating expenses which are subtracted from the effective gross income in order to determine the net operating income of a property. Property taxes differ from federal taxes which is primary focus of this chapter. Property taxes are typically assessed by the county or municipality where the subject property sits, with taxes being assessed based on the tax value of the land and improvements to real property. The municipality will typically assess taxes based on a stated rate per $100 or $1,000 of value. The local government will typically reassess the values of real estate very few years, on a regular schedule. The rate of tax assessment per dollar of value is typically referred to as the millage rate.

As an example of this concept, assume that the subject property has a land value of $200,000 and a building value of $1,800,000, with a stated assessment rate of $0.30 per $100. The property tax is calculated at $6,000. This amount would be used for the current year as the operating expense for property taxes. In the United States, most county governments have tax information for residential and investment property which is freely accessible via the internet. Websites for a given state's tax assessor's office or register of deeds can provide ownership information, whether a given property owner is current on their taxes, and the current tax value for land and improvements. While some readers outside of the United States might cringe at the ease of disclosure of this information, these websites are very helpful for investors and lenders alike!

7.2 Effects of Interest Expense and Non-Cash Expenses on Taxable Income

One of the key benefits of investing in real estate is the possible tax benefits. In the United States, along with other developed nations, property owners are allowed to deduct depreciation and interest expense from their taxable income. The deduction of these expenses allows the property owner to reduce their taxable income from an investment property. These differences in the pro-forma operating statements provide for the basis for calculation of the after tax internal rate of return (ATIRR).

In Chap. 4, we discussed the different forms of depreciable basis involved in the construction of the cost approach valuation. Terms such as physical deterioration and functional obsolescence, were introduced, along with the concept that land value does not depreciate. The terms discussed in Sect. 4.2.2 shed light on the definition of depreciation. Under Title 26, section 167 of the IRS tax code, depreciation is defined as "a reasonable allowance for the exhaustion, wear and tear (including a reasonable allowance for obsolescence) of property used in trade or business and of property held for the production of income" (IRS 2011A). As we discussed in Sect. 4.2.2, some portion of an asset's depreciation is curable, whereas others are not. From a tax perspective, such categorization of depreciation is not material. What is important is how the annual depreciation allowance is determined. The annual depreciation allowance is the amount of annual depreciation which can be deducted from taxable income. This represents the realization that real property is not immutable. Over time, even the best of properties will experience a decline in value owing to general wear and tear, obsolescence, or net absorption effects.

7.2 Effects of Interest Expense and Non-Cash Expenses on Taxable Income

Residential Rental Property	27.5 Years
Non-Residential Real Property	39 Years
Personal Property	From 3-15 Years

Fig. 7.1 Cost recovery periods for select assets

As newer properties come to market, the probability that a given property can command the highest leasing rates in a given area will deteriorate at some level. Thus net absorption effects have an impact on the value of a property over the course of time. The annual deduction for depreciation is the accounting and taxation realization of this concept. Annual depreciation is defined as the depreciable basis of the property divided by the cost recovery period. If only there was a rational analysis of how the cost recovery periods are determined. Like many things in the public sphere of influence, the cost recovery period is the end result of negotiating by all of the relevant vested interests. One only has to look at the current cost recovery periods to realize the implications of this process. Figure 7.1 summarizes the latest mandates (IRS 2011). The cost recovery periods are hardly intuitive.

As stated earlier, land does not depreciate. This makes intuitive sense as land should not decrease in value over time owing to general wear and tear. There are possibilities for amortization for land, but typically amortization expense is associated with intangible assets such as goodwill, copyrights, etc. If land has timber or other income producing elements on site which are used over time, then amortization may be allowable. For our purposes, we will not consider the value of land as a typical asset class allowing for depreciation or amortization.

Government regulations also allow for the owner of a property to deduct the total annual interest expense paid to a lender for loans secured by the subject investment property. Loan points, or origination fees, are also eligible for deduction for tax purposes, but the points paid are deducted equally over the length of the balloon period for the mortgage. Loan points are used to essentially reduce the mortgage amount at loan origination. The lender's effective return will typically increase by 0.25% for every two points charged (Brueggeman and Fisher 2010). If no points are charged, then the effective rate of interest is equal to the contract rate of interest. Loan points allow the lender to recoup their principal a bit quicker than if the entire loan amount was amortized. The benefit for the borrower is accrued interest savings for the amount essentially prepaid, and possible tax benefits.

For example, assume that an investor has a loan for $1,500,000 based on a 240 month amortization, priced at an interest rate of 6%. Assume that the bank has offered a 5 year loan term. Payments are based on a constant payment mortgage (CPM). In addition to the loan, the customer paid an origination fee of 1% of the loan balance at origination.

Fig. 7.2 shows the benefits from a tax perspective over the 5 year rate commitment by the bank. As you can see from the Fig. 7.2, the loan points are spread out equally over the 5 year rate commitment period. With the loan of $1.5 million, a 1% fee would be equal to $15,000, but is only deducted as $3,000 annually. The interest expense varies each year as well. The annual interest figure can be easily determined using a financial calculator. Once the inputs of 240 months, 6% interest, and starting loan amount of $1.5 million are input, and the monthly payment of $10,746.46 is determined, the annual interest expense can be viewed by pressing the "gold key" and then "Amort" on the HP 10 B II calculator.

In order to view the difference between investor cash flow and taxable income, we can extend our example to include depreciation effects. Assume that an investor owns an investment property where the depreciable basis of the real improvements is $1.8 million. If 39 years is taken as the cost recovery period, the annual depreciation allowance is $46,154.

Fig. 7.3 compares the investor cash flow with the taxable income. The depreciation, interest expense, and loan points serve as primary differences from the investor's calculated cash flow. In this example, the real estate, when coupled with property indebtedness, serves as a partial tax shelter as the taxable income is less than the investor's cash flow. Pyhrr and Cooper (1982) state that a tax shelter exists whenever the depreciation expense is greater than the amortization of principal. The game plan of the investor regarding tax shelter is to generate cash flow from operations before tax and at the same time to produce tax losses that are as high as possible for as long as possible (Pyhrr and Cooper 1982). There are various types of tax shelters ranging from an excess shelter, where a tax loss is reported, to a complete tax shelter where taxable income is zero, to a partial shelter as in our example above, and finally to no tax shelter whatsoever. In this case, the taxable income and the cash flow before tax to the investor are the same. A negative tax shelter could also exist where taxes exceed the cash flow of the property. Hopefully, based on some of the insights provided in the preceding chapters of this book, this situation is a rare occurrence!

The key to the tax shelter concept is depreciation working through the leverage magnification process. As discussed in Chap. 6, increased financial leverage does increase financial risk. Additionally, while depreciation is an artificial accounting loss, it could reveal actual economic loss if the property is not adequately maintained. As also mentioned in Chap. 6 on legislative risk, the structure upon which most tax shelters are built is very fragile, so investors are cautioned to keep up to date with possible legislative changes that might impact their returns.

Investors should also keep in mind that tax savings accounts for one of four cash flow benefits resulting from ownership of property. The four benefits are summarized below:

- Cash Flow from Operations
- Tax Savings
- Refinancing Proceeds
- Net Proceeds from Sale.

Should any investor focus too heavily on one of the four cash flow benefits, some mistakes can occur. We will review this concept when we discuss like-kind exchanges later in this chapter.

7.3 Introduction of After Tax Internal Rate of Return (ATIRR)

Deduction	Year 1	Year 2	Year 3	Year 4	Year 5
Interest	$88,910.61	$86,440.60	$83,818.24	$81,034.15	$78,078.33
Points	$ 3,000.00	$ 3,000.00	$ 3,000.00	$ 3,000.00	$ 3,000.00

Fig. 7.2 Tax benefits of indebtedness on investment property

7.3 Introduction of After Tax Internal Rate of Return (ATIRR)

In Chap. 3, we introduced the internal rate of return. The examples in that chapter were on a before tax basis. In this chapter, we will highlight the differences on an after tax basis. As mentioned earlier in the chapter, tax considerations represent one source of concern for investors. While it is not the only consideration, the after tax IRR can aid an investor in assessing the after tax implications for their investment property.

Assume that an investor wishes to estimate the after tax internal rate of return (ATIRR) for an investment property currently valued at $2 million. The investor has approached the bank for a loan of 75% of the value of the property, and has received a rate quote of 6% of 5 years based on a 240 month amortization period. The investor has projected a year one net operating income of $175,000, and has further projected that NOI will increase by 3% for each year of the rate commitment period. The investor plans on selling the investment at the end of the fifth year and would like to evaluate the after tax results of this transaction based on the assumed tax rate of 28%. For simplicity purposes, the interest expense and points calculations mirror those shown in Fig. 7.2, while the depreciation expense calculation mirrors that seen in Fig. 7.3. While the property is currently valued at $2 million, only $1.8 million is the building value, while the remainder is the value of the land.

Viewing the ATIRR analysis at a high level, the investor is concerned with including interest expense, depreciation, and any known points in the analysis in order to determine the taxable income for each year of the holding period. Our example in Fig. 7.3 represents the calculation of taxable income versus before tax (investor) income for the first year of the holding period. Additionally, the investor will calculate the capital gain or loss from selling the property at the end of the 5 year holding period. From a BTIRR standpoint, the calculation of before tax cash flow for each year would result from subtracting the annual debt service from the NOI as is shown in Fig.7.4.

The NOI is projected at $175,000 for year one, with annual increases of 3%. The debt service is based on the loan received from the bank, and does not change over the holding period as the rate is fixed for 5 years. In order to arrive at the BTIRR, assumptions must be made concerning the subsequent resale of the property. There are two primary methods of estimating the reversionary cash flow (Brueggeman and Fisher 2010). The first method is to simply assume a compound interest rate by which the initial property value of $2 million will appreciate over the holding period. The second method is to project the year *six* NOI, which based on the

	Investor Cash Flow	Taxable Income
Gross Potential Income	$ 272,000	$272,000
Less: Vacancy & Collection Loss	$ 27,000	$ 27,000
Effective Gross Income	$ 245,000	$245,000
Less: Operating Expenses	$ 70,000	$ 70,000
Net Operating Income	$ 175,000	$175,000
Less: Interest	$ 88,911	$ 88,911
Points	$ -	$ 3,000
Amortization of Principal	$ 40,047	$ -
Depreciation	$ -	$ 46,154
Cash Flow Before Tax	$ 46,042	$ -
Taxable Income (Loss)		$ 36,935

Fig. 7.3 Investor cash flow versus taxable income

End of Year	1	2	3	4	5
Net Operating Income	175,000	180,250	185,658	191,227	196,964
Debt Service	128,958	128,958	128,958	128,958	128,958
Before Tax Cash Flow	46,042	51,292	56,700	62,270	68,006

Fig. 7.4 Calculation of before tax cash flow over 5 year holding period

investment horizon of the investor is what the subsequent buyer of the property would achieve in their first year of their ownership of the property. The year six NOI is divided by a terminal cap rate or the projected cap rate at the end of the holding period.

For our example, we have assumed an NOI of $203M for year six, and have divided by a terminal cap rate of 9% in order to arrive at the projected sales price at the end of the fifth year of $2,255 million. This number has been slightly rounded as sales prices typically end in either zeroes or fives. An alternative method for determining the sales price is to multiply the asking price by an assumed property appreciation rate compounded over the holding period of 5 years. We have also assumed sales costs of 3%. Figure 7.5 itemizes the calculation.

As Fig. 7.5 ilustrates, the sales price is subtracted by the sales costs and the outstanding balance of the loan at the end of the fifth year. The balance of the loan at the end of the fifth year can be found using a financial calculator by finding the original payment for the $1.5 million mortgage amortized over 240 months at a rate of 6% ($10,746.47, which when annualized is equal to the debt service shown in Fig. 7.4 of $128,957.59). The outstanding balance can be determined by then solving for the present value with 180 months remaining on the mortgage. The before tax cash flow from sale of $913,856 is then added to the year five operating cash flow of $68,006 to reveal the BTIRR as is shown in Fig. 7.6.

7.3 Introduction of After Tax Internal Rate of Return (ATIRR)

Cash Flow from Sale in Year 5	
Sales Price	2,255,000
Sales Costs	67,650
Mortgage Balance	1,273,494
Before Tax Cash Flow	913,856

Fig. 7.5 Calculation of before tax cash flow from sale

Year	0	1	2	3	4	5
Before Tax Cash Flow	(500,000)	46,042	51,292	56,700	62,270	981,862
BTIRR on Equity	21.71%					

Fig. 7.6 Before tax internal rate of return on investment property

	End of Year	1	2	3	4	5
Net Operating Income		175,000	180,250	185,658	191,227	196,964
Less: Interest		88,911	86,441	83,818	81,034	78,078
Depreciation		46,154	46,154	46,154	46,154	46,154
Loan Points		3,000	3,000	3,000	3,000	3,000
Taxable Income		36,935	44,655	52,686	61,039	69,732
Tax (Savings)		10,342	12,503	14,752	17,091	19,525
After Tax Cash Flow		35,701	38,789	41,948	45,179	48,481

Fig. 7.7 Calculation of after tax cash flow over 5 year holding period

This analysis is a recap of what we introduced in Chap. 3. In order to view the property return on an after tax basis, NOI must be reduced by the allowable depreciation, interest expense, and any possible loan points. The interest expense and loan point calculations are the same as were shown in Fig. 7.2, and the depreciation is calculated by the same method alluded to in Fig. 7.3. Figure 7.7 reveals the after tax cash flow from operations for our subject property. The after tax cash flow is found by subtracting the before tax cash flow for a given year as shown in Fig. 7.4 by the tax shown in Fig. 7.7.

In order to complete the ATIRR analysis, the after tax cash flow from sale must be included. As shown in Fig. 7.5, the assumed selling price after 5 years is $2,255,000. This is determined by dividing the assumed year 6 NOI of $203,000 by the terminal cap rate of 9%. In order to determine the capital gain on the sale, the net sales proceeds (sales minus selling costs), are subtracted by the adjusted basis of the property. In our example, the beginning value of the land and building was $2 million. Since we are depreciating $46,154 annually ($1.8 million building value over a 39 year cost recovery period), after 5 years the building would have been

Year	0	1	2	3	4	5
After Tax Cash Flow	(500,000)	35,701	38,789	41,948	45,179	878,714
	17.52%					

Fig. 7.8 After tax internal rate of return on investment property

subject to accumulated depreciation of $230,769. This leads to an adjusted basis of the property of $1,769,231. The capital gain is calculated as $418,119.

Sales price	$2,255,000
Sales cost	(67,650)
Adjusted basis	(1,769,231)
Capital gain	418,119

The tax associated with the sale is $83,624, or 20% of the capital gain amount. In some cases, the capital gains tax can be equal to the marginal tax rate, but for our example we have assumed that the capital gains tax rate is slightly lower than the marginal tax rate of 28%. The after tax cash flow from sale is then determined in the following manner:

Sales price	$2,255,000
Sales cost	(67,650)
Mortgage balance	(1,273,494)
Tax from sale	(83,624)
After tax cash flow	830,232

The after tax cash flow is then added to the year five cash flow to compute the ATIRR. This is summarized in Fig. 7.8. The after tax IRR is 17.52%, which is compared to the before tax IRR of 21.71% for the same project as shown in figure 7.6. A spreadsheet which shows the entire analysis of the calculation of ATIRR is located in the appendix to this chapter.

7.4 Various Forms of Property Ownership

7.4.1 The Global View

As with any form of business venture, there are numerous forms of ownership structures available for owners of investment property. The ownership structure chosen may depend on how many equity partners, or shareholders, are involved in a given investment. Some property investors choose to make a series of investments alone or with just one or two partners. Other investors are part owners of a series of entities with various other investors. For this second type of investor, the financial picture can become very complex, especially for their CPA or lender, given all of the various partnership entities and the various different individuals involved in each entity. In the aftermath of the "Yes era," many financial institutions are now

7.4 Various Forms of Property Ownership

requiring a more "global" view of an applicant's financial situation. Banks have learned that simply viewing the potential default risk of an applicant based on the valuation of the property is not an accurate assessment of risk for investors owning multiple entities. If the bank has made a loan commitment for a property of sound financial condition, but subsequently other investor-owned properties experience financial difficulties, the investor might utilize excess cash flows from their stronger properties to supplement the debt service payments for their weaker properties.

Thus investors should expect more financial disclosure requirements going forward than seen in prior years. This is especially true for any corporate entities where the investor applying for financing has majority controlling interest. If the investor in question does not have controlling majority interest in the entity that owns the real estate, they may not be capable of moving cash flow from that entity to supplement payment problems experienced elsewhere. In evaluating risk for minority controlled enterprises, lenders will desire to at least verify that the monthly cash flows from the minority owned properties can support the monthly loan payments. Otherwise, the investor in question might experience a capital call, where owners are required to contribute equity to an underperforming investment property in order for the bank's payments to be made on an ongoing basis.

Given the increased scrutiny of related entities of an investor from a lending perspective, and the aforementioned desire for the investor of reducing tax liability, we will now discuss available ownership strategies for investment property, especially in light of the interplay between business and personal tax returns.

7.4.2 Choose or Lose

Earlier in the chapter, we discussed taxation as it concerns a specific investment property. The tax paid here is known as an effective tax rate. This involves reducing taxable income via interest expense, depreciation, and any other allowable deductions. Implicit in our discussion of Fig. 7.3 which itemized the differences between investor cash flow and taxable income was that net operating income had already been reduced by expenses included in both approaches. Real estate allows for significant tax deductions in the form of property taxes, insurance, repairs and maintenance, and utilities that are subtracted from revenue to determine NOI (IRS 2010).

In this section of the chapter, we now discuss the marginal tax rate. This is typically viewed as the tax bracket of the investor including all sources of income. If property is held in a real estate holding company based in the United States, and that company is classified as either an S-Corporation or as a Partnership, federal taxes are not paid at the corporate level, but are paid by the investors in a pass-through fashion. A few of the statements in the preceding sentence require some clarification. An S-Corporation is essentially a "small corporation" whereby the number of shareholders is limited to 100 (PW Coopers 2011). This form of corporation differs from a C-Corporation which does not have the shareholder limitation, and is typically reserved for larger companies which are taxed at the corporate level. Sub-Chapter S Corporations and partnerships are not taxed at the corporate level. Items such as net profits, interest and dividends earned on company checking accounts, and capital gains are pro-rated for each shareholder relative to their ownership interest in the entity. The resulting income is then taxed at the individual level. Thus the earnings for an S-Corp or for various partnership entities such as limited partnerships, limited liability companies, and others, are passed through the corporate tax returns to the personal tax returns. The interplay of the personal and business tax returns is shown in Fig. 7.9.

There are many forms of taxable income. The first is active income, where the investor actively participates in its creation. This would consist of wages, bonuses, commissions, and specific real estate activities which are considered to be active. Ownership and active daily operation of hotels and nursing homes are often considered as active sources of income. Otherwise, real estate falls into the second category of passive income. This is defined as any income or loss where the investor does not actively participate in the management of the company or activity. This might seem strange to readers, especially to property managers who have received calls in the middle of the night from enraged tenants due to equipment failure at the property. The third form of income classification is portfolio income. This final classification includes interest and dividends where the investor does not control majority ownership in the property or company (Ajami et al. 2006). For the most part, the consequences of these income classes are transparent to the investor. The discussion serves to highlight how pervasive is the taxation of rental property ownership.

Investors have the option of either owning an investment property in their own name, or incorporating via a holding company to own the asset. In the former case, the revenues, expenses, and capital gains will report on the personal tax returns. In the latter case, the same information would report on the corporate returns.

For individually owned real estate in the United States, revenues and expenses are reported on the first part of the schedule E of the personal tax return. On the first part of the schedule E, income from rental property and royalties are reported. Capital gains are typically reported on the schedule D of the personal tax return. Additionally, any interest income earned on property operating accounts will appear on the personal schedule B. The personal schedule B is where interest and dividend income is reported.

If an investor also provides management services for rental property, the fees associated with property management are often reported on the personal schedule C. The schedule C is essentially an income statement for sole proprietorships. Businesses which are owned by individuals utilize the schedule C to itemize annual revenues and expenses. Under a sole proprietorship, all taxation occurs at the personal level. Additionally, there is no "corporate veil" as any liability associated with running the business or operating the property is borne 100% by the individual owners.

Sometimes the personal schedule F has relevance for real estate investment. If land is owned personally and farming activity takes place on the land, the schedule F is where the profit and loss is reported.

7.4 Various Forms of Property Ownership

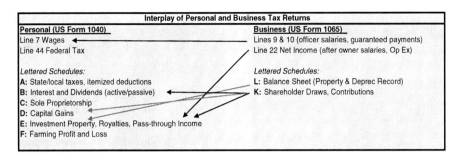

Fig. 7.9 Interplay of personal and business tax returns

Investors seeking shelter from aggregate liability typically opt for incorporation. As mentioned earlier, most real estate investment companies with fewer than 100 shareholders forgo incorporation as a C-Corporation owing to double taxation of profits and dividends. Small corporations such as S-Corporations and limited liability companies report financial performance via corporate tax returns, but the taxation occurs at the personal level. As shown in Fig. 7.9, all taxable income for these smaller companies flows to the personal tax returns of the shareholders on a pro-rata basis.

For example, net income from any S-Corporation will flow to the schedule E part II of the personal tax return. This is where income from pass-through entities is reported. Let's assume that Vic and Sal Tyan own Tyan Investments, LLC on an equal basis. If the last year's taxable income was $410,000, then each shareholder would receive a pass-through on their personal schedule E part II of $205,000. This figure would flow to the front page of the personal federal tax return and report on line 17 of the US form 1040 (IRS 2011). This may also flow to the personal schedule A, which itemizes state and local taxes as well as itemized deductions. This income, along with any other personal income, would serve as the basis for the determination of the marginal tax rate for each investor.

The primary point of interplay between the personal and business tax returns revolves around the schedule K. The schedule K aggregates the net income, equity contributions, equity distributions, interest and dividends, and capital gains during the calendar year. The schedule K-1 is prepared for each shareholder and serves as the basis of taxation at the personal level. Thus the financial performance of the corporation is pro-rated via the K-1 schedule. Returning to our previous example, if Tyan Investments' operating account produced annual interest income of $5,000, the schedule K-1 would itemize interest income of $2,500 for each shareholder, and this income would report on the personal schedule B. The shareholder will be taxed on the interest received from the company checking account as well as on any dividends received from the pass-through entity during the tax year.

From the lender's perspective, the income of $207,500 for each shareholder does not necessarily mean that the individuals actually received this income. The tax returns only indicate that the individuals were responsible for taxation at this amount owing to their ownership in the corporate entity. The schedule K-1 can be utilized to view actual partner draws and contributions during the calendar year.

The role of the small corporation, especially the limited liability company, has been criticized recently for creating an imbalance between risks and rewards given the partial shelter from personal liability for the shareholders (Sinn 2010). As the name implies, shareholders of a limited liability company are only responsible, or limited, to the amount of their ownership, or equity investment, should the company become entangled in legal or other disputes. The prevalence of limited liability companies in Germany (GMBH: Gesellschaft mit Beschränkter Haftung), France (SARL: Société à Responsabilité Limitée), England (LLP: limited liability partnership), and many other nations across the globe bodes well for its continued use.

7.5 The Good, the Bad and 1031 Exchange

The musician Frank Zappa reportedly said that the United States is a nation of laws, badly written and randomly enforced. While this statement of opinion may or may not be accurate, it contains at least a grain of truth in that sometimes seemingly well intended actions can have perverse consequences. Take, for example, like kind exchanges under IRS code section 1031. These exchanges, or 1031 exchanges as they are often termed, have the stated aim of deferring the capital gains taxes paid in qualifying situations when business or investment property is sold (IRS 2011B). Internal revenue code (IRC) section 1031 provides an exception for the payment of capital gains taxes on the sale of business or investment property at the time of sale. The concept of tax free exchanges under US tax law has its origins in the Revenue Act of 1921 (Jaffe and Sirmans 1982). This section identifies when a transaction may qualify as a 1031 exchange, introduces the economics of the exchange decision, and discusses the unintended consequences of this tax law. It is our contention that this simple potentially value enhancing transaction was yet another piece of coal in the financial meltdown furnace. Ultimately, what determines the value of a 1031 exchange is not the applicability of a particular property to arbitrary, governmental requirements, but rather, the impact the deal has on the investor's (i.e. the customer's) long-run returns (RMA, 2011).

7.5.1 Types of 1031 Exchanges

The simplest form of 1031 exchange applies to a simultaneous exchange of one qualified property for another. In this exchange scenario, there is no time interval between the closings of the relinquished and replacement property. In order to qualify for the 1031 exchange, the property sold *and* the acquired property must be classified for either a business or investment purpose. This is typically an easy hurdle to clear. Additionally, the property must previously have been held for a productive business or investment use, and it is up to the taxpayer to prove that the property being sold was not owned for only a short period of time. Consequently, "flipped" properties do not qualify for 1031 exchanges, but there are no official definitions for what qualifies as the required term of ownership. One rule of thumb is that the property should be held for at least 2 years. Two other requirements for all 1031 exchanges are that the property purchased must be owned in the same name as the property sold and both properties must be located either inside the United States or out of it. That is, if one property is located within the United States and the other is not, the transaction fails to meet the definition of a 1031 exchange (Brueggeman and Fisher 2010).

Because 1031 exchanges are also known as like kind exchanges, the latter term implies a similarity between the properties being exchanged. "Like kind," to someone unfamiliar with the process, could constitute something as rigid and specific as selling (relinquishing) one apartment building and buying (replacing it with) another apartment building, but this is not the intention of the "like kind" requirement. As long as each property involved in the exchange is classified generally as either for a business or an investment purpose, the matching of specific property types is of no consequence.

A second form of 1031 exchange is the delayed or deferred exchange. As the name implies, the delayed exchange applies to transactions that are not consummated on the same day. Deferred exchanges were first introduced in the late 1970s when a U.S. taxpayer named Starker challenged the requirement that exchanges had to be simultaneous. The IRS has strict rules regarding when the delayed exchange must be completed in order to gain the tax exemption. An investor has 45 days to identify possible properties, and has 180 days to close the transaction.

A deferred exchange is relatively common and typically requires the partnership of the investor selling the property with a QI as the first step. A QI is defined as a person or entity that is not the tax payer and is not an attorney, accountant, realtor, or other related party, and of course, the QI must enter into a written exchange agreement with the taxpayer desiring to facilitate the exchange. The QI will maintain the necessary paperwork used to validate the acceptability of the transactions and also will act as an agent for the taxpayer. The QI also can add value for an investor in the form of estate planning. Typically, a QI is a financial institution that acquires the relinquished property from the taxpayer (seller), transfers the property to the buyer and then acquires and transfers the replacement property to the tax payer. Obviously, the QI charges fees for this service and may also obtain interest on idle funds held while the 1031 clock ticks.

Another form of exchange is the reverse exchange. In this situation, the replacement property is purchased and closed before the relinquished property is sold. Typically, the QI will take title to the replacement property until the relinquished property is sold. Once the relinquished property is sold, the QI will then transfer the title of the replacement property back to the investor. Fig. 7.10 illustrates the required timeline for a qualified deferred 1031 exchange. If an investor fails to complete the reverse exchange within the mandated time limits, the investor will be subject to capital gains tax.

As Fig. 7.10 makes clear, once the relinquished property is sold, the investor has 5 days to identify a qualified intermediary (QI) that will facilitate the transaction.

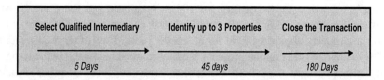

Fig. 7.10 Reverse 1031 exchange timeline

Subsequently, there is a 45 day inspection period in which the investor can select up to three properties for possible purchase as the replacement. Rules exist that allow for more than three properties, but most investors keep their choices limited, given the time pressures associated with closing the deal. Once the selection period has expired, the investor has 180 days in which to close the transaction. If the exchange of properties occurs within the mandated time limits, the investor does not have to pay taxes today on the capital gains associated with the sale of the initial property.

A final form of 1031 exchange is the improvement exchange. Under this scenario, the investor wants to acquire a property and arrange for improvements before it is received as the replacement. The improvements can consist of the construction of a building on a formerly unimproved lot, or simply can be improvements made to an existing structure. The improvements must occur prior to the conclusion of the 1031 exchange so that the value of the replacement property reflects the improvements at the completion of the exchange.

7.5.2 Exchange Economics

In this section, we will discuss how an investor might determine whether a deferred tax exchange is beneficial. While most investors would want to save taxes, the saving of taxes is not the only consideration. Once a 1031 exchange is completed, the gain that is not reportable in the disposition of the relinquished property becomes a basis adjustment in the replacement property. The basis in a property acquired via a 1031 exchange is known as a substitute basis. Since the goal of the 1031 exchange is to defer all capital gains taxes associated with the sale of the first property, let's review what happens in this scenario.

At the end of the discussion in Sect. 7.3, we provided an example where a property was sold for a market price of $2,255,000, with an adjusted basis after sales costs of $1,769,231. This resulted in a realized capital gain of $418,119. Under a 1031 exchange, the basis of the property acquired in the exchange is lowered by the amount of the deferred gain. This means that the lower basis is what is depreciated. Thus the capital gain of $418,119 would reduce the annual depreciation by $10,721, as we would divide the capital gain by the cost recovery period of 39 years. If the investor is in the 28% tax bracket, this would lead to $3,002 in less tax deductions each year due to depreciation (i.e. $10,721 × 0.28).

In order to determine if the capital gain is beneficial, investors should calculate whether the lower capital gains tax today, which in our earlier example was $83,624, exceeds the sum of the present value of reduced depreciation tax benefits

7.5 The Good, the Bad and 1031 Exchange

each year after the exchange occurs plus the higher capital gain taxes paid upon the eventual sale of the replacement property.

Once the replacement property is sold, the deferred gain of $418,119 will be recognized. Since the exchange results in less depreciation each year for the property, the basis of the property is lower than what it would have been had the 1031 exchange not occurred. Assuming a 10 year holding period, depreciation is lowered by $107,210 in total. This would offset the difference in capital gains now being realized as is shown below.

Deferred gain from exchange	$418,119
Less additional depreciation without exchange	107,210
Net amount of capital gain	$310,909
× Capital gains tax (20%)	62,182

We have now determined that at period zero, the time of exchange, the investor could save capital gains tax of $83,624, and that this would lead to less tax savings for depreciation of $3,002 each year. We have also calculated that we have $62,182 in additional taxes at the time the second property is sold given the greater capital gains. If we view the effective interest rate assuming a 10 year holding period, a present value of ($83,624), an annual payment of $3,002, and a future value of $62,182, our borrowing rate is 1.16%. In essence, the investor is borrowing $83,624 and repaying $3,002 per year for 10 years and a lump sum payment of $62,182 in the tenth year. Since the low borrowing rate would be attractive in most economic environments, the 1031 exchange is most likely beneficial in this case.

7.5.3 Exchanges at High Noon

The goal of a tax deferred exchange typically is tax minimization. Nonetheless, the goal of deferring all of the capital gains taxes is achieved only when the following conditions are met:
- All cash from the sale is used to purchase the replacement property.
- The price of the replacement property is at least as much as the relinquished property.
- The investor obtains the same financing on the replacement property that was held on the relinquished property.
- Both properties meet the like-kind definitions and time frames.

Taxes will be assessed when these conditions are not met and if the equity is not "balanced" between the properties used in the exchange. Differences in the property values usually require the equities to be balanced with what is known as boot – an old English term meaning *an addition*. Boot can take the form of money, debt relief, or the fair market value of "other property" received by the taxpayer to equalize the equity of properties in an exchange. Because it can take several forms, boot does not qualify as like-kind property, and is thus subject to taxation.

For example, assume investor A is selling an office property to investor B with a market value of $2 million and that the property is encumbered by $1.5 million of

Fig. 7.11 Balancing the equity in a 1031 exchange

	Investor A Office	Investor B Retail
Market Value	2,000,000	1,900,000
Loans (-)	1,500,000	1,500,000
Equity (=)	500,000	400,000
Difference		100,000
Balance	500,000	500,000

outstanding debt. Thus, investor A has $500,000 of equity in the property. Assume that investor B is selling retail property to investor A with a current market value of $1,900,000 and the same level of indebtedness.

Figure 7.11 reveals that Investor B must provide some additional equity to consummate the exchange, but there are numerous alternatives available to both parties. Investor A might take $100,000 in cash out of the office property and replace it with debt to equilibrate the financing of the two properties. Of course, investor B could add $100,000 in cash or other security in order to balance the equities; or, investor B could pay down the debt on the retail property by $100,000. Ultimately, the two parties could undertake any combination of these actions in order to balance the equities being transferred.

Even if investors A and B come to agreement regarding a method to balance the equity of the properties being transferred, taxes will still apply because any of the alternatives represents "boot." The tax payer receiving boot is subject to taxation. Based on the example illustrated in Fig. 7.11, investor A is subject to tax on the $100,000 received from investor B. Consequently, to avoid capital gains tax on boot, investors should never purchase a property that is selling for a price that is less than that which they are trying to replace via a 1031 exchange.

7.5.4 The Good: A Fistful of Dollars

From a taxation standpoint, the benefits of 1031 exchange can be substantial. For one thing, the deferment of the capital gains tax can greatly improve an investor's internal rate of return. In fact, if we return to our ATIRR example from earlier in the chapter, the after tax internal rate of return for the property in question moves from 17.52% to 19.43% simply from the elimination of the $83,624 in capital gains tax. Of course, the investor is not technically eliminating the tax but is deferring it to the next investment transaction.

Additionally, investors can employ a 1031 exchange to acquire properties that are appreciating faster than those relinquished, or consolidate assets by exchanging many properties for one that is equal in value to that of the combined properties. IRS regulations allow for up to three properties to be selected irrespective of the aggregate fair market value, but a greater number of properties can be employed as replacements in a 1031 exchange, as long as at least one of two provisions are met. The first is that the aggregate fair market value of the replacement properties does not exceed 200% of the aggregate fair market value of all exchanged properties as of the initial transfer date. This 200% rule allows an investor to diversify, or consolidate, in an effort to enhance the risk and return characteristics of the portfolio of properties. Alternatively, the 95% rule allows for the selection of any number of properties, as long as at least 95% of the aggregate fair market value of the properties identified is eventually selected. Given the advantages, the decision to undertake a 1031 exchange seems simple, but these transactions do have some problems.

7.5.5 The Bad: Unintended Consequences (For a Few Dollars More)

1031 exchange was very popular during the "Yes era." For many investors, the idea of saving on capital gains *today* seemed to outweigh any losses associated with depreciation and higher capital gains *tomorrow*. In the *rock, paper, scissors* world of real estate investor decision analysis, tax savings today would appear dominant. What should become apparent from our discussion of tax deferred exchanges is that the speed, size, and the leverage of the transaction are all crucial elements of its success.

If the avoidance of boot taxation is accomplished by continuously purchasing properties with higher and higher market value relative to what is being replaced, and if the general requirement of maintaining *at least* the current level of property indebtedness is satisfied, the 1031 exchange fits in well with the property bubble mentality of the years preceding the recent financial crisis. Furthermore, when coupled with lenders justifying loan-to-value and debt coverage ratios by requiring additional equity investment into a new loan via the exchange, the result is that 1031 exchanges at least partially maintained inflated property values. Finally, adding the ticking clock of the 1031 exchange timeline, market psychology comes to the forefront of the decision process.

As the 1031 clock shows that a deal's deadline approaches, it is usually the case that the investor has spent a considerable amount of time and resources in the attempt to defer taxes. Rusbult's investment model states that the equity calculation of any relationship should not only include the costs and benefits of the current interactions and viable alternatives, but should also include previous investments in the relationship, where the commitment to the relationship considers the past, present and future (Raab et al. 2010). Because the 1031 exchange process entails an investor receiving and relinquishing at least two properties, and because many investors roll the capital gains over on more than one occasion, the consideration of banking customer relationships is relevant to the 1031 exchange decision.

In an era of cheap money, inflated and unsustainable property values, and inexperienced investors, the 1031 exchange was ripe to be overused. In this context, a myopic investor is more concerned with saving on the capital gains tax than on undertaking the due diligence that likely would occur in the absence of government imposed rules and artificial time constraints. It is better to purchase the right investment at the right price, than to make a hasty decision to save on the capital gains tax today. Minimizing tax obligations is certainly a worthwhile endeavor, but it should not be the only goal. Bankers and financial institutions that market themselves on the basis of "relationship" should never lose sight of the fact that partnering with customers via 1031 exchanges has potential benefits for all parties.

Nonetheless, this is true only when the properties are being financed and purchased at reasonable, economically justifiable values.

An investor that hastily enters into a 1031 exchange will likely drop a QI when pursuing a subsequent transaction if he is not satisfied with the eventual return on an investment. In the final analysis, bankers are cautioned to work in partnership with their clients in an effort to find the right properties to finance via the 1031 exchange mechanism. If the bank can approve the loan only on the basis of a high equity position, the easy loan decision today may beget an unsatisfied client tomorrow when it is eventually discovered that the property was purchased at too high a price. Understanding the nuances of the process helps to navigate the good and the bad of 1031 exchange.

Mini-Case: Sensitivity Analysis Revisited

In conclusion to this chapter on taxation, it will be beneficial to return to our example of sensitivity analysis which concluded Chap. 6. In this chapter we have introduced the inclusion of interest and depreciation expense to lower the taxable income for the investor. If you recall from our example in Chap. 6, the net operating income of the grocery anchored retail property was calculated at $381,152. As is shown in Fig. 7.12 below, the inclusion of interest and depreciation expenses will significantly reduce the net operating income to arrive at taxable income.

The taxable income as shown in Fig. 7.12 is $77,761, which represents an 80% decline when comparing investor cash flow with taxable income. While American revolutionary Benjamin Franklin has said that the only certainty in life are death and taxes, it appears that owning real estate as an investment makes the taxation part of life a little more bearable.

Effect of Interest and Depreciation Deductions on first year Net Operating Income from Ch. 6 Sensitivity Example

Tenant	Sq. Ft.	Annual Rent		Depreciation Assumptions		
Winn Dixie	22,000	$ 462,000				
Dollar General	6,000	$ 123,000		Original NOI	$ 443,502	
Franco's Pizzeria	2,500	$ 60,000		Cap Rate	9%	
Uncle Jimmy's Toy & Hobby	4,000	$ 92,000		Value	$4,900,000	
Total Gross Revenue		**$ 737,000**		Cost Recovery Period	39	Years
				Depreciation	125,641	
Market Vacancy Factor	10%	$ (73,700)				
Effective Gross Income		**$ 663,300**		**Interest Expense Assumptions**		
				LTV	70%	
Property Taxes		$ 75,000		Loan	$3,430,000	
Insurance		$ 27,000		Loan Interest Rate	7.00%	
Repairs & Maintenance		$ 55,000		Interest Expense	240,100	
Utilities		$ 18,000				
Management Fee	5%	$ 33,165				
Other Expenses		$ 5,000				
Replacement Reserve	1%	$ 6,633				
Depreciation		$ 125,641				
Interest Expense		$ 240,100				
Total Operating Expenses		**$ 585,539**				
Taxable Income		**$ 77,761**				

Fig. 7.12 Sensitivity analysis case revisited

7.5 The Good, the Bad and 1031 Exchange

Questions for Discussion
1. Explain the various benefits from a taxation standpoint of investing in real estate.
2. Differentiate between the various forms of like-kind exchanges.
3. Itemize the various "lettered schedules" of the U.S. personal tax return and describe what is reported on each schedule, and how it may interplay with the U.S. corporation tax return.
4. Describe the differences in calculation of the before tax internal rate of return (BTIRR) and the after tax internal rate of return (ATIRR).
5. Where might an investor find the historical property taxes assessed on a specific property and determine how future property taxes will be calculated?

Problems
1. Broad Street Investors is a limited liability corporation that owns numerous investment properties. The partnership was formed years ago by doctors seeking to diversify their asset portfolios by investing in investment real estate. There are currently six shareholders who own a percentage of the company relative to their equity investment. The owners and the percentage of ownership in the LLC are as follows: Dr. Raab (25%), Dr. Ajami (15%), Dr. Gargeya (12%), Dr. Winkler (16%), Dr. Clarke (20%), and Dr. Tyan (12%).
   (a) Assuming net income of $325,000 for the LLC, how much would each investor report on their personal tax return as income? On which schedule of the U.S. Personal tax return would this be shown?
   (b) Assuming that the company maintains an operating account of $100,000 at their local bank, and that this account earns interest of 1% annually, how much would each investor report on their personal tax return? On which schedule of the U.S. personal tax return would this be shown?
2. Broad Street Investors is considering the purchase of a self-storage facility for $2 million. Broad Street Investors is in the 32% income tax bracket and capital gains will be taxed at 20%. The property is to be depreciated over 39 years and has a building to value ratio of 80%. Rents are expected to be $275,000 for the first year, increasing by 3% annually thereafter. Vacancies are expected at 10% of gross rents, while operating expenses are expected to be 35% of EGI. The property is expected to appreciate in value by 3% each year, and is expected to be sold at the end of 5 years. Financing can be obtained at 70% of purchase price, with terms of 8% interest based on a 20 year amortization.
   (a) Calculate the BTIRR and the ATIRR for this prospective investment.
   (b) What is the expected debt coverage ratio in the first year?
   (c) What is the terminal cap rate for this investment?
   (d) What is the present value of the property at a 12% discount rate?
3. An investor is considering selling a property that has an adjusted basis of $1.5 million for $2 million. The property has a loan balance of $1.75 million. All capital gains would be taxed at 20% and ordinary income would be taxed at 32%.

The investor is considering doing a tax-deferred exchange. He would acquire a second property for $3 million and assume a loan of $2,750,000. The building to value ratio is 80%, and the second property is planned for resale at the end of 5 years. The replacement property would be depreciated over 30 years. Is the exchange strategy better than selling the property for cash and then purchasing the second property?

4. An office building has three floors of rentable space with a single tenant on each floor. Each floor has 12,000 in rentable space. The first floor tenant is renting the space for $10 per square foot with an expense stop of $5 per square foot. There are 4 years remaining on the lease. The second floor tenant is renting the space at $12 per square foot with an expense stop of $6 per square foot in place. There are 3 years remaining on this lease. The third floor tenant is renting the space for $13 per square foot with an expense stop of $6 per square foot. There is only 1 year remaining on this lease. The owners have estimated that expenses per square foot are to be $5 per square foot for the next year, excluding management fees of 5% of effective gross income. Annual lease rates are expected to increase with the CPI, which is estimated at 3% for the next 5 years. Operating expenses are expected to increase by 4% annually for the next 5 years.
   (a) Project out EGI for the next 5 years.
   (b) Project out expense reimbursements for the next 5 years.
   (c) Project out NOI for the next 5 years.
   (d) Assuming that the property is purchase for $2.4 million, what is the going in cap rate?

5. A borrower and a lender negotiate a $7,500,000 interest only loan at 8% interest for a term of 5 years. There is a lockout period of 3 years. Should the borrower choose to prepay the loan at any time after the end of the third year, a yield maintenance fee will be charged. The yield maintenance fee will be calculated as follows: a treasury security with a maturity equal to the number of months remaining on the loan will be selected, to which a spread of 400 basis points will be added to determine the lender's reinvestment rate. The penalty will be determined as the present value of the difference between the original loan rate and the lender's reinvestment rate.
   (a) How much will the yield maintenance fee be if the loan is repaid at the end of year two when the 2 year treasury rates are 2%?
   (b) How much yield maintenance fee would be charged if 2 year treasury rates are 5%?

Appendix: Complete ATIRR Example Spreadsheet

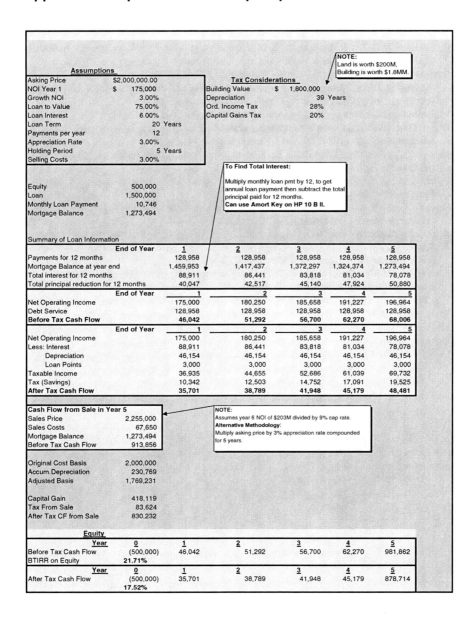

References

Ajami, R., Khambata, D., Cool, K., & Goddard, G. J. (2006). *International business theory and practice* (2nd ed.). Armonk, NY: M.E. Sharpe.

Brueggeman, W. B., & Fisher, J. (2010). *Real estate finance and investments* (14th ed.). New York: McGraw-Hill.

Internal Revenue Service (IRS). (2010). Publication 527: Residential rental property (including vacation and rental homes), 2010 edition, Publication 527 category 15052W, IRS website, http://www.irs.gov/pub/irs-pdf/p527.pdf. Accessed March 7, 2011.

IRS website. (2011A). www.irs.gov. Accessed February 24, 2011

IRS website. (2011B). Like-kind exchanges under IRC code section 1031. www.irs.gov/newsroom/article/0,,id=179801,00.html. Accessed March 10, 2011.

Jaffe, A. J., & Sirmans, C. F. (1982). *Real estate investment decision making*. Englewood Cliffs, NJ: Prentice Hall.

Pyhrr, S. S., & Cooper, J. R. (1982). *Real estate investment: Strategy, analysis, decisions*. New York: Wiley.

Raab, G., Goddard, G. J., Ajami, R. A., & Unger, A. (2010). *The psychology of marketing: Cross-cultural perspectives*. Aldershot, UK: Gower.

Risk Management Association (RMA) Journal. (2011). The good, the bad, and the 1031 exchange, September 2011 (pp. 48–51). Excerpts from section 7.5 were taken from the RMA Journal article authored by Goddard and Marcum.

Sinn, H. W. (2010). *Casino capitalism: How the financial crisis came about and what needs to be done now*. Oxford: Oxford University Press.

U.S. Master Tax Guide. (2011). Price Waterhouse Coopers, CCH a Wolters Kluwer Business, Washington, DC.

Investing in Residential Apartment Projects 8

All rising to great place is by a winding stair.
Francis Bacon

Contents

8.1	Property Life Cycle Pyramid	164
8.2	Types of Residential Apartment Investment Projects	165
8.3	Valuing Residential Apartment Projects	167
8.4	Jasmine Court Apartments Case	169
8.5	Mixed Use Properties as an Investment Alternative	172
8.6	Off-Campus Student Housing as an Investment Alternative	173
	8.6.1 Location, Location, Location	173
	8.6.2 If You Build It, Will They Come?	174
	8.6.3 Market Analysis in Student Housing	175
	8.6.4 Class Project: Off-Campus Student Housing Occupancy Study	176
References		181

The next three chapters of the book will elaborate on characteristics specific to the "four food groups" of investment property as was defined in Chap. 1. Chap. 8 will discuss multi-family (or residential apartment) projects, Chap. 9 will discuss retail and office properties, and Chap. 10 will discuss the various types of industrial and warehouse properties. Each chapter will conclude with case studies with corresponding answer keys located on *the Springer website*.

Regardless of the specific type of investment property selected, the bulk of the analysis from a financial standpoint is consistent with the prior chapters in the book. Specific differences by property type will be itemized over the next three chapters, but each property type will be evaluated based on the "QQD" framework presented in Chap. 1.

In the following sections, specific types of apartment complexes will be defined. This will be followed by a discussion of valuation considerations which are specific to multi-family investment properties. Off-campus student housing properties will be highlighted, and the chapter will conclude with case studies as well as an exercise recommended for university instructors utilizing this book as a course text.

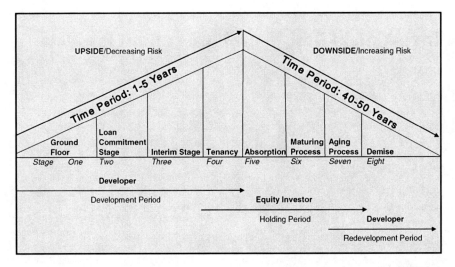

Fig. 8.1 Property life cycle pyramid (Pyhrr and Cooper 1982)

8.1 Property Life Cycle Pyramid

Pyhrr and Cooper (1982) have utilized the property life cycle pyramid to illustrate that regardless of property type, investment real estate tends to exhibit certain known characteristics from an investment timeline perspective. As shown in Fig. 8.1, the pyramid has an upside period where risk starts high and decreases, and a downside period where risk increases as the property ages over time. Developers are the typical investors during the first four stages of the pyramid. Developers will typically ascertain how an existing parcel of land (or an underutilized building) can be developed so that the entrepreneurial profit can be achieved. This process is shown as the "ground floor" first stage in Fig. 8.1.

Once plans have been developed and the investor feels assured that the idea of property development has merit, the second stage can begin. The investment assurance is typically achieved via the completion of a feasibility analysis. The feasibility analysis focuses on the development of new projects and explicitly considers non-financial objectives of the investor. This analysis would consider the marketing, legal, political, physical, and social dimensions of a project, as well as the financial dimensions.

The second stage of the pyramid would typically involve pursuing a loan commitment from a financial institution. Once financing has been secured, the investor will invest further equity into the project so that construction (or renovation) can commence. This construction or renovation of the building is shown as the interim stage of the pyramid. Once construction of the building, or of any major renovations for an existing property, is complete, a certificate of occupancy (C/O) will be issued. Investors who heeded the advice in Chap. 1 and opted against speculative investment should transition more smoothly into the fourth stage referred to as tenancy. During this stage, tenants occupy the property in the hope of first achieving a break-even occupancy rate, where the gross revenue is equal to the sum of the annual operating expenses and annual debt service requirements. As occupancy rises above the break-even rate, cash flow begins to accrue to the investor. This stage represents the "top of the pyramid" and offers the best chance for the initial investors to achieve profit by selling the property to equity investors. In the aftermath of the "Yes era", many projects had trouble moving from the tenancy stage to the absorption stage, as break-even occupancy proved to be a chimera. The absorption stage is defined as when a now "stabilized" property is subject to the vagaries of the market in terms of vacancy and rental rates. As a property ages, any premiums achieved in rent and occupancy given novelty or superior condition will decrease. Conservative equity investors thus may find the absorption and maturing process stages to be more compelling periods for investment than the earlier stages. It should be noted that for existing structures, each time a new investor enters the picture, the loan commitment stage will more than likely be revisited. For illustration purposes, Fig. 8.1 assumes that the third stage is the initial construction or major renovation of the property for its highest and best use.

The absorption stage begins the long downward slide from a property value perspective. Given the time period of 40–50 years, numerous equity investors could buy and sell the property prior to the need for redevelopment at the end of the economic useful life of the property. The property life cycle pyramid is useful to illustrate how different stages appeal to different investors. The illustration also suggests that there is a trade-off between risk and return over the property life cycle.

If the property life cycle is fused with the life cycle of the specific investor, most young investors have limited property management experience and limited equity capital. For these investors, the entry point into investment real estate is typically via residential houses, duplexes, or small multi-family property which is why the preceding discussion appears in this chapter.

8.2 Types of Residential Apartment Investment Projects

The first form of residential investment is the single family residence. Many investors start out by purchasing these homes with the intention of leasing to third party tenants. Sometimes the home was purchased for owner occupancy, but the family was forced to move for a job and the owners could not sell the property. This represents a situation where the property is leased as a defensive strategy rather than when an investor purchases the home for investment purposes. Typically, single family residential houses and duplexes are viewed separately from the categorizations that follow given the lack of market intelligence concerning rental rates for single family homes, the lack of diversification in income since the home is either vacant or leased, and since the landlord will typically manage the upkeep of the property in a similar vein to owner occupied housing.

Apartment projects can come in many shapes and sizes. The number of units can range from one to hundreds, but traditionally multi-family projects are viewed as consisting of over four dwellings. The dwelling itself can be considered an efficiency (or studio) apartment. In this type of dwelling, the apartment itself does not have a separate bedroom. A garden unit is an apartment unit located all on one floor, while a townhouse consists of a unit on two floors. Amenities are another point of distinction for apartments. Amenities can include specific features in a particular unit, such as vaulted ceilings and fireplaces, or could consist of things offered in the complex as a whole, such as tennis courts, swimming pools, or laundry facilities.

Garden apartments are typically housed in two or three story buildings and have a wide range of units per building. Garden apartments provide tenants with access to a semi-public front yard and a semi-private backyard, so tenants can tend to their gardens as the name implies. Garden apartments are typically located in suburban areas rather than in densely populated urban environments. This has traditionally been the case since the unit to land ratio is lower in suburban environments. Garden apartments also rely on surface parking rather than on parking decks for similar reasons. A target tenant to parking space ratio of 1.50:1.00 is typical, although as the prevalence of two-income households increases, the ratio might be higher. The rule of thumb is that the number of parking spaces should equate to the number of working adults living in the units. The Americans with Disabilities Act (ADA) requires a certain number of handicap accessible parking spaces depending on the number of units. For apartments of under 100 units, a handicapped accessible parking spot is required for every 25 parking spaces (ADA 2011).

Townhouse complexes consist of row houses with party walls, which are walls shared between neighboring tenants. Young families or single professionals are typically attracted to this form of multi-family property. Townhouses have separate entrances and private patios which help to create the feeling of a detached primary residence for the occupants. Typically, upkeep for interiors is left to the occupants, and this form of property may also be owner-occupied in the form of condominiums. Whether the townhouses are leased or owned by the occupants, property management consists of maintenance of the common areas and repairs to any structural deficiencies of the buildings. Condominiums will require that the owners pay a monthly fee associated with managing the homeowners association (HOA). The HOA is responsible for utilizing the proceeds from monthly assessments to compensate those that provide landscaping and other services at the property. Sometimes special assessments are necessary in order for the HOA to pay for less common services such as rooftop replacement and other large expense items that will occur over the property's life cycle. These special assessments are levied based on the owners of the units voting in favor of the need for these repairs. For example, monthly assessment fees should be sufficient for paying for yard and swimming pool maintenance, but replacing rooftops or resurfacing parking lots could require a special assessment. The make-up of the HOA board and the age and condition of the buildings go a long way in determining the expenses to be shared by the condo owners over the holding period of the investment. For older properties with deferred maintenance, or for HOA boards served by individuals seeking short term property appreciation for resale, special assessments could be higher than for properties of sound condition with HOA boards served by owners with a longer term investment horizon.

As apartment complexes rise in the number of floors, the properties are classified as walk-up (low-rise), mid-rise, and high-rise. Walk-up apartments typically consist of from three to five floors. Depending on the municipality and the situation, the Americans with Disabilities Act (ADA) may require an elevator on site for apartments of over three floors, and older properties may not comply with handicap access requirements under this legislation (ADA 2011). In terms of amenities, walk-up apartments will typically provide a laundry center, basic storage facilities, and possibly a game room. Tenants on the first floor will typically have a patio, while those on the higher floors will typically have a balcony.

Mid-rise apartments typically consist of from six to nine stories in height. These properties most assuredly have an elevator, so keeping the elevators serviced is the responsibility of property management. Units are typically separately metered for electric and basic utilities. The property manager is responsible for general maintenance of the building as well as replacement and repair of carpeting and appliances in the units. Amenities vary to include exercise facilities, swimming pools, tennis and basketball courts (or volleyball courts for warmer climate locations), as well as laundry and storage facilities.

High-rise apartments are much taller buildings typically located in dense urban environments. One rule of thumb is that there should be an elevator for every eight floors (Phyrr and Cooper 1982). The same basic services provided in mid-rise apartments are provided for high-rise properties, but the greater size and scope of high-rise apartments require some additions. Amenities could include valet parking, on-site security, doormen, garage attendants, retail outlets, or other conveniences.

Another form of investment class under the heading of multi-family property is off-campus student housing. While the property types can be any of the types thus far discussed, off-campus student housing is different and popular enough to require its own section later in this chapter.

8.3 Valuing Residential Apartment Projects

In this section, we will discuss the valuation considerations specific to apartment projects. The capitalization effect looms large in apartment valuation. Location can contribute to or reduce the value of a property. Apartments near schools and employment, as well as the quality of the schools and employment opportunities, will affect the demand for a given property's units. This will directly influence the occupancy rates for an apartment project which will thus impact property value. Whether it is the proximity to schools, employment, or recreational areas which drives value for a given property will depend on the target market of the tenant base. Tenants who are primarily comprised of single adults may not value the nearness to a school, while the opposite may be true for properties which cater to occupants consisting of young families with children.

The crime rate of a subject area will also influence value. If regular visits by the local law enforcement agency are a requirement of property management, occupancy rates should suffer. Additionally, investors evaluating apartment complexes should note if operating expenses associated with evictions and court costs materialize. These expenses can imply underlying problems with the current tenant base. Problems with the credit quality of tenants can be viewed in terms of market vacancy and rental rates. If market rental and vacancy rates are superior to the subject property's performance, investors (and lenders) should explore further to elucidate the cause of the trouble. Simply assuming that the problem lies with the seller or existing property manager may prove to be an erroneous and costly assumption.

The initial drive-by assessment of the property by the investor (or lender) should help to alleviate some concerns about negative implications on property value owing to poor location and tenant mix considerations.

The quality of the construction and condition of the apartment property can also be ascertained at a high level via the property inspection. For example, an apartment complex of brick construction is typically of higher quality than one of wood construction. Central air conditioning is preferable to window air conditioning units. An inspection of the interior of the units will reveal if washer and dryer hookups are available, and what amenities are on offer in the subject property. Investors might also consider the benefits of engaging a qualified third party to perform a study concerning the structural quality as well as environmental and legal (given the aforementioned ADA concerns) acceptability of the subject property. Deferred maintenance which goes undetected prior to purchase may have serious implications on the investor's actual return should higher than anticipated renovation expenses be required during the investment holding period. Since you cannot always judge a book by its cover, it is recommended that that property inspection include visitation to random apartment units, rather than those specifically recommended by the seller's agent. There have been cases where required improvements for the entire subject property were underestimated by the investor based on viewing the condition of units and appliances in units thought to be randomly selected. Rather than focusing on the nicest units, inspection of the average or even worst conditioned units may allow for more accurate expense projections.

The calculation of replacement reserve is typically different for multi-family apartments relative to other forms of commercial real estate as was discussed in the mini-case on the subject at the end of Chap. 4. Rather than estimating a replacement reserve on an annual basis on a cost per square foot, or via a percentage of effective gross income (EGI) basis, apartment reserves are typically calculated on a per unit basis. A typical range for replacement reserves is from $100 to $300 per unit. Newer properties will obviously require less from a replacement reserve perspective, while older properties would tend to require a higher per unit allocation. While the replacement reserve is not an actually incurred expense, its inclusion in the direct capitalization and discounted cash flow (DCF) analysis stems from the origins of investment property lending. Banks historically required, as a normal course of business, that an investor set up an escrow account whereby a portion of the net cash flow from the investment property be retained in the account in case something should go bump in the night, and an unforeseen expense item would materialize. Nothing is less desired than the phone call at 3 A.M. from a tenant whose heat or air conditioning has stopped working. In today's marketplace, having liquidity for unforeseen expenses is certainly a valid concern. Lenders will evaluate the extent of unencumbered liquidity held by an investor or corporate borrowing entity to help mitigate the lack of an escrow account being established. Most lenders will require that the property operating account be held at their bank in order to assess this issue in real time. Regardless of the extent of the sponsor's strength from a liquidity standpoint, the concept of property valuation is based on the perspective of the average investor, which requires the inclusion of the replacement reserve.

8.4 Jasmine Court Apartments Case

Jasmine Court Apartments consists of 2 two story and 1 three story buildings which consists of brick exterior and wood frame construction. The roofs are pitched with architectural shingles, metal gutters and down spouts. The property was constructed in 1973 but is considered to be in good condition. Each of the 45 units are furnished with carpet flooring, and vinyl in the kitchens and bathrooms. All public utilities are available on site including electricity, telephone, water and sewer, cable television and natural gas. There is a paved asphalt parking lot on site with a total of 80 marked parking spaces. The site contains a coin operated laundry service in one building and there is a children's playground on site. The property is nestled in a residential wooded area and sits on three acres of land. There are three grocery anchored retail centers within 2 miles of the complex, and the elementary school that services the apartment's district is one half mile away. Historically, occupancy rates have been around 95%, with current occupancy at 100%. The rent roll depicted in Fig. 8.2 details the size of the units at Jasmine Court.

As you can see from the rent roll, the current lease rates at Jasmine Court are above the market averages. This could be owing to the strong property management or the general favorability of the location. In any event, utilizing the direct capitalization valuation technique will achieve divergent values based on the first projected year of operating performance. Should the rents remain at the current levels, the value will be $240,000 higher than if the rents revert back to the market averages. This is shown in Fig. 8.3 by comparing the estimated values shown in "projection 1" and "projection 2".

If the property is for sale for $1,400,000, a loan of 75% of the sales price would be $1,050,000. A loan amortized for 20 years at a fixed rate of 7% would produce debt coverage ratios based on the four scenarios shown in Fig. 8.4.

For situations like this, investors may find that the discounted cash flow (DCF) valuation technique is a more effective approach given the uncertainty concerning future income and vacancy levels. One key valuation consideration for Jasmine

Units	Description	Sq Ft	Actual Rent	Income	Rent/ Sq Ft	Market	Market Income	Rent/Sq Ft
15	1 Bed, 1 Bath w/ Den	830	$ 370	$ 66,600	$ 0.45	$ 330	$ 59,400	$ 0.40
10	2 Bed, 1.5 Bath w/ Den	990	$ 565	$ 67,800	$ 0.57	$ 445	$ 53,400	$ 0.45
16	2 Bed, 1 Bath w/ Den	950	$ 450	$ 86,400	$ 0.47	$ 445	$ 85,440	$ 0.47
4	3 Bed, 2 Bath, w/ Den	1100	$ 595	$ 28,560	$ 0.54	$ 600	$ 28,800	$ 0.55
45	Totals			$ 249,360			$ 227,040	

Fig. 8.2 Rent roll for Jasmine Court Apartments

	Prior Actual	Last Actual	Projected 1	Projected 2
Gross Revenue	$ 202,613	$ 234,680	$ 249,360	$ 227,040
Other Income	$ 918	$ 918	$ 946	$ 946
Total GPI	$ 203,531	$ 235,598	$ 250,306	$ 227,986
Vacancy/Collection (5%)	$ -	$ -	$ 12,515	$ 11,399
EGI	$ 203,531	$ 235,598	$ 237,791	$ 216,587
Property Taxes	$ 8,359	$ 7,300	$ 7,300	$ 7,300
Insurance	$ 4,580	$ 5,496	$ 5,660	$ 5,660
Management	$ 4,800	$ 4,800	$ 11,800	$ 11,800
Utilities	$ 27,865	$ 30,995	$ 31,963	$ 31,963
Advertising	$ 300	$ 300	$ 330	$ 330
Repairs & Maintenance	$ 38,734	$ 40,648	$ 41,867	$ 41,867
Replacement Reserves ($250)	$ -	$ -	$ 11,250	$ 11,250
Total Op Ex	$ 84,638	$ 89,539	$ 110,170	$ 110,170
Net Operating Income	$ 118,893	$ 146,059	$ 127,621	$ 106,417
Cap Rate	9%	9%	9%	9%
Estimated Value	1,320,000	1,620,000	1,420,000	1,180,000

Fig. 8.3 Direct capitalization at various scenarios for Jasmine Court Apartments

	Prior Actual	Last Actual	Projected 1	Projected 2
Net Operating Income	$ 118,893	$ 146,059	$ 127,621	$ 106,417
Annual Debt Service	$ 97,688	$ 97,688	$ 97,688	$ 97,688
Debt Coverage Ratio	1.22	1.50	1.31	1.09

Fig. 8.4 Debt coverage ratios for Jasmine Court Apartments

Court Apartments is whether the rental rates can sustain above market performance. Over the long term, the assumption of above market lease rate sustainability may not prove to be wise. It is therefore recommended that the investor run numerous iterations of the DCF. One trial may assume that the value of the subject property is based solely on market lease rates. This would be a conservative case given where the lease rates are currently in relation to market averages. Another DCF trial might assume that the above market lease rates do prove to be sustainable. This would be considered a best case scenario.

A most likely case scenario for the DCF might consist of higher than market rental rates for the first 2 years, with the remaining years of the holding period assumed at market lease rates. Another point of contention in the valuation for Jasmine Court is the vacancy factor assumption. Since the property is currently fully occupied, and assuming that occupancy rates have achieved 95% for the last

8.4 Jasmine Court Apartments Case

	Year 0	Year 1	Year 2	Year 3	Year 4	Year 5	Year 6	Year 7	Year 8	Year 9	Year 10	Reversion	Year 11	Comments
Gross Potential Income		250,306	250,306	227,986	233,686	239,528	245,516	251,654	257,945	264,394	271,004		277,779	2.5% Annual Increase after year 3
Vacancy Factor		12,515	12,515	11,399	11,684	11,976	24,552	25,165	25,795	26,439	27,100		27,778	5% years 1-5, then 10%
Effective Gross Income		237,791	237,791	216,587	222,001	227,551	220,964	226,488	232,151	237,954	243,903		250,001	
Taxes		7,300	7,300	7,300	7,483	7,670	7,861	8,058	8,259	8,466	8,677		8,894	Grow at 2.5% for all after year 3
Insurance		5,660	5,660	5,660	5,802	5,947	6,095	6,248	6,404	6,564	6,728		6,896	
Repairs & Maintenance		41,867	41,867	41,867	42,914	43,987	45,086	46,213	47,369	48,553	49,767		51,011	
Management		11,800	11,800	11,800	12,095	12,397	12,707	13,025	13,351	13,684	14,026		14,377	
Utilities		31,963	31,963	31,963	32,762	33,581	34,421	35,281	36,163	37,067	37,994		38,944	
Other		330	330	330	338	347	355	364	373	383	392		402	
Replacement Reserves		11,250	11,250	11,250	11,250	11,250	11,250	11,250	11,250	11,250	11,250		11,250	$250 per unit
Total Expenses		110,170	110,170	110,170	112,643	115,178	117,776	120,439	123,169	125,967	128,835		131,774	
Net Operating Income		127,621	127,621	106,417	109,358	112,374	103,188	106,049	108,982	111,988	115,069	1,313,628	118,227	9% Terminal Cap Rate
Annual Debt Service		97,688	97,688	97,688	97,688	97,688	97,688	97,688	97,688	97,688	97,688			
Tenant Improvements		-	-	-	-	-	-	-	-	-	-			
Leasing Commissions		-	-	-	-	-	-	-	-	-	-			
Debt Coverage Ratio		1.31	1.31	1.09	1.12	1.15	1.06	1.09	1.12	1.15	1.18			
Net Cash Flows	(350,000)	29,933	29,933	8,729	11,671	14,686	5,501	8,362	11,294	14,300	529,725			
BTIRR On Equity	7.84%													
Property Value	$ 1,210,000.00													

Fig. 8.5 Discounted cash flow analysis for Jasmine Court Apartments

few years, the DCF might assume a 5% vacancy factor for the first 5 years and a 10% vacancy factor as the property reverts to mean performance over the last 5 years of the investment holding period. Fig. 8.5 illustrates the DCF valuation under the just described most likely case scenario.

Based on the DCF analysis in Fig. 8.5, the estimated property value was $1,210,000. This value is in between the two direct capitalization values, which makes intuitive sense given that our DCF was a blended assumption concerning the market rental and vacancy rates.

Since the current sales price of Jasmine Court Apartments is $1.4 million, whether this is a good investment at the price depends on whether the above market leasing situation can endure, and if the before tax internal rate of return (BTIRR) of 7.84% is a reasonable return for the risk to the investor. The BTIRR will increase depending on the optimism shown in the financial projections over the term of the loan. As the rental income increases relative to operating expenses, the percentage of the income coming from the reversionary cash flow at the end of year 10 will hopefully decrease. Based on the DCF case shown in Fig. 8.5, approximately 75% of the net cash flow is projected from the reversion.

What should be apparent from the preceding discussion is the importance of a strong property manager. The property manager is responsible for screening out lower quality tenants who may create immediate problems such as non-payment of rent, annoyance of other tenants, and illegal or other such activities which serve to drive away more stable tenants. While some investors may prefer to serve as their own property manager, professional management may become essential. Rates charged for professional management vary widely, but a range of from 3% to 7% of effective gross income (EGI) is a general rule of thumb. Management fees should vary relative to the effective gross income, i.e. revenue after vacancy and collection losses. In this way, property managers are properly motivated to keep the subject property leased to a sufficient degree at rates deemed comparable in the market (or perhaps better).

Phyrr and Cooper (1982) have itemized the services offered by professional property managers. These eight key areas of assistance by professional management are shown below.

1. Selection and training of on-site personnel
2. Tenant selection and leasing
3. Rental collection and evictions
4. Timely maintenance and repairs
5. Cleanliness of improvements and grounds
6. Provision and supervision of tenant rules and regulations
7. Budget and expense control
8. Record keeping and reporting to the owner

Returning to Jasmine Court Apartments, the eight keys for an effective property manager may help the complex maintain the existing above market lease rates. Given the age of the apartments, new construction may not necessarily compete with Jasmine Court for occupants, as newer, more luxurious apartments do not typically compete with standard apartments. Barring an influx of new construction of competing units, the occupancy rates for Jasmine Court should not suffer materially over the holding period. Timely maintenance, quality tenant selection, and quality on-site personnel should help keep rental rates up as well. Retention of qualified professional management may make these conjectures more of an economic certainty.

8.5 Mixed Use Properties as an Investment Alternative

Earlier in this book, we mentioned that multifamily housing often represents a typical steppingstone for an investor who may begin owning their own home, to then owing a few rental homes, to then eventually owning an apartment complex. A similar case could be made for investors who own apartment complexes utilizing mixed use properties as a step toward diversifying their real estate investment portfolios. As the name implies, mixed use properties consist of a blend of one or more of the traditional categories of investment real estate.

Sometimes mixed use properties can blend retail and apartment units into one investment property. In this case, the retail units would be constructed on the bottom floor, with apartment units constructed on the higher floors. The draw of this type of mixed use property is the presence of pedestrian foot traffic near the property, making the retail units on the ground floor more appealing. These types of properties are typically found in more urban environments.

Apartment, retail, and office units can also be packaged together in mixed use properties. If you envision a three story property, apartment units could occupy the third floor, office units could occupy the second floor, and retail units could occupy the ground floor. The UK grocer Tesco has expanded its operations in China via large lifestyle centers which also contain the Tesco grocery chain as the primary tenant (Financial Times 2010). The benefits of the Tesco version of the lifestyle center in China is that residential, office, and retail services can be provided all under one very large rooftop.

While the diversification by property type within one investment complex has benefits, there are also some disadvantages. The primary disadvantage is that the investor must obtain rental and vacancy market information for each of the sectors represented in the property in order to estimate future demand. Additionally, expense projections will consist of blended averages for the presented property sectors, at least until the subject property experiences a few years of operating performance. Lastly, comparable properties may be more difficult depending on the level of diversification in the property. Being the first to market with an innovative property can be advantageous, but only if the market is ready to accept the new offering. If the mixed use alternative is too much of a trail blazer, the property may be slow to lease.

8.6 Off-Campus Student Housing as an Investment Alternative

Many investors prefer to invest in apartments which cater to college students. Since college dormitories are located on the campuses of the colleges and universities that they serve, on-campus student housing is not typically open to individual investors. State owned universities utilize annual budget appropriations to construct new on-campus student housing as needs arise and as funding allows. Private universities utilize contributions from alumni and other sources for on-campus housing construction and development.

As student enrollment increases over time, on-campus housing availability may not necessarily be able to keep pace with the demand for student housing. Thus off-campus student housing becomes a viable investment alternative.

8.6.1 Location, Location, Location

Off-campus student housing certainly is a good example of the "three most important considerations in real estate" as is itemized in the heading to this section. Investors rush to obtain the closest available land near the subject university, in order for students having the option of walking to campus. After years of polling our students, it is with some degree of confidence that we label "location" as an important consideration for off-campus housing projects. Other important considerations include the amenities on offer, the quality of construction, and the monthly rental rate per bed in comparison with the market averages.

In many off-campus student housing markets, the existing supply consists of older, well-positioned units, and newer units with a more modern appearance and amenities such as fitness centers, swimming pools, and tennis courts. Some markets supply fully furnished units, whereby students can move in easier than in years past. The fully furnished units are typically more expensive than non-furnished units, but those in occupancy justify the additional cost given the convenience provided.

If there is one constant in off-campus student housing, it is the continual raising of the bar concerning what students expect in an apartment offering. Depending on the competitive situation of the particular market, some property owners provide internet, cable, and other services as part of the monthly rental assessment, while other properties do not have these services included in rents. Ceteris paribus, students typically flock to the most up-to-date units at the expense of those properties where the owners do not keep up with the rising tide of expectations.

8.6.2 If You Build It, Will They Come?

During the last 20 years, much construction of off-campus student housing has occurred, especially in the southeastern United States where land availability is high. A typical modus operandi of an investor would be to purchase various small tracts of land in proximity to the target university. The goal of such purchases is to acquire enough contiguous land parcels to subsequently construct a multi-family project catering to college students, and the parents that guarantee their leases! Given the boom in college enrollment over the last few decades, student housing projects were seen as recession-proof, or at least more immune to down cycles than other investment property options. This idea stems from the fact that more people return to school during recessionary times than when the economic climate is more vibrant. As the preceding chapters of this book make clear, nothing is immune to the business cycle, or to the possibility of imbalance between supply and demand.

In the prior section, we discussed the possibility of an investor's property not keeping up with the expectations of students for what a quality off-campus living arrangement should provide. Let's refer to this as the units being *too old* to maintain profitable occupancy rates amongst the sea of newer properties in the market. As students seek out the newest available units, older property vacancy rates are the first to suffer once the market equilibrium point is surpassed.

Another risk that may impact a property's occupancy rate is that of being *too late* to market. In the United States, the fall semester is the beginning of the academic year which ends upon the completion of the spring semester in early May. The fall semester typically begins in late August or early September. A new property to market which is not complete and ready for occupancy by the end of July may miss an entire academic year of potential occupancy. Unless there is a dire need for the newly constructed units, most students will not wait for a property to be complete before moving in during the fall semester. As most college students will attest, moving to a new complex during Christmas break is only to be considered in the most-dire of circumstances. Thus, being "too late" to market has the potential of leaving a property vacant or under-occupied for an entire academic year. This risk is furthered by the typical annual lease agreements, which compensate property owners for the lower occupancy rates experienced in the summer months when the students are away from school.

An additional "if you build it risk" is that of being *too new*. This involves a property owner being the last property to market in a saturated environment. For those who are the last to market, full occupancy might be tough to achieve, and it may only be achieved via rent concessions or by expanding the target market of the tenants. Typically, off-campus student housing should be restricted to students, rather than expanding the target market to include families or the general public. Such "mixed occupancy" properties are typically not successful as students and families live on different schedules and generally do not mix well in the same property. Investors seeking to mix occupancy between students and other groups are cautioned to consider keeping the student housing for the students, and not falling into the *too broad* target market trap. All of the risks outlined in this section make the analysis of supply and demand all the more important.

8.6.3 Market Analysis in Student Housing

Market analysis for off-campus student housing has similarities and differences to traditional apartments. Traditional market analysis such as the ingress/egress of the property, the distance to schools, shopping, recreation, and employment pertain to both traditional apartments and student apartments. Other factors such as vacancy rates, competition from rental houses, and the specific geographic boundaries of the comparable properties require a more targeted focus when evaluating student housing.

The first step in evaluating the market is to delineate between student housing and traditional apartments. There may be some properties catering to both students and traditional occupants, but the primary focus should be on those catering specifically to students. Once that the existing stock has been identified, regional or city planning commissions should be contacted in order to identify any planned demolition, renovation, or construction in the subject market in the foreseeable future. If the existing supply of housing units is adjusted based on the projected amount of units in the market, a reasonable projection of housing availability can be made for the next academic year. When this information is coupled with the traditional vacancy rate for off-campus properties in the market, a conservative estimate of the supply of units can be determined. In validating the investor's projections, local market professionals such as appraisers, property managers, and brokers can be engaged.

Projecting the future demand for student housing involves a similar process to estimating supply. The starting point would be to verify the projected enrollment at the target university for the next academic year. This figure would constitute the most optimistic view of the off-campus housing demand as some students may live on campus while others may commute from areas further from the university than the submarket being evaluated. University officials involved with campus housing may represent another viable information source for the off-campus housing investor. University officials may also be able to share future on-campus housing plans so that future demand projections can include expected changes in the on-campus housing environment. While this process is not an exact science, estimating the supply and demand for student housing is crucial in order to understand future value implications on movements in rents and vacancy over the investment holding period.

When in doubt, investors might consider engaging an appraiser for completion of an occupancy study, or might consult additional third party sources for data in the subject market to serve as confirmation or refutation of their market assumptions. As was made clear in the *too late* discussion earlier, market equilibrium analysis is only good for one academic year, thus it requires periodic revision. Prudent investors will keep a continued focus on economic and demographic changes in their markets which may impact the demand for off-campus student housing both directly and indirectly.

8.6.4 Class Project: Off-Campus Student Housing Occupancy Study

Given our discussion of off-campus student housing projects, readers utilizing this book for a college course in commercial real estate might consider an interesting project whereby the students take part in an occupancy study during the semester. The purposes of this occupancy study are twofold: (1) allow students to physically inspect an off-campus student housing project near their campus (2) interview a property manager in order to obtain market data required to assess the occupancy and rental rates of the off-campus student housing market during the current semester.

Instructors should compile a first-come, first-served listing of the off-campus housing projects selected by the students which can be shared with the students so the same property is not selected numerous times. The basic idea is for each student to select a unique property as their subject property, and then to complete the "off-campus student housing occupancy study" which follows this section, along with a short paper where the student comments on anything observed while on-site which is deemed useful when collecting the market data from the property manager. Students can comment on the management ability of the property manager, the quality of the landscaping and the general aesthetic appeal of the property, and the age and condition of the units. The student may also discuss any difficulties in obtaining information from the property managers, and if anything was observed on-site which directly relates to topics covered in the course in terms of property management or tenancy issues.

If the available properties considered to be the primary off-campus student housing product have all been selected, instructors may allow students to study general apartment properties in the market so that a comparison can be made between rents and occupancy rates for general apartments as compared with off-campus student housing. Students should be instructed to contact the property manager to set up an agreeable interview time, rather than either dropping in randomly, or conducting the interview over the phone. The student should obtain the brochure of the property as well as any other handouts deemed useful in verifying that the data obtained from the property manager is accurate.

Once the instructor has received all of the completed occupancy study worksheets and papers, the results can be compiled to share the total number of properties and units involved in the study, the average age of the units in the survey, the current vacancy and rental rates, and the average vacancy rate for the last 3 years. The results can be shared with the students on the evening where the class evaluates apartment projects. In fact the market rental and vacancy rate information can be utilized to represent the current market information, and this information can be used in class discussion reflecting how this information might change the analysis of the apartment projects being discussed. Each student will thus contribute a small part in the creation of the class occupancy study, and it will provide students with real world experience for some of the actions taken by an appraiser in order to determine market vacancy and rental rates.

The chapter thus ends with the publication of the one page questionnaire necessary for the off-campus student housing assignment, and concludes with two apartment case studies. As similar to the Jasmine Court Apartments case study, students are instructed to compile the direct capitalization and discounted cash flow analysis for each case study.

Questions for Discussion

1. Describe the property life cycle pyramid. How is this framework relevant for determining when an equity investor or a developer may be interested in a given investment property?
2. Given that most residential leases are for less than 1 year, how might an investor become comfortable with the "durability" of the income stream of a multi-family property?
3. Describe the differences between multifamily housing and off-campus student housing properties. Does it make sense to include both the general public and students in the same residential housing complex?
4. Elaborate on the four risks associated with new off-campus student housing construction.
5. Provide some examples of successful mixed use developments and some examples of when combining property types might not be appropriate.

Mini-Case: Property Manager Interviews for Off-Campus Student Housing

Off-Campus Student Housing Occupancy Study

Student Name _____
Date of Study _____
Name of Housing Complex _____
Address (street, city) _____
Property Manager Contact _____
Property Management Company _____
Age of Complex (Year constructed & last major renovation*) _____
Number of Floors in Complex _____
Number of Buildings in Complex _____
Type of Construction (brick, etc.) _____
Total Number of Units in Complex _____
Total Number of Bedrooms per Unit _____
Total Number of Bathrooms per Unit _____
Total Square Footage per Unit _____
Current Occupancy _____%
Occupancy Rates for the last 3 years ____% ____% ____%
Number of new units added over same period _____
Rent per Student per month $_____
Rent per unit per month $_____
Any expenses paid by landlord? _____
If yes above, what and for how much monthly? _____
List amenities in complex _____
Washer/Dryer Units or Laundry Room? _____
Any additional comments*: _____
_____
_____

*If major renovations have been undertaken within last 5 years, please describe in the section above or on additional pages. Additionally, please itemize any other concerns or observations that you feel are important and relevant to the class as related to the subject property.

Names of other complexes considered comparable by Property Manager:

Apartment Case Study #1

Mr. & Mrs. Robbins have recently sold their Ace Hardware business, and want to invest $150,000.00 to buy a small apartment complex. The Robbins' will operate the day-to-day management of the apartment complex. This represents the couple's first foray into investment real estate.

Some specifics of the apartment complex are as follows:
- The apartments were built in 1989 (it looks its age)
- There are two 8 unit buildings
- All units are two bedroom/two bathroom and are 1,000 square feet
- The property has no amenities
- The apartments are located in a small to medium sized town. There are no anticipated changes to the base employment in the town, and no material changes in economic growth are projected.

The investors' market research reveals the following:
- Current rents are $675.00 per month per unit and these rents are within the market averages.
- Hundred competing units will be available in 30 days. There are currently 75 additional units in the market. Both of these properties have amenities including swimming pools, tennis courts, and a fitness center.
- Current occupancy is 100%, though vacancy has fluctuated up to 10%.
- The market vacancy is 7% for apartments.

Total acquisition costs for this property are $800,000.00. Mr. & Mrs. Robbins will invest $150,000.00 in personal equity, and are seeking financing from a financial institution for a loan for the remainder.

Historical operating expenses are:

Taxes	$17,595
Insurance	$1,102
Repairs and maintenance	$1,749
Utilities	$1,260
Other expenses	$7,162

1. Determine the net operating income for this property using the direct capitalization spreadsheet.
2. Determine the first year debt service coverage ratio for a loan of $650,000.00, at an interest rate of 8%, for 20 years.
3. Using a cap rate of 9%, what is the estimated value for this property?
4. Determine the breakeven occupancy rate and interest rate for this property.

Apartment Case Study #2

Overlook, LLC was formed to purchase Overlook Apartments. The purchase price is $2,500,000.00. The investors would like to finance as much of the purchase price as is supported by the net operating income of the property. The apartments were built in 1997 and are in good condition. The investors have provided you with a rent roll for the subject property, as well as the income and expense statements from the most recent 2 years for the subject property. Market vacancy is stable at 10%, and the location of the complex is considered good based on your recent tour of the property and the surrounding market. The investor's equity into this purchase is coming from the personal savings of Mr. and Mrs. Ernest Smith. The investors have requested financing of 90% of the purchase price, but traditional bank guidelines allow for financing of only up to 80% for this type of property. The investors have supplied a copy of the most recent rent roll as follows. The current occupancy in the apartments is 95%.

# of units	Layout	Sq. ft.	Rent rate	Tenant history
24	1 Bed and 1.5 Bath	800	$650	Avg. occupancy is 3 years
24	2 Bed and 2 Bath	980	$700	Avg. occupancy is 4 years

The most recent 2 years profit and loss statements for the subject property are summarized below:

	Prior year	Most recent
Gross rents	$375,000	$385,000
Expenses		
Advertising	$4,000	$5,000
Cleaning and maintenance	$6,000	$8,000
Insurance	$7,500	$8,000
Repairs	$15,000	$14,000
Taxes	$37,000	$40,000
Utilities	$2,300	$2,450
Depreciation	$15,000	$26,000
Landscaping	$3,000	$3,000
Total expenses	$89,800	$106,450
Net income	$285,200	$278,550

1. Determine the net operating income for this property using a direct capitalization spreadsheet.
2. Determine the first year debt service coverage ratio for a loan equal to 80% and 90% of the purchase price, at an interest rate of 8%, for 20 years.
3. Using a cap rate of 9.50%, what is the estimated value for this property? What does the loan to estimated value look like at 80% of purchase price? What about at 90% of purchase price?

4. Determine the breakeven occupancy rate and interest rate for this property at both loan amounts.
5. Under what circumstances would you consider financing over 80% (and moving outside of the typical financing guidelines for your bank)?
6. Calculate a Discounted Cash Flow for this property. Assume that your direct cap numbers represent year 1, and project a 2.5% increase in income and expenses over the holding period of 10 years. Assume further that the property will appreciate 3% annually, with 1% selling expenses. Further assume that management expenses will be 5% of EGI and that replacement reserves will be $100 per unit each year. Discount the NOI at 9%, and find the BTIRR for the 80% loan case. Is there a material difference in value between the direct cap and the DCF in this case? Why is there (or when might there be) a material difference?

References

Americans with Disabilities website, http://www.access-board.gov/adaag/html/adaag.htm?bcsi_scan_2EC956E02D10C4E6=MMHU2QPoN9O6AvTKdpgK1WnsvWsVAAAAKumeFw==&bcsi_scan_filename=adaag.htm#4.9. Accessed 25 March 2011.

Financial Times, Flested, Andrea (2010). Supermarkets: Tesco finds thriving market and launch pad for China 10 Nov 2010, Financial Times web edition, at http://www.ft.com/cms/s/0/c0ab4740-eb8a-11df-bbb5-00144feab49a.html#axzz1LyVcvPId. Accessed 10 May 2011.

Phyrr, S. S., & Cooper, J. R. (1982). *Real estate investment: Strategy, analysis, decisions.* New York: Wiley.

Investing in Retail and Office Property 9

Keep thy shop and thy shop will keep thee. Light gains make heavy purses.
George Chapman

Contents

9.1	Subtle Differences Between the Property Types	184
	9.1.1 Destination Versus Spontaneous Orientation	184
	9.1.2 Projected Demand for Space	185
9.2	Classifying Retail Properties	185
	9.2.1 Outparcels and Single Tenant Properties	185
	9.2.2 Multi-Tenant Retail Properties	186
9.3	Evaluating Retail Property Projects	188
	9.3.1 Retail Property Location Considerations	188
	9.3.2 Retail Tenant Considerations	188
9.4	Classifying Office Investment Properties	190
	9.4.1 Office Property Delineation by Size and Class	190
	9.4.2 Office Property Categorization by Use and Design	191
9.5	Evaluating Office Properties	192
9.6	Office Condominiums as an Investment Alternative	193
9.7	Case Studies in Retail and Office Property	193
9.8	Conclusion	195
References		202

During the height of the "Yes Era", retail and office property was constructed based on the growth in new housing starts in various markets across the United States. Speculative home construction may have led to speculative retail and office properties being constructed nearby. While the level of speculative construction was not as prevalent during the "Yes Era" as was the case in prior real estate-led recessions, the "if you build it, they will come" philosophy is an alluring concept during economic expansion.

In this chapter, we will explore the valuation considerations specific to retail and office properties. While each is considered separate classes of real estate investment, the two property types have enough in common to offer a joint analysis in Chap. 9. Before we cover the valuation issues for each property type, it is helpful to differentiate subtle differences between retail and office properties. Then, retail and office valuation considerations will be discussed, followed by a discussion of office condominiums as an investment alternative. The chapter ends with case studies on retail and office properties, including a construction case which involves a mixed use property hypothetically located in Heidelberg, Germany.

9.1 Subtle Differences Between the Property Types

9.1.1 Destination Versus Spontaneous Orientation

A primary distinction between retail and office property is the motivation by which the tenants customers decide to visit the property. The motivation of the customer is defined as the reason for behavior which helps to explain the direction, intensity, and form of behavioral patterns which is leading the customer to venture into the retail or office tenant's place of business (Raab et al. 2010). Retail customers are driven by more spontaneous motivations relative to office clientele. If a customer is driving down a busy thoroughfare during lunchtime and they decide that they feel hungry, they are more likely to turn their car toward a restaurant location than a dentist's office! If the same customer also determines that their car requires refueling, the prevalence of a gas station will also influence which way they turn. Given the spontaneity of the decision to visit the retail property, the ease of entering and exiting the property is an important consideration. If the traveling customer is located in an area with many competing restaurants near gas stations, how easy it is to arrive and exit may determine their order of preference.

Office property is by definition more destination-oriented than retail property. Office tenants will typically set appointments for their clientele rather than letting the visit depend on the whims of chance. Medical office tenants are more typical of the "appointment oriented" schedule than other office tenants such as insurance or attorney firms, but the distinct lack of whim site visitations is a subtlety to distinguish office from retail properties. While ingress/egress is important for office properties, it is crucial for retail properties.

Other categorizations allow for retail and office properties to occupy the same chapter of this book. The demand for retail and office property is directly and indirectly affected by general economic conditions. When the economy is vibrant, jobs are typically up and consumer and investor confidence is high. During these times, demand for office and retail property is higher as consumers are more likely to be employed and spending money. Higher levels of unemployment negatively affect the demand for office and retail space. As consumer spending is a primary means of assessment of economic conditions, retail spending is seen as being directly influenced by the general economy. The demand for space is thus indirectly tied to consumer and business expenditure.

9.1.2 Projected Demand for Space

In order to project the demand for a retail site, the primary trading area should be established. The trading area is a measurement for the potential revenue of the tenants in a given property from the geographic boundary where 60–80% of the sales in a given property originate (Appraisal Institute 2008). This can be estimated based on the size of the subject property, the types of goods on offer, population densities in the area, and transportation facilities (Phyrr and Cooper 1982). The higher the level of competition in a given market, the lower is the possible revenue of the specific retail property. Since retail property success is defined on a sales-per-square foot basis at a given property, the trading area is of importance in helping to assess the sales potential of a property relative to where it is located in a competitive market.

In order to project demand for office properties, the types of service businesses must be evaluated in light of local and regional competition. Various third party resources analyze metropolitan areas in order of preference for many of the variables which comprise a favorable business climate. The availability of employment for the various positions sought in an area and the balance between the average educational levels of the local population relative to the kinds of office jobs available will also help to assess the derived demand for office space.

Now that we have itemized some general differences between retail and office property, we will discuss important characteristics for a successful investment property by specific property type.

9.2 Classifying Retail Properties

9.2.1 Outparcels and Single Tenant Properties

Retail properties can come in all shapes and sizes. The size and number of tenants in occupancy in a building can vary from a small single tenant to a larger super-regional center. In many cases, retail properties can consist of numerous buildings on-site to include an outparcel tenant. Outparcels are classified as retail pads and are typically free-standing buildings with one to a few tenants located in the front of the primary property. Outparcels generally enjoy superior street exposure. Nationally or regionally known tenants typically occupy these properties, such as pharmacies, auto parts stores, or restaurants. Outparcels may be sold by investors or ground leases can be established. In this form of ownership, the tenant owns the building but pays for access to the land. As discussed in Chap. 5, at the end of the lease, real improvements typically revert back to the owner of the land, so there is some risk for an investor who does not own the "dirt" where the property resides. Given the title transfer at the end of ground leases, financial institutions are typically less willing to finance loans secured with only the buildings involved in a ground lease scenario.

Free-standing restaurants are another form of retail investment property. Typically, single tenant restaurants are properties designed and constructed to the tenant's specifications, and are also known as build-to-suit properties. These by-design restaurants are typically located in a cluster of similar properties and command higher rents per square foot then alternative retail tenants in the market. The higher rents per square foot are typically required to reimburse the property owner for the leasehold and structural improvements required by the tenant. Single tenant retail property is by its very nature risky given the binary cash flow (either the property is leased or vacant) and the possibility of the property valuation being based on an above market lease rate. Even if a vacant property is subsequently leased, if the above market lease rate of the original tenant is not achieved, the property owner may have a hard time in paying the monthly debt service requirements or in achieving their original internal rate of return expectations.

9.2.2 Multi-Tenant Retail Properties

A strip center is an attached row of stores or service outlets which are managed as a unified retail entity. Parking is typically located in front of the stores. Open canopies may connect the store fronts, but there are typically not enclosed walkways which link the various stores in the complex. A strip center is also known as a convenience center, and it may be configured in a straight line or it may have an "L" or "U" shape. The strip center is among the smallest multi-tenant retail configurations, and the tenants typically provide a narrow mix of goods and personal services catering to a very limited trade area. These properties are also known as unanchored retail as there is not typically one primary tenant which serves to draw customers into the property.

A somewhat larger multi-tenant property is the neighborhood center. These properties are typically catering to convenience, and typically have an anchor tenant, usually a supermarket, as well as other nationally or regionally known tenants. General size estimates are from between 30,000 and 150,000 sq. ft. of gross leasing area (GLA), and the properties are situated on from between three and five acres of land (Appraisal Institute 2004). The primary trade area is typically within three miles and the anchor ratio is often between 30% and 50%. The anchor ratio is the percentage of the center's total square footage that is attributable to the anchor tenants.

Fashion or specialty centers are slightly larger multi-tenant retail properties which cater to higher end and often fashion-oriented tenants and anchors. Given the higher-end and specialty nature of the property, the primary trade area is wider than for the neighborhood center, ranging from 5 to 15 miles. **Outlet centers**, while larger in size than specialty centers, contain manufacturer's outlet stores and thus cast a wide net in terms of the primary trade area. Some customers are quite willing to drive long distances to cash-in on the discount prices.

Community centers typically provide general merchandise encompassing from 100,000 to 350,000 sq. ft. of GLA. These properties will typically have two or more

9.2 Classifying Retail Properties

Type of Shopping Center	Size (Sq. Ft.)	Anchor Ratio	Primary Trade Area
Neighborhood Center	30,000-150,000	30-50%	3 miles
Fashion/Specialty	80,000-250,000	30-50%	5-15 miles
Community Center	100,000-350,000	40-60%	3-6 miles
Regional Center	400,000-800,000	50-70%	5-10 miles
Power Center	250,000-600,000	70-90%	5-10 miles
Super-Regional Center (Mall)	800,000 +	50-70%	5-25 miles

Fig. 9.1 Characteristics of Various Shopping Centers (Appraisal Institute, 2004)

anchors, which may consist of a supermarket, drugstore, home improvement, or discount department store. The anchor ratio for community centers is often between 40% and 60%, with a primary trade area of three to six miles. Anchor tenants are defined as those tenants that represent the primary draw for customers to the property. In-line tenants are supporting players relative to the anchor tenants.

As properties become larger than community centers, they are often priced outside of the affordable price range for most small investors. Primary buyers include real estate investment trusts and larger corporate entities. Regional centers reflect a general merchandise or fashion orientation and are typically enclosed malls with two or more anchor tenants. Total GLA ranges from 400,000 to 800,000 sq. ft. Anchors here consist of department stores, mass merchants, fashion apparel, and discount department stores. The anchor ratio is often between 50% and 70%, with primary trade areas ranging from five to ten miles. Power centers will typically have three or more anchors, with a higher anchor ratio than is seen in regional centers. Super-regional centers are also known as malls. They are similar to regional centers, but are larger and have more variety in tenant offerings. Super-regional centers usually consist of over 800,000 sq. ft. of GLA, with anchor ratios of from between 50% and 70%, and primary trade areas of up to 25 miles. Fig. 9.1 summarizes the discussion concerning typical characteristics of various types of shopping centers.

In recent years, the performance of enclosed malls have generally declined given the convenience offered by smaller, open air shopping areas that allow customers to park near their store of choice rather than walking inside in order to locate their chosen destinations. When considering retail property for investment, prospective owners should consider how the current and predicted demand will be influenced by convenience factors in order to assess if an existing retail configuration is likely to stand the test of time.

Phyrr and Cooper (1982) have identified factors which can cause changes in shopping center returns over time. This list is reproduced as a good summary of the preceding section and an introduction to what follows. Specifically, key factors contributing to change in shopping center returns can be defined as:
- Competition entering the market
- Outdated design and layout
- Changes in trade area income levels
- Changes in population density
- Changes in highway construction (or traffic patterns).

9.3 Evaluating Retail Property Projects

9.3.1 Retail Property Location Considerations

Given the spontaneity of the demand pattern for retail properties, convenience is an important element in determining the successful retail property location. Retail property should be in proximity to residential neighborhoods such that the target consumer does not have to travel long distances in order to reach the property. The department of transportation can provide general statistics for a given area's periodic traffic count while ingress and egress to primary access routes must also be relatively smooth (Buchanan 1988). The proximity to major highways helps for both consumers and for product deliveries to the location. The back of the retail property should provide sufficient clearance for delivery trucks, while road ways in front of the property should ideally have a stoplight. At a minimum, the frontage of the property should be near to an intersection. In order to aid in the customer's view of the property, retail construction should be facing the road rather than being perpendicular to the road. This principle is often violated by property developers wishing to increase the rent per square foot for an oddly shaped parcel, but in the long run, parallel construction relative to the road is often considered to be more favorable.

Traffic patterns can change over time as evidenced by many a retail property becoming vacant at the expense of a newer, better located property. Retail properties with grocery stores should be located on the "going home" side of the road, as customers are more likely to shop for groceries on the way home then on the way into work in the morning. Contrarily, coffee and bakery shops would rather be located on the "going to work" side of the street to provide that morning enhancement that we often need when going to work. As mentioned at the close of the prior section, changes in highway networks can effect change on which side of the street represents the optimal location given a property's tenant mix.

Other considerations for retail property location include the parking availability and the parking location relative to the tenants. If large parking lots are available on site, investors may consider constructing outparcels on the outer edges to generate additional revenue. A final consideration for retail location involves a comparative analysis of the amenities and tenant product and service lines offered on site relative to the competition in the area. If the primary trade area has been invaded by properties offering similar merchandise, the rent per square foot achieved by the subject property's tenants may provide hints concerning the probability of those tenants renewing their leases if they are expiring in the near term. This information in conjunction with a comparison of the subject property' lease rates relative to market can offer additional guidance as to whether tenants will remain in occupancy at the subject property, and if they will desire a decrease in their rents in order to be motivated to sign the lease extension agreement.

9.3.2 Retail Tenant Considerations

Investors considering the purchase of a multi-tenant retail property should consider the implications of tenant mix for the subject property. At a basic level, the list of tenants should be reviewed in light of whether the tenants offer complimentary services and if the general mix makes sense from a customer-draw standpoint. For example, a property that contains stores that cater to the needs of children might not fit well with stores catering to adult nightlife activities. The property might have the potential for "twenty-four hour income" but more likely the pairing of incompatible tenants will not be successful. The tenant mix should be considered in relation to the characteristics of the potential consumers in the area. In this regard, high fashion oriented retail would be misplaced in a working-class neighborhood. Another consideration of the tenant mix analysis is whether a property contains too much of a certain type of tenant. For example, a six bay retail property that has all fast food oriented restaurants in occupancy may be too highly dependent on one service industry. Nearby competing properties should also be reviewed to assess if duplication in tenant offerings is feasible. While competition is a good thing, too much of a good thing can have deleterious effects on the sales per square foot for all involved.

In many markets across the United States, a veritable fight to the death is being borne out before our eyes. On the corner of any reasonably travelled intersection, two or three national drugstore chains are often positioned in opposition to each other. Over the last decade, consolidation in the industry has occurred as weaker rivals have been literally consumed by the apparent victors. Students of our real estate courses should be familiar with the classification of the "there can be only one" slogan from the *Highlander* television and movie series. In that series, immortals fight to determine who will be the final standing victor, with each winner being comparatively stronger as they defeat the weaker rivals. While the competitive fervor of the series is similar to the drug store wars, the ironic twist is that by acquiring the store locations and extra debt of the defeated rivals, the victors in the drugstore wars become weaker over time. This same competitive struggle is repeated in other retail branches such as grocery stores, discount stores, and apparel outlets. The message for investors should be that the presence of a long term lease to a nationally known tenant does not guarantee success. As with most things in life, circumstances change over time. Investors should require at a minimum store specific annual sales figures (historical sales figures should be requested of anchors when considering purchasing the property) and complete annual financial statements for privately held firms occupying large percentages of the gross leasing area of the subject property. For publicly traded tenants, savvy investors will keep a watchful eye on credit rating agency analysis and quarterly earnings results so if the tenant in the subject property is not destined to be "the one" left standing, property owners can utilize available market data to read the tea leaves ahead of potential problems.

Another telling variable for the viability of retail property is the average length of time it takes to release a vacant space in the subject market. Appraisers and brokers can be utilized to determine the average time to release a property. The longer it takes to fill a vacant space, the less income the investor will receive during the year.

A final consideration in retail tenancy pertains to buildings with more than one floor. Typically, second and third floor space is difficult to lease to retail tenants, unless the property is situated in an urban environment with much pedestrian traffic.

In the absence of urban settings, customers typically do not want to climb stairs or ride elevators to retail locations. Often, what was originally conceived as second story retail space is converted into office space or possibly into apartment space. In the next section, we begin to discuss office property.

9.4 Classifying Office Investment Properties

There are various sizes and uses for office property. Given the appointment based orientation for clientele of office tenants, the ingress and egress are not as crucial for success as is the case for retail property. Location is an important consideration however, as clients will not want to be inconvenienced by travelling too far for a scheduled meeting. As will be discussed in the following section, location considerations reflect the primary services offered by the tenants of the subject property.

9.4.1 Office Property Delineation by Size and Class

A primary method of distinction for office property is by the number of stories, or size. Low rise office buildings are typically from one to six stories, with mid-rise office buildings typically ranging from 7 to 25 stories, and high rise office buildings are typically over 25 stories in height (Appraisal Institute 2008). The number of stories is typically utilized to differentiate office properties rather than measures such as the gross building area or net rentable area. The gross building area is an aggregate measure of the building according to exterior dimensions, while the net rentable area is the total area available for occupancy by tenants excluding stairwells, common areas, and mechanical areas.

A second method of distinction for office property is by the quality of the construction, or class. Marshall and Swift Valuation services and other third party providers specify features which can be utilized to segment office property by class. The types of construction materials used, floor and roof structures, the amount of fireproofing, the quality of interior finishing, amenities at the property and general location considerations are all variables utilized (Marshall and Swift 2011). These specific components are viewed in order to classify property as either A, B, C, or D. For practical purposes, real estate industry participants reference these same class distinctions, but may not specifically cite the academic components of the categorization (Appraisal Institute 1993). To the layman practitioner, a "class A property" represents the "best" product available in the market, thus a class A building in Williamston, North Carolina would differ substantially from a class A building in Manhattan, New York City, New York.

Class A is typically of sky scraper quality. The buildings consist of reinforced concrete framing, are in excellent condition, and have professional management. Access to these properties is considered strong and the location is considered superior when compared to other market locations. Class B property is typically located in suburban settings, and consists of a brick façade. These properties are professionally landscaped and are more likely to be midrise level properties. Class C property is typically older, often between 15 and 25 years old. Age alone is not the only consideration, as there can be plenty of buildings of much older origination which stand the test of time given numerous renovations. Thus the remaining economic useful life is a more appropriate benchmark than simply the year of construction. Class C properties consist of block wall construction, and often reveal notable physical deterioration. These properties may also be converted from retail use. Class D is typically below investment grade and consists of property in need of significant renovation or otherwise is subject to significant obsolescence when compared to the existing supply of property in a given market.

9.4.2 Office Property Categorization by Use and Design

Office property may or may not contain tenants whose ownership differs from the owners of the building. Some office buildings contain both owner occupants and third party tenants. For our purposes, we will discuss competitive buildings, or those that offer space to the public. Office property can be leased exclusively to medical industry tenants, can be leased to the general office market, or can be targeted to professionals such as attorneys, accountants, financial services, and doctors. The latter building is known as a professional building, as it is used exclusively by professionals. Professional buildings should be located in an office park, or in neighborhoods of similar quality and utility. For example, a suburban setting for an office park, which consists of two or more office buildings in a coordinated development, is more likely to be successful if the properties are located in proximity to the households who utilize the office properties for employment or as clientele. Office or business parks provide a campus-like setting with extensive landscaping, underground utilities, and architectural standards which promote a harmonious and attractive working environment (Appraisal Institute 2008).

Medical office buildings are exclusively leased to tenants engaged in healthcare services. These properties should be located in close proximity to a hospital, as many of the tenants are offering complementary services supplementing hospital demand. Medical office buildings are typically located in clusters of medical office space. These properties often require significant leasehold improvements in order to ready the space for the provision of medical care. Property owners may require that the improvements are funded by the tenants, or that the costs of the leasehold improvements are repaid by the tenants over the term of the lease. Additionally, as mentioned in Chap. 5, property owners should consider which expenses are key cost drivers for the tenants in occupancy, and require that those expense items are borne exclusively by the tenants, or that the tenants agree to pay for expenses above a predetermined cost per square foot. Such expense stops in leases allow for property owners to control their liability for tenant expenses so that proper incentives are in place where tenants pay for expenses which make up a large portion of the tenants' operating expenses. For medical office space, the utilities to include electricity and water use are examples of key cost drivers. Additionally, medical office users are typically parking intensive, requiring typical parking of five spaces per one thousand square feet (Appraisal Institute 2008).

9.5 Evaluating Office Properties

Office buildings can either be located in suburban locations or in downtown urban environments. For suburban office property, successful locations include such factors as proximity to major highways which are central to the city. This will allow the appointment oriented customers ease of access, as well as for office employees. Suburban office property should also be centrally located relative to major business centers such as shopping malls, banks, restaurants, and other new office buildings or parks. For office property located in downtown or urban settings, proximity to main railroad terminals, mass transit stations, and proximity to governmental buildings and financial districts are considered location enhancements. Regardless of whether an office property is located in a suburban or urban location, the absence of heavy industrial buildings or other aesthetically unpleasing activities should be avoided. If the office property caters to professional tenants, some consideration should be given to the distance from the subject property to the existing "hub" of professional space in the community. If the property is in close proximity to the existing hub, the probability of finding replacement tenants quickly is more likely than if the property is in a pioneering or trend-setting location for the market.

From a property specific standpoint, asbestos has been a major problem for office properties in the past. This represents a significant environmental concern which can translate into significant renovation expenses for property owners of older office properties. The age and condition of the elevators is also a cause of concern for owners of office properties, as is the age and efficiency of the heating, ventilation and air conditioning (HVAC), roof and utilities on site. Since tenants expect good elevator service, temperature control, ventilation and lighting, deviations from the standards available in the most recently developed buildings will be penalized by lower rental rates, unless these factors are offset by a superior location. Even if the current property is up to the standards in a given market, property investors should consider the feasibility of modernization at some future date. If the elevator or HVAC cannot be easily renovated or replaced, this might produce a large expense item or otherwise impede the sale of the property in the future at a desirable price for the current property owner. Another property specific consideration is the availability of parking. If the property is located in a downtown environment, the availability of parking near the property is a very important subjective component of property value. This subjective factor may express itself objectively in lease rates as compared with other properties in the market.

Given the derived demand for office property, general economic trends can be reviewed in order to assess the future demand for office space. Unemployment, area vacancy rates, and projected job growth are all leading indicators of office space demand. Additionally, investors should segment the market vacancy rates by location, age and features so that the effects of economic variables on properties considered in direct competition to the subject property can be realized. Investors should also pay close attention to new office space under construction or development which might alter the demand for the subject property in the future. Net absorption rates can be obtained from local appraisers which disclose the net additional space leased when considering new construction and demolition of older properties in the market.

Finally, tenant rent rolls can be viewed in light of the quantity, quality, and durability of the income as was introduced in Chap. 1. Office tenants sign leases of from 3 to 5 years typically, so an analysis of the tenant rollover risk in any one year is a key component of risk assessment for office properties. As with all property types, the current leasing scenario should be compared with the market averages for occupancy and rental rates so that if a property is over-performing or under-performing relative to the market, assumptions can be made for properties returning to the mean at some point during the investment holding period.

9.6 Office Condominiums as an Investment Alternative

Condominiums are defined as multi-unit buildings in which ownership of individual units can be separately held by individual unit owners (Appraisal Institute 2008). Common areas in condominiums are undivided and shared in common from the standpoint of ownership. Condominium ownership can exist for most property types, and the similarity for all is that the common areas are typically managed by an association whereby monthly or annual assessments are charged to individual condominium owners for the upkeep of the common areas.

Investors purchasing condominiums are attracted by the relatively low cost of obtaining the investment, as in the case of office properties, the investor is purchasing one unit in the building. This situation results in fractured ownership, as there are numerous owners for the units under one roof. From a lending perspective, fractured ownership presents a problem, as the existing condominium owner's association agreement may make foreclosure of individual units problematic. From an investor standpoint, understanding how expenses will be shared among the units on an ongoing basis, and how assessments will be made for unforeseen costs is a key component of the investment decision. The monthly assessment fees are an additional expense borne by the property owner, and thus should be included in any income approach valuation exercise.

9.7 Case Studies in Retail and Office Property

This chapter concludes with five case studies. The first two cases are retail oriented, where the first case is considered at beginner's level, and the second case is more advanced. The third and fourth cases are office oriented. The final case is a mixed use property. It is recommended that students complete a direct capitalization underwriting for each of the first four cases, and that a discounted cash flow analysis be completed for second retail case, the second office case, and the final mixed use case.

Regarding the discounted cash flow specifically, multi-tenant retail and office projects (as well as multi-tenant warehouse and industrial property to be discussed in Chap. 10) introduce expense items not included in the apartment analyses of Chap. 8. During this chapter, we have discussed the property owner's periodic needs concerning tenant improvements to specific units in the subject property. As retail and office tenants vacate the subject property, owners will typically need to make improvements to the vacant units once a replacement tenant has been located. Sometimes, when existing tenants renew their leases, part of the lease negotiation involves making certain improvements to the space. Thus, some level of tenant improvements expense associated with the renewal or new leasing of space should be included in the discounted cash flow analysis. If a new tenant is required, the assumed cost to the property owner is higher than if an existing tenant renews their lease. The cost to the owner is expressed in terms of the price per square foot of the space being improved. For example, assume that a tenant space consisting of 10,000 sq. ft. comes up for renewal sometime during the fifth year of the projected holding period. The investor should assess the probability of the tenant renewing or vacating their space. This can be based on the level of rent paid per square foot relative to market averages, the sales per square foot for the tenant's business relative to a corporate benchmark, as well as by the amount of equity that a tenant has invested in the location. If it is assumed that the tenant will renew with 70% probability, and that costs to the owner for tenant improvements for renewing tenants is $3 per square foot, and if it is assumed that the costs to the owner for tenant improvements for a new tenant are equal to $10 per square foot, the following expense would appear in the year 5 DCF for tenant improvements:

$$10,000 \text{ square feet multiplied by } ((0.70)^*(\$3) + (0.30)^*(\$10)) = \$51,000$$

Spreadsheet programs such as Argus also consider inflation effects in the analysis. Thus if inflation is assumed to be 3% for the first 5 years, the $51,000 would be multiplied by $(1.03)^5$ to arrive at the final DCF expense for tenant improvements in year 5 of $59,123.

Whether an existing tenant in the subject property renews their lease or decides to vacate the premises, there will typically be a cost to the owner in the form of leasing commissions. The property owner may require the services of a commercial broker to either find a replacement tenant or to help negotiate the lease renewal for an existing tenant. Thus, some level of leasing commission expense associated with the renewal or new leasing of space should be included in the discounted cash flow analysis. If a new tenant is required, the assumed cost to the property owner is higher than if an existing tenant renews their lease. The cost to the owner is expressed in percentage terms relative to the effective gross income in the year the cost is incurred. For example, assume that the same tenant space of 10,000 sq. ft. coming up for renewal sometime during the fifth year also requires leasing commissions to be paid. Further assume that the space represents 20% of the total space leased in the subject property. If it is assumed that the tenant will renew with 70% probability, and that costs to the owner for tenant improvements for renewing tenants is 2% of effective gross income in year 5, and if it is assumed that the costs to the owner for tenant improvements for a new tenant are equal to 4% of EGI in year 5, the following expense would appear assuming a year 5 EGI of $150,000:

$150,000 EGI multiplied by 0.20 equals $30,000 multiplied by ((0.70)* (2%) + (0.30) * (4%)) = $780×5 year lease = $3,900.

Since leasing commissions are paid upfront for the length of the lease, a 5 year lease would result in a year 5 DCF expense for leasing commissions of $3,900.00. Since EGI in year 5 already includes inflation effects, no inflation is required for LC cost assumptions.

While the projections for tenant improvements and leasing commissions require a lot of assumptions, it is recommended that some assumption be made for these items rather than ignoring the costs. The assumptions for LC and TI costs should be based on the local market averages wherever possible. The above analysis only considers the known rollover risk for stated expirations, and does not include consideration of costs associated for replacing tenants who vacate prior to the formal end of their lease agreement. In order to account for these uncertainties, the vacancy factor or the probability of a tenant not renewing their lease could be increased.

9.8 Conclusion

This chapter has combined the evaluation considerations of retail and office property. In Chap. 10, we will discuss warehouse and industrial property specifically, along with the impact on commercial real estate owing to the outsourcing and off shoring of manufacturing and service sector jobs from developed countries to the developing world. Chapter 10 will also discuss self-storage properties, a hybrid form of investment which combines business operation and third party leasing into one income producing property valuation.

Questions for Discussion
1. Describe the difference between retail and office properties in terms of the tenants' decision to visit a given property.
2. Itemize some important location considerations for various types of retail and office property.
3. Provide examples of good and bad tenant mix possibilities.
4. Describe the process of estimating tenant improvement and leasing commission expenses in a discounted cash flow analysis.
5. Describe the strengths and weaknesses associated with condominium ownership.

Retail Case Study #1

Strip Center Properties, LLC is looking to purchase a five bay retail strip center for $900,000.00. The investors plan to inject equity of $100,000.00 and seek bank financing for a loan of $800,000.00. The property is a typical retail strip center that contains local small businesses. The property has 12,000 sq. ft. of leasable space, and the strip center was constructed in 1989. It is in fair condition, but it looks its age based on your recent inspection of the property. Professional inspectors could not find any deferred maintenance at the property. The property would be considered average when compared to similar properties in the surrounding area. All leases are on a gross basis, but each call for common area maintenance (CAM) expense reimbursements of $1.50 per square foot leased for each tenant which also covers taxes and insurance for the property. The property is located near the area mall and has had a historical vacancy rate of from 8% to 10%, though it is currently fully leased. The market vacancy rate is 10% for the local area. Based on local averages, standard management fees for this type of space are 5% of the effective gross income (EGI).

The investors have provided a current rent roll for the subject property as follows:

Unit	Tenants	Monthly rent	Size	Lease maturity
1	Nails are us	$1,650.00	1,800	1 Year
2	Hardware plus	$2,625.00	3,000	MTM
3	Junior's Deli	$2,500.00	2,500	2 Years
4	Hospital florists	$2,083.33	2,000	Next month
5	Mike's hobbies	$2,362.50	2,700	3 Years

Also included in the financial package submitted to the bank for financing consideration was the following breakdown of the annual operating expenses for the subject property:

Taxes	$8,000
Insurance	$2,000
CAM	$8,000
Recurring repairs	$9,000
Utilities	$2,000

1. Determine the net operating income for this property using the direct capitalization spreadsheet.
2. Determine the first year debt service coverage ratio for a loan of $800,000.00, at an interest rate of 8%, for 20 years.
3. Using a cap rate of 9%, what is the estimated value for this property?
4. Determine the breakeven occupancy rate and interest rate for this property.
5. What questions would you ask concerning the rent roll information?
6. Given the tenant mix, what would be some ideas for replacement tenants that would be comparable to those currently in the property?

9.8 Conclusion

Retail Case Study #2

Safin Properties has an opportunity to purchase a mixed use retail property located in the City Market section of Charleston, SC. The property is located on North Market Street, in a neighborhood bound by Meeting Street to the west and Concord Street to the east, and is generally known as the most active tourist district in Charleston, SC. The investors feel that this property would make an excellent investment as Charleston attracts over five million visitors per year and the properties in this submarket have some of the highest rents found anywhere on the peninsula. The property was originally constructed in 1880, but was completely renovated over the last 3 years. The building is constructed on a concrete foundation and has solid masonry and frame exterior with painted stucco and wood side covering. The interior consists of a wood frame structural system, with one main masonry wall, and a built-up roof. The building is two stories, with a currently vacant loft apartment on the second floor, along with a small portion leased to Joe's Bar and Grill (also a first floor tenant) and 4,099 sq. ft. of space currently vacant that was previously occupied by the Port City Inn. The buyers (who also owned the Port City Inn) feel that there is an opportunity to convert this space to mixed use retail, with a Market Street front entrance. The buyers have stated that they are willing to sign a master lease of $25.00 per square foot for 5 years if this is required to make the deal viable.

The purchase price of the property is $6,000,000, and Safin properties and its owners are seeking financing in the amount of $3,000,000. The remaining proceeds are to come from a 1031 Exchange. The property has the following rent roll:

Tenants	Sq. ft.	Annual rent/sq. ft.	Lease expiration
Joe's Bar and Grill	4,559	58.13	1 Year
Praline Candy Company	1,772	44.02	10 Years
Trinkets and Baubles	724	93.73	2 Years
Port City Inn	4,099	25.00	
Spender's Coffee	1,915	49.21	3 Years
Loft Apartment	900	29.33	

You have market intelligence that rental rates in the area are between $45 and $55 per square foot, with vacancies in the area typically falling between 5% and 10%. The leases are structured as triple net. The only expenses incurred by the prior owners for the last 3 years are shown below.

	Year 1	Year 2	Year 3
Taxes	$ -	$ -	$8,000
Insurance	$ -	$ -	$2,000
Repairs and maintenance	$3,000	$5,000	$7,500
Management	$12,000	$15,000	$17,000
Other	$10,000	$15,000	$25,000

Questions for Discussion
1. Determine the net operating income for this property using a direct capitalization framework. Be sure to justify what annual rental rates you are using in your analysis.
2. What is a master lease? Does the property underwrite without the master lease? What issues do you have with the proposed master lease?
3. Why were there expenses for taxes and insurance in year three?
4. Using a cap rate of 8%, what is the value for this property? Prepare a DCF for this property to determine its value on a multi-period basis.
5. What questions would you ask about the rent roll and property historical performance?
6. Determine the debt service coverage ratio for the first year, assuming loan terms of 8% interest for 20 years.

Office Case Study #1

Office Investors, Inc. completed the construction of an office building in 2003, and achieved 100% occupancy in the subject property by late 2004. They would like to refinance the existing $2,500,000.00 note, and would like another $500,000.00 to replenish their personal cash holdings. The building is housed on 3.1 acres of land in a favorable location in Good Sized Town, Ga. The building size is 52,000 sq. ft. gross, but has only 45,000 sq. ft. of rentable space. The building consists of three stories of net rentable area, 15,000 sq. ft. each. All leases are on a gross basis. Market vacancy rates are between 10% and 15%.

The current rent roll is as follows:

Unit	Tenants	Monthly rent	Size	Lease maturity
1A	Doctor	$10,000	7,500	2 Years
1B	Insurance	$5,000	5,000	Next month
1C	Temp staffing	$2,708	2,500	6 Months
2A	Court reporter	$4,583	5,000	MTM
2B	Vacant		10,000	
3	Attorney	$18,750	15,000	5 Years

The location of the subject property is on a busy street in the downtown section of Good Sized Town, GA. The market rents for medical space are in the $15.00–$16.50 per square foot range, while the market rents for standard office space is $10.50 per square foot. The standard management fee in the market for this type of space is 5% of effective gross income (EGI).

9.8 Conclusion

The owners of the property have supplied the following information regarding historical operating expenses at the subject property:

	Year 1	Year 2	Year 3
Taxes	$12,500	$14,750	$16,000
Insurance	$2,500	$3,250	$4,000
Janitorial	$8,455	$10,152	$12,000
Utilities	$6,500	$7,250	$8,000
Automobile	$4,000	$5,000	$5,000
Leasing commissions	$7,500	$8,500	$9,000
Landscaping	$9,875	$11,530	$11,000
Repairs and maintenance	$8,000	$14,000	$12,000

1. Determine the net operating income for this property using the direct capitalization spreadsheet.
2. Determine the first year debt service coverage ratio for a loan of $3,000,000.00, at an interest rate of 8%, for 20 years.
3. Using a cap rate of 9%, what is the estimated value for this property?
4. Determine the breakeven occupancy rate and interest rate for this property.
5. What questions would you ask concerning the rent roll and expense information?
6. Given the tenant mix, what would be some questions concerning the property's location?

Office Case #2

You have been presented with an opportunity to finance a purchase of an existing office building in Irvine, CA. The property is located in Old Town Irvine, located in Orange County California, one of the most heavily urban counties in California today. The property is a six tenant office building and is currently on the market for $5,000,000. The applicants are currently banking at Bank of America and live in Hasbrouck Heights, NJ. They have recently created a real estate holding company, California Dreamin', LLC, which will borrow the funds, while the principals will guarantee the debt. They have requested a loan amount of $3,000,000 from your organization, which is a primary competitor to B of A. The prospects have over $10 million in personal liquidity, primarily in the form of small cap stocks less than $10 a share (as a side note, when meeting with the prospects they mentioned that they prefer high risk and high return investments, and that this is their first foray into investment real estate.) Given the recent drop in property values in California, they felt that now was a good time to venture into real estate investment on the west coast.

The property rent roll is as follows:

Tenants	Sq. ft.	Annual rent	Lease expiry
Interactive communication	12,500	168,316	6 Months
Jung architecture	1,500	38,160	2 Years
Tong and Lang management	7,500	92,700	2 Years
The learning center	3,500	68,291	5 Years
Alliance Francaise	5,000	117,000	10 Years
AT&T wireless (ground lease)		39,700	9 Years
Totals	30,000	524,167	

The property was constructed in 1925, and was renovated from its original use as a grain storage building, into an office building in 1985. The property has a wood exterior, but was completely renovated in 1985 for office space. Given its prior use, the property sits on 2.73 acres. The building is listed in the California Register of Historic Places, and the area has vacancy rates of 15%, which is up from 10% last year due primarily to the problem in the mortgage industry in California. Office rental rates in the area are between $2.25 and $2.50 per month.

Property operating expenses for the last 3 years are as follows:

	Year 1	Year 2	Year 3
Taxes	35,000	35,000	35,000
Insurance	14,000	14,000	15,000
Repairs and maintenance	24,250	35,685	27,500
Management	13,250	14,000	14,153
Utilities	17,750	18,000	18,250
Janitorial	6,800	7,200	7,500
Other	7,500	2,500	3,500
Totals	120,555	128,391	122,910

A few of the selling points for the investors were the consistency of the operating expenses, and historical nature of the property. Since this is your first opportunity to speak to these potential clients, you want to make a good impression, prove that you can be a valued advisor for their financial portfolio, and hopefully to win their entire banking relationship.

Questions for Discussion
1. What are some issues that you have with the management situation in the case?
2. What are some issues that you see with the rent roll?
3. Assuming an interest rate of 8% and a 20 year amortization, what is the debt coverage ratio? What is the value of property assuming a cap rate of 7%?
4. Does this property support 20 year financing? What questions would you ask to better understand this issue?
5. Is there a gap between the loan request and what the property supports for traditional bank financing? How far would you stretch to win this business, and to possibly win the relationship of these clients?
6. What financial advice would you offer these potential customers?

9.8 Conclusion

DCF Mini-Case

Construction Project in Heidelberg, Germany

You have recently joined an investment team that operates a well-diversified portfolio of investment properties located in tourist destinations throughout the world. The group's portfolio of assets consists of office, retail, and residential properties located primarily in North America and Europe. The group is considering the construction of a mixed use office-apartment building located in Heidelberg, Germany. The site has recently been razed, and the prior usage was two residential properties that were owned by professors from Heidelberg University. The site is located on Neuenheim Landstrasse, a primary corridor in the city which has a scenic view of the Neckar River. The site is on the opposite side of the river from the famous castle in Heidelberg, and is near "Philosopher's Way", a trail made famous by German philosophers such as Hegel and Nietzsche. In years past, philosophers such as these would traverse the trail, contemplating philosophical questions. Now the path is occupied primarily by tourists, hoping for similar deep insights.

Given the scenic view from the proposed new mixed use property, the project will be known as "King's View Park", and will consist of a two story building on site where the two residential houses once stood. The top floor of the building will contain three apartment units (two bedroom, one bath), while the bottom floor will consist of four office units.

Based on your firm's market research in conjunction with local real estate professionals, it has been determined that the apartment vacancy in the city of Heidelberg is essentially zero at the present time. Some competing units are being constructed currently, but none with the scenic view of the subject property. Your firm has also determined that the existing office market vacancy rate in the Rhein-Neckar metropolitan region is approximately 10%.

Given the strong level of occupancy in the Heidelberg apartment market, you have been able to obtain letters of intent for each of the three apartments, as well as three of the four office units.

Each apartment is 1,400 sq. ft., and will rent for $4,250 per month. The letters of intent have been signed for 1 year, which will begin once construction is complete and the property is ready for occupancy.

The rent roll for the office spaces is shown below:

Tenants	Sq. ft.	Term	Monthly rent
Prof. Dr. Raffeisen	750	2 Years	2,500
Heidelberg Univ.	1,000	4 Years	4,000
Heidelberg Univ.	1,225	4 Years	6,000
Vacant	1,225		

One of the office units will be occupied by a retired professor, who will use the space as his personal office while he writes his memoirs and updates his previously published textbooks which are in various stages of completion. He has signed a pre-lease agreement for a 2 year term. Two other office suites will be occupied by Heidelberg University. Since the University has grown in recent years, these offices will be occupied by university staff. They have signed a letter of intent for 4 years. The final office space has yet to be leased, but given that 85% of the total square footage of the property has been pre-leased prior to commencement of construction, your firm believes that this property will be fully and quickly absorbed by the market prior to completion.

Your firm typically assigns a replacement reserve of 1% of EGI annually for projects of this age and condition. Additionally, you have contracted with Dahler & Co. GmbH, a well known real estate brokerage and management firm, for a management contract of 3 years, with compensation equal to 3% of the effective gross income of the property.

Additionally, you have estimated the following expenses for the first 3 years for the project:

Operating expenses	Year 1	Year 2	Year 3
Taxes	25,000	25,500	26,000
Insurance	12,500	13,000	13,500
Repairs and maintenance	8,575	9,900	12,375
Utilities	10,125	11,600	14,500
Other	–	2,500	1,500

Total project construction costs are estimated at $3,500,000, and you have requested a loan to cost of 75%. Your firm already owns the land. A local bank has offered 25 year financing at 7% interest.

Your assignment is to develop a discounted cash flow analysis (DCF) for this property. Assume a 10 year holding period, which is consistent with your firm's investment horizon for this project. Given the strong location, a 5% annual appreciation rate is expected for this property, and selling costs run 3.5% in this market. Your projections should include a most likely case which assumes annual GPI increases of 2.5%, and a best case which assumes annual GPI increases of 5%. Assume a 9% discount rate, and a terminal cap rate of 8%. Your DCF should take the market intelligence, your projections of income and expense associated with this project, and lease rollover risk into account.

References

Appraisal Institute. (1993). *The dictionary of real estate appraisal* (3rd ed.). Chicago: Appraisal Institute.
Appraisal Institute. (2008). *The appraisal of real estate* (13th ed.). Chicago: Appraisal Institute.
Appraisal Institute. (2004). *Appraisal Institute Commercial Data Standards, version 2.0*, May 14, 2004, web site http://www.redi-net.com/adn_db/AI_standards.pdf. Accessed 25 May 2011.
Buchanan, M. R. (1988). *Real estate finance* (2nd ed.). Washington, DC: American Bankers Association.

References

Marshall & Swift Valuation Services website. (2011). http://www.marshallswift.com/FAQ.aspx. Accessed 27 May 2011.

Phyrr, S. S., & Cooper, J. R. (1982). *Real estate investment: Strategy, analysis, decisions.* New York: Wiley.

Raab, G., Goddard, G. J., Ajami, R. A., & Unger, A. (2010). *The psychology of marketing: Cross-cultural perspectives.* Aldershot, UK: Gower.

10. Investing in Warehouse and Industrial Property

All you need is a place for your stuff.
George Carlin

Contents

10.1	Industrial and Warehouse Investment Properties	206
10.2	Evaluating Industrial and Warehouse Property	208
	10.2.1 Location Considerations	208
	10.2.2 Projecting Demand for Industrial Space	208
10.3	Outsourcing and Its Effect on Commercial Real Estate	209
10.4	Self-Storage Facilities as an Investment Alternative	211
	10.4.1 Brief History of the Self-Storage Industry	211
	10.4.2 Delineation by Size and Class	213
	10.4.3 Projecting Demand for Self Storage Space	213
	10.4.4 Know Thy Customer	214
	10.4.5 The Ayatollah of Climate Controllah	216
10.5	Conclusion	217
References		223

In Chap. 10, we complete our analysis of the primary investment property sector considerations with a discussion of industrial and warehouse property. These properties are generally classified as industrial property given that their function relates to manufacturing, production, or storage of industrial products. A relatively new subsection of industrial property is known as self-storage, or alternatively as mini-storage. As will be discussed later in this chapter, self-storage property has widened the scope of demand for storage space to include the general public.

Chapter 10 begins with a discussion of differences in industrial property types. Then a discussion of the projected demand for space is viewed in light of outsourcing and off-shoring initiatives in the product and services markets. After the impacts are understood in a commercial real estate perspective, we then discuss self-storage property specifically, with the chapter ending with case studies.

10.1 Industrial and Warehouse Investment Properties

Industrial property consists of numerous uses and designs. Any line of manufacture, research and development, and storage of goods can be designed in construction for a specific user. Generally, the more that the construction and interior composition of a property is geared for a specific user, the more likely that the property will be considered special purpose property, and will be occupied by one primary tenant. Additionally, the more specific that a building is constructed based upon the needs of a particular tenant, the higher the probability that the primary occupant business will own the property. Thus, certain industrial properties are more likely to be owner-occupied and not serve as possible investment by equity investors. Furthering the probability of owner occupancy would be cases where the commercial activity conducted on-site involves a high risk of environmental site contamination.

Industrial activities such as the manufacture of products or the storage of products where hazardous materials are involved would generally require owner-occupants rather than equity investors seeking a return from the eventual resale of the property at the end of the investment holding period. Industrial processing plants where chemical products are created on-site, as well as facilities utilized in converting raw materials such as coal, oil, and salt into various products, would more than likely not be a typical candidate for investment property. Water treatment, waste treatment, and minerals processing facilities are also more likely to be owner-occupied properties. Additionally, environmental liability risks associated with heavy manufacturing such as pollution, fumes, noise, and vibration dictate that these properties should be owned by the operating entities present on-site. These risks were further elaborated in Sect. 6.1.7 of this book.

Light and high tech manufacturing sites are potential candidates for investment property given the lower environmental risk and the lower special use orientation of the property. Generally, if the risks associated with on-site production are considered low-to-average, and the potential of releasing the space without considerable specialized (and costly) leasehold improvements is considered reasonable, industrial properties may be seen as a viable investment alternative.

Business parks consist of master planned developments which encompass a group of predominantly industrial buildings on large acreage tracts (Appraisal Institute 2004). Higher quality parks typically feature a campus-like setting with extensive landscaping which promotes an attractive working environment. Some business parks cater specifically to high technology industries, and may contain a hybrid of office, manufacturing, and warehouse space. Business parks can be both owner-occupied or investment property depending on the degree of specialization and environmental risk contained on-site. Prospective investors should request copies of environmental reports to assess whether the property contained any "at-risk" manufacturing on-site in the past. Future considerations should be made for any current tenant engaged in environmentally suspect manufacturing or processing on site. Investors contemplating purchasing such a property should consider obtaining an updated environmental assessment by an expert third party.

10.1 Industrial and Warehouse Investment Properties

A warehouse is a structure that is designed and utilized for the storage of wares, goods, and merchandise. Warehouses are generally considered a subsection of industrial property as they are used in conjunction with the manufacturing industry. Distribution warehouses are storage buildings designed to promote the logistical movement of goods. Construction considerations involve providing adequate loading facilities and ensuring that trucks have the ease of ingress and egress to and from the property. Modern distribution buildings feature a minimum clear height of 24 ft and contain one or more dock-high loading doors for every 10,000 sq. ft. of space (Appraisal Institute 2004). Ceiling height and loading capacity are very important for distribution warehouses as the goals of the tenant are to stack as many products as possible in the property and that valuable time in transit is not wasted on properties of insufficient design. Given the "stack them high" aims of the tenants, these properties are also known as bulk warehouses. Bulk warehouses generally contain single tenants. Prudent investors would want to determine the financial health of the prospective tenant before signing them to a long term lease, or purchasing the property under an existing lease arrangement.

Storage warehouses are structures which are designed for the storage of wares, goods, and merchandise where obsolescence has occurred such that the characteristics of the building are no longer considered competitive with distribution warehouses. Ceiling heights may be lower than is desirable and or loading capacity may not be up to the current standards. These buildings are used to store inventory with low turnover where time is not as crucial.

Service warehouses typically contain more office space than in distribution or storage warehouses. Service warehouses also have lower ceiling heights than bulk warehouses, typically coming in around 18 ft. As the percentage of office space increases, these properties are then classified as flex space. Flex space generally contains glass on three of the outer walls and has additional parking relative to other warehouse properties (Appraisal Institute 2004). An 80/20 split of warehouse and office space is generally enough to earn the flex space classification, although as the name implies, any composition up to a 50/50 designation is possible for flex space. These properties are designed to allow conversion of industrial units to a higher percentage of office space, if so desired. Flex properties allow tenants to provide mixed uses in a single location in order to reduce the need for multiple sites and the redundancy in staffing. Another advantage of flex space is that the lease rates are typically lower than for regular office space. In soft leasing markets, flex properties often become occupied by any tenant who wishes to pay a reasonable lease rate per square foot.

Office showrooms are similar to flex/office in terms of basic construction. Approximately 50% of the interior is finished and there is a higher amount of parking as compared with other warehouse properties. Office showroom space is typically located along a freeway or other major thoroughfare where the traffic exposure of the property can be exploited for direct retail sales. The interior build out typically favors sales floor orientation rather than office design, with the balance of the space devoted to warehouse and stock.

Office warehouses have much better curb appeal relative to bulk warehouses. This higher curb appeal typically translates into higher rents per square foot relative to general warehouse space. These properties are typically multi-tenant facilities, with office space ranging from 25% to 50% of the total space. The warehouse ceiling clearance for office warehouses is lower than that of bulk and service warehouses.

10.2 Evaluating Industrial and Warehouse Property

10.2.1 Location Considerations

Given the uses associated with industrial property, there are certain features of a site location which prove to be favorable to investors as well as to tenants. The industrial property should have easy access to primary highways especially if the purpose of use is storage and distribution. Bulk warehouses are often located in industrial parks which contain wide streets facilitating easy access for large trucks. Proximity to railroads and airports is also desirable. In some communities, special economic zones have been created near airports and other key transportation hubs to help encourage the export of products at favorable customs rates. The prevalence of these preferential trading areas or other favorable features of the state or local tax structure, make distribution and other industrial facilities a better investment based on this capitalization effect of legislation targeting a specific area to the detriment of other areas.

Generally, industrial services expand to meet the needs of the dominant industry in an area in order to attract other industries that require similar services. This results in an agglomeration or clustering tendency where one type of industrial property tends to be concentrated in one place. Industrial users seek least cost locations. The location selected will depend on the location of competitors, the importance of proximity and direct contact with customers, the extent of the market area, and costs including the price of land, labor, materials, equipment, and transportation (Phyrr and Cooper 1982). Deglomeration forces are factors that detract from an otherwise favorable location. Increased demand for industrial space in a given market serves to increase the cost of land and improvements with an eventual effect on the future demand for industrial space. Other detractions are plant obsolescence in the dominant industry, or the basic shift of an area from a manufacturing base to a service economy.

10.2.2 Projecting Demand for Industrial Space

Demand for industrial space will come from new firms moving to a given area, from relocations within a given area, and from expansion of existing establishments in an area. The first characteristic requires projections of the current number and productivity of manufacturing and industrial workers in a given area. Once this is estimated, future industrial employment by type of industry and size of firm should be projected. As industries continue to modernize, future employment densities should be projected, taking into consideration the trend toward a greater output per square foot of floor space.

Future land requirements can be estimated by multiplying the projected area required per worker, known as industrial density, by the expected industrial employment for each use. These estimates must take into account the subject market's land use regulations. The estimated demand for industrial land should then be compared with the estimate of future supply of industrial land. As should be apparent from this overview of demand analysis, obtaining accurate projections is often difficult, especially given the increasing interdependence of domestic and international markets.

The uncertainty surrounding future expected industrial land use makes the analysis of the current tenants financial condition all the more important. Tenants with improving financial trends may have more of a need to expand their operations, but this alone does not rule out the possibility that an existing tenant will not renew their lease in a given property. An existing tenant may find a suitable location in market that meets their objectives more than the current location. For this reason, investors should keep abreast of area trends in industrial space requirements and follow their existing tenants' financial condition so they do not wind up with an empty facility.

Concomitant with the projection of demand for a given industrial property is the consideration of the building characteristics. During our earlier itemization of the various types of industrial properties, the various demands for ceiling heights were discussed. Other factors in the building analysis generally include verifying the age and condition of the electrical, gas, water, and sewage installations. The existing fire protection, sprinkler system, heating, light and ventilation should also be evaluated prior to purchase. Finally, considerations of the quality of the security systems and any elevators or conveyors on site should be conducted. As with any investment property, special attention should be made for any improvements which might be expected to be made over the holding period of the investment.

10.3 Outsourcing and Its Effect on Commercial Real Estate

Now that we have covered the basic four "food groups" of investment real estate discussed in this book, it is time to quantify some general risks in the global economy that can threaten occupancy levels for investment property. The risks identified in this section are specifically the outsourcing and off shoring of business services and goods manufacture. Outsourcing is defined as "engaging a third party company to operate and administer a specific set of business processes", while off shoring is defined as "repositioning of operations in global markets" (Bhagwati et al. 2004; Ajami et al. 2006). Both of these processes can affect and have affected occupancy levels in various property types over the last decade.

While outsourcing and off shoring is nothing new, both have been the subject of spirited debate in recent years. While the financial crisis may have slowed the speed of the off shoring of goods and services overseas, the past and future movement can be classified in three primary waves (Goddard and Ajami 2008).

The first wave had the aim of lowering costs and primarily involved the outsourcing and off shoring of manufacturing processes and jobs to China. US industries which were adversely impacted were furniture, textiles, and agribusiness processing. While many manufacturing operations are classified as owner-occupied property, as the manufacturing industry experiences a decline in domestic production, complementary investment property such as distribution warehouses may also experience decreased occupancy rates.

The second wave of outsourcing and off shoring had the primary aim of restoring the quality of the goods manufactured and also involved the inclusion of service sector business processes. Occupancy rates for office buildings thus began to experience decline in many metro markets given the off shoring of backroom office functions to locations such as India. Retail operators enjoyed the lower cost of products manufactured overseas, but did not enjoy dampened sales volumes which materialized in cities where job losses were high given clusters of jobs relocated abroad.

The third wave of outsourcing and off shoring has the focus of transformational relationships. In this wave, global service providers help the onshore firm with strategic planning for the future with a long term relationship orientation between the parties. When firms are considering locations for various business processes, the selected global service provider will represent the "best shoring" alternative of the highest quality at the most reasonable cost. These service providers are often located in clusters such as Bangalore, where there is an ample supply of well-educated workers benefitting from knowledge transfer within the cluster of employment.

Investment property owners should consider how global economic forces may impact the overall demand for goods and services on offer by the existing and prospective tenants in the subject property and submarket. Each of the four food groups are either directly or indirectly influenced by the outsourcing and off shoring of business operations. Industrial property is directly affected by manufacturing locations being moved overseas, and is indirectly impacted by lower demand for warehouse space in a given market. Office property occupancy is directly affected by the relocation of business processes as space demand falls when less office activity exists in a given market location. Retail property is both positively and negatively affected by outsourcing and off shoring as was noted earlier, while apartment unit demand is indirectly impacted by lower office occupancy rates in a given market.

Investors should view the operations conducted on premises by the tenants in light of the future impact on occupancy given future relocation decisions. Future candidates for off shoring include positions that are separable from the customer, in departments which are not imbedded in the firm, and where local knowledge is not essential for the job function being considered (Sakthivel 2007). Whether a department in an organization is embedded depends on the amount of interaction that the employees of a given department have with others in the organization. If these considerations are evaluated in advance of lease execution or property purchase, property owners may find the occupancy in their investment properties to be of a more durable nature in the future.

The preceding discussion of outsourcing and off shoring is an example of a global economic issue that is worth following by owners of investment property. While real estate is inherently local, the increased integration of global markets reveals that only having a local orientation may come at a price.

10.4 Self-Storage Facilities as an Investment Alternative

The comedian George Carlin sometimes wondered why the other guy's stuff is junk while our junk is stuff (Carlin 2009). Of course, the difference is that stuff is for keeping and junk is for throwing away. Ultimately, people need a place to keep their stuff. Owners of self-storage facilities seem to have taken this to heart, given the dramatic rise in self-storage facilities across the United States over the last few decades. Perhaps to some extent the growth of the industry has been fueled by the perception that it is recession proof, or even counter cyclical. In Sect. 10.4, we examine the self-storage industry in an attempt to verify whether it is immune to the vagaries of the business cycle.

Self-storage property is defined as real property designed and used for the purpose of renting or leasing individual storage spaces to customers for the purpose of storing and removing personal property on a self-service basis (Self Storage Association 2002a). Self storage property is alternatively known as mini-storage, and is comprised of small units ranging in size from 20 to 500 sq. ft. The small units are typically leased by customers on a monthly basis, and some of the units may be climate controlled. Public storage is a facility that includes an operating business that provides warehousing storage and service for multiple customers (Appraisal Institute 2004). Self storage is typically either categorized as a type of industrial warehouse property, or is considered a special purpose facility. Self storage property may also contain storage yards which generally consist of a plot of land enclosed by a high fence or other screening for the purpose of external storage of equipment, parts, vehicles, or materials (Appraisal Institute 2004).

10.4.1 Brief History of the Self-Storage Industry

Personal storage has its origins in ancient China 6,000 years ago where people stored their belongings in clay pots in public underground storage pits (Tropical Storage website 2011). The conventional concept of personal storage began in England in the nineteenth century. British banks were asked to safeguard valuables for clients embarking on extended voyages, and overcrowded vaults led them to seek storage in lofts (Self Storage Association 2002b). During the Great Depression, moving and furniture companies began sideline storage operations for large appliances that were repossessed or unclaimed (Hess 1973). The initial development of self storage facilities in the United States occurred in the late 1950s in Texas to meet the needs of migrant oil workers. In the early 1960s in the western and Sunbelt states self storage demand arose to serve a transient population moving to seek new jobs and retirees moving in search of better climate (Correll 2003). In the late 1970s and early 1980s, increased self storage construction occurred along the eastern coast of the United States, as well as in Canada, Europe, Australia and Japan (Self Storage Association 2002b). By the mid 1980s, there were fewer than 7,000 self storage facilities in the United States. Today, there are approximately 50,000 such facilities in the United States, with an additional 58,000 facilities in the rest of the world. The second and third largest volume of self storage property is found in Canada and Australia at 3,000 and 1,000 respectively, with Europe and Asia seen as future growth areas. The United States may be more culturally receptive to self storage given consumption habits and land availability (Chappell 2011a). Self storage property in Japan is known as "rental storage" and consists of either indoor storage, similar to the US model, or containers which look like metal shipping containers stored on ocean-going freighters (SSA Globe 2011). Containerization of storage in the United States consists of "portable on-demand storage" (PODS) where customers can request storage units to be transported to their chosen destination and can then subsequently move them to an enclosed storage facility (PODS 2011). This variation of self storage truly blends business operation and property investment given the need for transportation vehicles for the portable storage units.

In the aftermath of the financial crisis, self storage occupancy generally increased owing to the movement of individuals from homes to apartments (Wall Street Journal 2011). Self-storage property is considered an attractive investment option, because the rental fees per square foot are among the highest of any real estate investment alternative. Such fees are possible because tenants are not concerned with comfort or living space and regularly lease units as small as 5'x5', allowing extraordinary occupancy density. Over the last two decades, however, competition has intensified and saturation points have been reached in many metro markets. Today, the industry is highly fragmented with most investors owning two or three facilities (Chappell 2011b).

During the nascent period of the self storage industry, properties were often referred to as mini-warehouses. This moniker was derived from community planners viewing the property type as small conventional warehouse space. The label served to prevent zoning officials from viewing self storage as sufficiently separable from warehouses (Self Storage Association 2002b). The Self Storage Association, the oldest non-profit organization devoted to owners, operators, and suppliers in the self-storage industry, has offered commentary on differences between warehouses and self storage properties as follows:

- Warehouses have employees, Self Storage has tenants
- Manufacturers use warehouses, families and businesses use self storage
- Self storage properties generate low traffic and can be designed to conform to the aesthetics of the neighborhood, both are contrary to warehouse property

As perceptions changed, many communities have begun to allow self storage in retail areas and in higher density residential districts. Today, a self storage facility is seen as being an operating business that reflects the design needs of the neighborhood that it serves.

10.4.2 Delineation by Size and Class

A self storage facility is an operating business that reflects the design needs of the neighborhood that it serves. Depending on the needs of the general population and the size of the land parcel, self storage facilities can vary from as small as a ten thousand square foot building to a site with numerous buildings with size upwards of one hundred thousand square feet. Self storage property construction can vary from block concrete buildings surrounded by barbed wire fencing, to modern metal and block buildings complete with keyless entry, security cameras, and climate controlled unit offerings. Many investors choose to build self storage facilities in phases. This strategy is well positioned with the utilization of the facilities, as tomorrow's next phase construction is today's vehicle and large equipment storage yard. Successful self storage properties have the desirable combination of low construction costs relative to high income concentration per square foot.

While the self storage industry is highly fragmented in terms of ownership, typically the larger, better capitalized firms construct more modern facilities with climate control capability and a small office on site which is constructed of similar design to other properties in the area for aesthetic appeal. Small investors typically specialize in the acquisition of older facilities that do not come equipped with more modern features. These older properties are typically located in less urban settings and may represent the "proto-type" first entrant of self storage in the area. When small investors have property near the larger players, the better capital position of the larger firms allows for the temporary lowering of rental rates in an attempt to drive out smaller rivals and to increase market share for the larger firm. Thus, how much storage space a given market can support is a crucial issue to be resolved for investors and lenders alike.

10.4.3 Projecting Demand for Self Storage Space

Given the relatively short history of modern self storage facilities relative to the other property types explored in this book, obtaining reliable data for market equilibrium points for self storage space is often difficult. Demographic information alone is not particularly telling, as the presence of consumers does not necessarily dictate that they will utilize the storage facilities in a given market. With this caveat now in place, the primary competitive market area for a given self storage property is typically considered to be within a three to five mile radius from the subject property (Correll 2003). In order to measure potential demand for space, the investor should obtain an estimate of the average rentable square feet of storage space per person relative to the current population in the primary competitive market area. This estimate should then be compared with the existing supply of units. Square footage per capita is not infallible, as some communities can absorb more storage space than others given the tendency to own recreational equipment, and the lack of basements or enclosed garages for residential property in a given market area. Other demographic factors such as the age of the population, the mobility rate of the population, the employment status and the income level of the population, are other important considerations (Self Storage Association 2002b).

Unless there are third party service providers in a given market that concentrate on self storage occupancy, rental and expense rate information, obtaining current market estimates is often problematic. Investors, appraisers, and lenders often find themselves calling market participants in order to survey the current situation in a given area. The absence of quality third party information sources for self storage, especially for properties outside of metro markets, has led to lenders showing caution in approving new, "out of the ground" self storage projects. If an investor is adding a second or third phase to a given facility, lenders may be more willing to provide financing if the existing units provide at least break-even debt coverage ability, where existing NOI is equal to debt service on the loan.

A break even debt coverage ratio is often possible as the costs of constructing subsequent phases for an existing, successful project are often lower on a per square foot basis as the foundations and other land infrastructure improvements are typically funded with the initial phase. Short of the existing facility providing the lender a break even debt coverage ratio, strong market occupancy rates and quick prior market absorption of completed units may not allow for new loan financing, especially in the presence of strong market competition. Lenders typically want to see a successful track record of operating performance before lending in the self storage property sector.

The difference between an under-performing self storage property and a successful property is often based on the quality of on-site management. The presence of an experienced, full-time, on-site manager can do wonders for a self storage property's occupancy rates. On-site management can also provide cross-selling potential as many self storage properties generate additional income from the sales of moving supplies, locks, packaging materials, and from the collection of late fees and insufficient funds (NSF) charges. On-site management can also improve the return on advertising dollars spent for the self storage property. Without daily on-site management, knowing why a customer utilizes the self storage facility is difficult. The more that management understands who their customers are and why they are storing items at the property, the more targeted the advertising effort can be to effect increased occupancy rates.

10.4.4 Know Thy Customer

As should be apparent from the prior discussion, the primary customer for self storage property is adults in transition. Families in the process of moving represented the first storage opportunity for moving companies in the early twentieth century (Hess 1973). Consumers who are in the process of downsizing or increasing their living quarters will often utilize self storage on a short term basis. In the aftermath of the subprime mortgage crisis, self storage occupancy increased as families moved from houses into apartment units. Those families that are staying in their current home may also utilize self storage space to house recreational vehicles or collectibles. During the "Yes era", the concept of making money from "house flipping" was pervasive in the media and in practice. In the wake of the financial crisis, television programs featuring house flipping techniques have been replaced by bottom-fishers who purchase the contents of self storage units sold via an auction when the former owner has defaulted on their monthly rental payments. Storage facility contents investors look for profitable items to resell in order to profit from the speculative purchase. For those who have seen television programs highlighting storage auctions, collectibles are a common item found in storage units, along with furniture and other bulky items.

A second large category of storage is the seasonal user. Students may not wish to lug furniture and other belongings back home during the summer, and may utilize self storage property as a temporary storage solution during the summer months. Recreational users exhibit similar behaviour exhibit similar behavior, as recreational items are stored in facilities located near vacation destinations rather than storing items in the rental houses nearby or back home during the off-season. Seasonal visitors to resort communities may also utilize self storage during the off-season for the same convenience reasons.

A third category of self storage user is legal professionals. These users store property for a period of time in relation to an estate in transition due to death, litigation, or some other reason. Self storage again provides a temporary solution for these customers until such time that ownership or other issues are resolved.

A fourth category of self storage user can exhibit either a temporary or an ongoing need for storage space. Some small businesses utilize storage facilities for inventory warehousing as well as storage of records, supplies, and equipment. For example, a pharmaceutical representative may utilize climate controlled storage space to house their excess inventory. Generally, business needs for storage space can be temporary, as in the case of seasonal display signage, or can be longer term in nature given the expansion or contraction of the firm over time.

A final primary customer for self storage facilities is military personnel. Over the last two decades, much construction of self storage space has occurred just outside of military bases in order to fill the need that many soldiers have for storing their belongings during active military operations, or for a general desire for low cost storage space. In this fashion, self storage facilities present a counter cyclical orientation as the occupancies generally increase during deployments, which negatively affect other businesses in a military-based community.

10.4.5 The Ayatollah of Climate Controllah

We got the title of this section from a sign for a self-storage facility located in the Winston-Salem, NC area several years ago. Of course, the owner was trying to convince passing drivers that his units were reasonably impervious to the whims of nature. Evidently, all he really managed to do was irritate people. The next day the sign was replaced with "Come on in, we're cool".

Obviously, we were looking for a way to work that sign into this article. Here's the hook: some investors exhibit an almost religious fervor when it comes to the belief that self-storage properties are "recession proof" – impervious to the whims of the economy. While not offended, we would nonetheless, like that sign to be taken down.

The belief in the resilience of self-storage investments is nothing new and can be traced back to the Great Depression when the moving and storage industry was thought be a safe haven in the storm (Hess 1973). Today, the thought is that people will always need storage space and this is true in a deep recession. During the bad times, people become more transient – in search of work or even displaced – and need a place to "keep their stuff".

In fact, self-storage facilities have fared relatively well in the aftermath of the financial crisis, but occupancy rates and operating performance have faltered, especially in contrast to the boom years. Many metropolitan areas currently have a glut of self-storage space (Cushman and Wakefield 2011). All in all, while a case can be made for some degree of counter-cyclicality, many markets are suffering, suggesting that the investment return performance of these properties is cross-sectionally variable and anything but "a sure thing". In the face of the optimism surrounding self-storage properties, we stand by the belief that what determines value is the quality, quantity and durability (QQD) of the cash flows of a potential investment. This means that the key to success is due diligence – gaining an understanding of the current and expected future market conditions is paramount.

Self-storage projects are similar to apartment projects in that leases typically are written on a monthly basis and operating expenses are evaluated on a per unit basis. Naturally, climate controlled units generate a premium rental rate, but the expenses associated with their maintenance also come at a premium. Per unit comparables for occupancy rates, rental rates, and expenses should be viewed in light of the quality of the property's construction, the percentage of units that are climate controlled, and the implications of an increase in the supply of units in a given market to the operating performance of the property. Given the issues noted earlier regarding difficulty in compiling aggregate market information, self storage properties with a long track record of success should be viewed in a more favorable light than the "Johnny come lately" variety. New construction, or properties without a track record of success, can be more negatively influenced by the *phantom menace* of new construction and an uncertain future.

Some simple, anecdotal evidence refuting the notion that self-storage property is recession-proof is readily available on popular television and in short books. These outlets typically focus on the speculative purchase of storage unit contents via a rapid-fire auction process. While the white hot spotlight of reality TV has contributed to the relatively recent, popular awareness of this cottage industry, these stressed sale auctions have actually been around since the late 1970s.

In the late 1970s, a number of states passed laws allowing self-storage proprietors to place a lien on the contents of units that were arrears in rent. In fact, a few states allow proprietors to seize the contents of a unit as quickly as 40–50 days after the unit goes into arrears. Regardless of their rights, however, many self-storage property owners typically wait 90–120 days before pursuing the auction process, which begins with a mandatory classified legal notice in the local newspaper at least twice within fourteen days of the auction. Of course, the recent surge of publicity of storage unit content auctions has increased the number of bidders at the auctions. Inexperienced participants often unwittingly increase bid prices, allowing the property owner to acquire an amount that is closer to the outstanding rent on the unit. Unfortunately, at least for the property owner, a self-storage auction rarely generates enough cash to cover the outstanding debt; in fact, it is much more common that less than half of the balance will be collected via the auction process.

Interestingly, any funds collected in excess of the cumulative, outstanding rent via the auction, must be returned to the original owner of the unit's contents. Consequently, the auction is more of a "debt recovery event" than a profit generating alternative for self-storage facility owners. In most cases, the renters whose units end up in an auction have not voluntarily walked away from their property – they typically face significant economic challenges. Regardless of the circumstances, communication between the owner of the unit's contents and the facility owner are severed. Ironically, the death of a unit's renter does not usually result in the liquidation of a unit's contents by auction; family members or estate settlement procedures usually handle the situation.

The existence of *self-storage contents speculators* suggests that changing economic circumstances affect the renters of storage units, which, in turn, affects the owners of the units. While self-storage investments may enjoy an element of counter cyclicality, they are not immune to the turbulence of a declining market.

10.5 Conclusion

In this chapter, the differences and similarities of various industrial and warehouse properties has been discussed. This concludes the presentation of property specific differences which began in Chap. 8. By way of conclusion to this chapter, please review the following industrial, warehouse, and self storage property case studies and attempt to value the properties using what you have learned from the discussion in this book.

The final two chapters of this book will discuss the securitization of real estate as well as portfolio considerations by way of a discussion on Real Estate Investment Trusts (REIT).

Questions for Discussion
1. Elaborate on some specific location considerations for the various forms of warehouse and industrial property.
2. Discuss the four primary users of self-storage property.
3. What are the arguments for and against the case of self-storage being immune to the business cycle?
4. Itemize for which industrial property types ceiling heights are crucial and why.
5. Elaborate how outsourcing and off-shoring of manufacturing and service jobs impact investment real estate.

Warehouse Case Study #1

Industrial Properties, LLC is looking to purchase a warehouse that was leased to Ford Motors 2 years ago. The investors are looking for a loan on the subject property of $3,000,000.00. The purchase price of the subject property is $3,800,000.00, with the remaining equity to come from personal investments and 1031 Exchange dollars.

The subject property is a distribution warehouse located in Rural Hall, NC and sits on 4.5 acres of land. The warehouse has 125,000 square feet of rentable space, of which 20% is office space. The ceiling heights in the warehouse are 20 ft. Based on your market research prior to receiving this request, it was determined that this level of ceiling height is acceptable for most tenants seeking this type of tenant space. The warehouse is 15 years old but is in good condition. Ford Motors has 15 years remaining on their lease and is leasing the entire space in the subject property. The lease terms are triple net at $4.00 per square foot, with a scheduled rent increase, after each 5 year interval, of 5%. Per the lease, the landlord is responsible for the upkeep of the exterior building, but the tenant is responsible for the heating, air conditioning and other internal repairs and maintenance. You have estimated the annual replacement reserves should be 1% of effective gross income.

You have hired Messick Properties to manage the property for you, and you have agreed to 5% compensation annually based on 85% of the gross potential income of the property. At the time the lease was signed, Ford Motors was considered a credit tenant, but in subsequent years their rating with Moody's and Standard & Poor's has fallen close to junk bond status. The location of the subject property is excellent, as it is located near a major highway. Local appraisers note, however, that should Ford vacate the premises prior to the expiration of their lease, that 125,000 sq. ft. is a lot of empty space for this market. Additionally, other space is being built in response to Ford moving into the area. You also wonder how the recent announcement by Dell that they are closing their Winston-Salem based facility will affect market vacancy rates in the area. Based on your conversation with local appraisers, the rent being paid by Ford is within the market averages for this type of space, and the overall market vacancy rates for this type of space have run from 10% to 15% in the recent past.

1. Determine the net operating income for this property using the direct capitalization spreadsheet.
2. Determine the first year debt service coverage ratio for a loan of $3,000,000.00, at an interest rate of 8%, for 20 years.
3. Using a cap rate of 9%, what is the estimated value for this property? Would the value be materially different if you used the DCF valuation technique?
4. What happens to the DCF value as your discount rate increases from 9% to 10%?
5. Determine the breakeven occupancy rate and interest rate for this property.
6. What questions would you ask concerning the financial strength of the tenant?
7. What questions would you have for the borrowers concerning this single tenant facility?

Warehouse Case #2

Chapman & Loftis Company are seeking to diversify their real estate investment portfolio with a small industrial property. Since diversity is the goal, the investors desire to obtain a multi-tenant industrial property rather than a single tenant facility. While perusing the current commercial property listings, the investors discovered a property which appears to fit their needs located on Division Avenue in Garfield, Bergen County, New Jersey. Bergen County is located in northeastern New Jersey and has been known as a bedroom county to New York City given its proximity to the metropolis. Bergen County has historically been among the wealthiest and most economically vibrant counties in New Jersey. A key to the economic vibrancy of Bergen County was the construction of the George Washington Bridge which connects the county to New York City by traversing the Hudson River. Transportation in the county is considered excellent, as the county is crossed by a well organized network of highways connecting the New York region with the northeast and the rest of the country. Interstate Route 95 (the New Jersey Turnpike extension), and Route 80 are the most important, followed by Routes 1–9, Route 46, the Garden State Parkway, State Routes 4 and 17, and numerous county roads.

Six rail lines extend through Bergen County, with three carrying freight, and the other three carrying freight as well as passengers. Newark International

Airport in Essex County, which is within 30 min travel from Bergen County, is one of the most comprehensive airfreight systems on the east coast of the United States. The city of Garfield is a small, semi-urban area located in southwestern Bergen County, and is in close proximity to the area's main road systems. The city covers an area of 2.13 square miles and has an estimated population of 30,000 people. Approximately 75% of Garfield's land parcels are utilized for residential and multi-family purposes, while approximately 10% is utilized for industrial activity.

The subject property is situated on a 0.75 acre parcel located on Division Avenue, and was constructed in 1950. Access is considered excellent, as Interstate Route 80, US Route 46, and State Route 17 are within minutes of the subject property. The immediate area of the subject is improved with several light industrial uses, lumber yards, small retail and commercial uses, and residential uses. Vacancies in the area are scarce owing to a declining number of similar small and medium light industrial buildings in the area. The subject is located near the Dundee Dam, which is a 450 foot dam extending across the Passaic River from Garfield to Clifton, New Jersey.

The subject property consists of a 30,000 square foot building currently 100% occupied by four tenants. The building is of a concrete block foundation, a masonry and wood frame, and an exterior comprised of brick, concrete block, and vinyl siding. Doors consist of metal frame glass for office space, and metal clad overhead bay doors. The office area located on the second floor is warmed and cooled by a roof mounted gas-fired HVAC system, while the shop area is warmed by space heaters. The overall quality of the facility is considered average for the neighborhood and the property age.

As part of your due diligence, you have obtained a copy of the current rent roll as is shown below:

Tenants	Sq. ft.	Rent psf	Lease term
La Voie Lighting Co.	7,500	$5.50	5 Years
Tsang Furniture Restoration	8,500	$4.75	3 Years
Saade Garden Supplies	9,000	$6.50	1.5 Years
Gallo Lock and Key	5,000	$4.80	3 Years

Based on your review of the leases, the tenants are responsible for their own utilities, repairs, and maintenance. The landlord is responsible for structural repairs to the building, insurance, and the property management fee. The tenant's reimburse the landlord for annual property taxes on a pro-rata basis. Your research indicates that typical management fees associated with similar properties are approximately 5% of effective gross income, while replacement reserves run from $0.25 to $0.50 per square foot on an annual basis. The subject property is currently listed at $1.2 million. Typical bank lending requirements for comparable properties indicate a 75% loan to value ratio, an interest rate of 6%, a 20 year amortization, and a desired debt coverage ratio of $1.30\times$. Market rental rates are between $5.50 and $6.00 per square foot annually, and the market vacancy rate has fluctuated between 5% and 10% in recent years.

10.5 Conclusion

The seller's agent has supplied the following expense information from the last 3 years for the subject property:

	Year 1	Year 2	Year 3
Taxes	$21,500	$22,000	$23,000
Insurance	$5,250	$5,750	$6,300
Other	$2,000	$1,000	$1,500

1. Assuming a cap rate of 8%, what is the estimated property value based on the direct capitalization approach?
2. What is the net operating income and debt coverage ratio for the first year?
3. Determine the break-even occupancy and interest rate for this property for the first year.
4. Prepare a DCF assuming your direct capitalization assumptions for the first year. Assume a 10 year holding period, a 10% discount rate, a terminal cap rate of 9%, and an increase in income and expenses of 3% each year.
5. What is the estimated value using the DCF approach?
6. What questions do you have about the quantity, quality, and durability of this investment opportunity?
7. What questions might you have about the condition and use of the property?

Self-Storage Case Study

Camden Yards Storage, LLC has requested 80% of purchase price for an existing self-storage facility located in Fayetteville, NC. The purchase price for this 725 unit multiple building self-storage complex is $3,900,000. The original phase of this project was constructed in 2003, and the final 175 units were constructed in 2005 and they contain climate-controlled capability. These are the only such units on-site and there are very limited climate controlled units in the Fayetteville market. The property consists of 6.70 acres, and all buildings were constructed on concrete slabs. All of the buildings have exterior walls that are a combination of pre-engineered metal and brick. They have masonry firewalls and metal louvered roll-up doors for access by the customers. A 6-foot chain link fence surrounds the property, and there are driveways in between each building which allow enough room for the largest of personal vehicles to enter into the storage area. There is 24-hour access by way of a keypad entry system, and there are security cameras on-site and inside of the buildings containing climate-controlled units. There is an office on-site which serves as housing for the onsite property manager. Relative to the competition, this is one of the nicer mini-storage facilities in all of Fayetteville. The investor has stated that they intend to pay $60,000 in annual property management fees for the subject property.

Fayetteville is located in Cumberland County, NC, and is the home to Fort Bragg and Pope Air Force Base. Both have a major impact on the economy and are considered to be the primary employers in the area. The military bases have received personnel from other facilities that have been closed in recent years, and these bases are expected to remain open. The primary customers for this property are military personnel as well as small business owners in the area who need storage space for their products. The area to the west, north, and east of the subject is primarily residential, and the area to the south is primarily light industrial. The southeastern area, the Hope Mills sub-market, is one of the fastest growing sub-markets in the region.

The investors have supplied the following current rent roll for the subject property:

Unit type	# of units	Sq. ft./Unit	Rent/Month
5×5	6	25	27.00
5×5	23	50	30.00
5×10	168	50	44.15
10×10	172	100	74.03
5×20	2	100	65.00
10×15	57	150	90.42
10×20	130	200	96.37
10×25	75	250	128.86
15×20	1	300	125.00
15×20	1	300	180.00
15×25	2	375	145.00
5×15	24	75	54.10
New 5×5 cc	6	25	40.00
New 5×10	4	50	65.00
New 10×10 cc	33	100	95.00
New 10×15 cc	10	150	120.00
New 10×20 cc	11	100	145.00

The current occupancy of the complex is 90%, as the newest units leased quickly after construction was complete. Most units are leased out on a monthly basis, although some of the climate-controlled (cc) units have 6-month lease terms. This is standard for this type of property in this market.

The investors also supplied the following historical operating expenses. Market averages dictate that replacement reserves should be equal to $25 per unit on an annual basis.

Taxes	25,000
Insurance	15,000
Repairs and maintenance	30,000
Utilities	25,000
Other	30,000

1. Determine the net operating income for this property using the direct capitalization spreadsheet.
2. Determine the first year debt service coverage ratio for a loan of $3,120,000.00, at an interest rate of 8%, for 20 years.
3. Using a cap rate of 10%, what is the estimated value for this property?
4. Determine the breakeven occupancy rate and interest rate for this property.
5. What questions would you ask and to whom concerning the future supply and demand of self-storage space in the area?
6. Who might you ask to get an idea of what the market rental rates and expense rates per unit are for comparable space in the market?

References

Ajami, R., Cool, K., Goddard, G. J., & Khambata, D. (2006). *International business: Theory and practice* (2nd ed.). Armonk: M.E. Sharpe.

Appraisal Institute. (2004). Appraisal Institute Commercial Data Standards, version 2.0, May 14, 2004, web site http://www.redi-net.com/adn_db/AI_standards.pdf. Accessed 25 May 2011.

Bhagwati, J., Panagariya, A., & Srinivason, T. N. (2004). The muddles over outsourcing. *Journal of Economic Perspectives, 18*(4), 94–114.

Carlin, G. (2009). *Last words*. New York: Free Press.

Chappell, C. (2011a). REIT real estate investment today. Consolidation continues in self storage sector, May/June 2011 (pp. 46–48).

Chappell, C. (2011b). REIT website, "Self storage REITs building momentum in 2011", http://REIT.com/Articles/Self-Storage-Building-Momentum-In-2011.aspx?p=1. Accessed 7 June 2011.

Correll, R. (2003). *Market analysis and valuation of self storage facilities*. Chicago: Appraisal Institute Press.

Cushman & Wakefield Self Storage Industry Group website (2011). Market equilibrium by core based statistical area based on the 2011 self storage almanac. Accessed December 10, 2011, http://www.selfstorageeconomics.com/pdf/2011cbsa.pdf

Goddard, G. J., & Ajami, R. A. (2008). Outsourcing: Which way forward? An essay. *Journal of Asia-Pacific Business, 9*(2), 105–120.

Hess, J. (1973). *The mobile society: A history of the moving and storage industry*. New York: McGraw-Hill.

Phyrr, S. S., & Cooper, J. R. (1982). *Real estate investment: Strategy, analysis, decisions*. New York: Wiley.

Portable on Demand Storage (PODS) website. (2011). http://www.pods.com/About-PODS.aspx. Accessed 7 June 2011.

Sakthivel, S. (2007). Managing risk in offshore systems development. *Communications of the ACM, 50*(4), 69–75.

Self Storage Association. (2002a). *Building codes and the self storage industry*. Springfield: Self Storage Association.

Self Storage Association. (2002b). *Self storage standards and the modern community*. Springfield: Self Storage Association.

SSA Globe Magazine. (2011). Surviving and thriving: Japan's self storage industry is poised for growth as the nation recovers. June 2011 edition, pp. 9–14. Web site http://www.ssaglobe.org/pdfissues/June2011/japan.pdf. Accessed 7 June 2011.

Tropical Storage Website. http://www.tropicalstorage.com/a-history-of-self-storage. Accessed 7 June 2011.

Wall Street Journal. (2011). REITs make about face, March 30, 2011, pg. c11 US print edition.

Securitization of Real Estate Assets 11

The rule is that financial operations do not lend themselves to innovation... The world of finance hails the invention of the wheel over and over again, often in a slightly more unstable version.

John Kenneth Galbraith, Economist.

Contents

11.1	Origins of Securitization: The Development of the Secondary Mortgage Market	226
11.2	The Agencies and Their Function	229
11.3	The Structure of Agency Guaranteed MBS	231
	11.3.1 Mortgage Conformability	232
	11.3.2 Residential Mortgage Prepayments	233
	11.3.3 The Problem with Prepayments	233
11.4	CMOs and REMICs	235
	11.4.1 Types of CMOs or REMICs	235
11.5	Securitization Without Agency Guarantees	238
11.6	Replacing an Agency Guarantee with Credit Enhancements	239
	11.6.1 External Credit Enhancements	240
	11.6.2 Internal Credit Enhancements	241
	11.6.3 Structural Credit Enhancements	242
11.7	Commercial Mortgage-Backed Securities	242
	11.7.1 A Peculiar Arrangement: Servicing the Loans of a CMBS	243
	11.7.2 Struggling to Recover: The CMBS Market	244
	11.7.3 Reviving the CMBS Market Through Quantity, Quality and Durability (QQD)	246
11.8	Financial Markets and the Problems of Securitization	247
	11.8.1 Meltdown	247
	11.8.2 Who's to Blame	249
	11.8.3 Originate to Distribute	249
	11.8.4 The Rating Agencies	250
11.9	The Upside of Securitization	250
References		252

One of the most significant financial innovations of the twentieth century was the introduction of securitization. Securitization involves pooling individual, usually illiquid, assets and using the pool as collateral for the issuance of an entirely new set of financial securities. People that invest in the new securities are promised a proportionate share of the cash flows produced by the pool of assets.

The origin of securitization is inexorably linked to the development of the secondary mortgage market in the United States and the technology was first applied to residential mortgages in 1970. Securitized assets now represent a significant segment of international financial markets and are an important source of funds for financial institutions and corporations. At the end of the first quarter of 2011, the estimated outstanding balance of securitized assets originating in the US was about $10.1 trillion, with mortgages comprising over $8 trillion (afme/esf 2011). Investors can now access securities backed by pools of home-equity loans, commercial mortgages, automobile loans, credit card receivables, student loans, equipment leases and other emerging asset classes.

It is difficult to overstate the impact securitization has had on mortgage borrowing and lending. There is evidence suggesting that it has lowered borrowing rates (Hendershott and Shilling 1989; Sirmans and Benjamin 1990) and increased the availability of loans (Sabry and Okongwu 2009). Conversely, critics assert that securitization lies at the root of the credit crisis that began in 2007/2008 (Oberg 2010) and that it can facilitate financial market instability (Shin 2009).

Chapter 11 provides the historical justification for the development of securitization and, as a result, focuses on the market for residential mortgages and mortgage-backed securities (MBS) because mortgage debt represents a huge portion of U.S. debt and MBS are a large part of the bond and securitization markets. Fortunately, whether the assets being securitized are residential or commercial mortgages, auto loans or student loans, the process is similar if not identical.

11.1 Origins of Securitization: The Development of the Secondary Mortgage Market

The mortgage market in the U.S. has played a historically important role in defining the 'American dream' as the ownership of a home. Because of its significance as a conduit to the acquisition of real property by individuals, the mortgage debt market has been the focal point of public policy and private initiatives since the Great Depression of the 1930s. This effort has been directed toward enhancing the accessibility of mortgage loans to individuals and establishing a market in which the loans are viewed as viable investment grade assets.

The Federal government began to foster modern mortgage finance with the creation of the Federal Housing Administration (FHA) in 1934. The FHA was established during a period when real estate markets were enduring the brunt of the overall economic depression in the U.S. Foreclosure rates were peaking and investors had seen real estate values drop to 50% of their 1927–1928 levels (Bartlett 1989). The slumping real estate market and high foreclosure rates also had forced many mortgage banks out of business. The collapse of the mortgage banking industry effectively eliminated the only form of financing available to prospective home buyers – short term contracts that required loan to value ratios of 40% and semiannual interest payments, with a balloon payment due at the end of the term of the loan.

Congress charged the FHA with the difficult task of rehabilitating the housing industry. The primary function of the agency was to insure mortgages against default in an effort to provide financial intermediaries and other potential investors a federal guarantee. To qualify for FHA-insurance, houses and the mortgages used to finance them had to conform to mandated standards. The subsequent improvement in the quality and condition of houses financed with the insured mortgages was substantial, but FHA conformability requirements had a broader impact on the housing finance market. The FHA's underwriting requirements promoted a standardized loan contract that was both long-term and self-amortizing – essentially the mortgage familiar to most borrowers today (Fabozzi and Modigliani 1992).

The default insurance and contract standardization initiated by the FHA, along with federal tax incentives, transformed mortgages into practicable investments for thrift institutions and insurance companies. Nonetheless, the absence of a secondary market and the resultant lack of liquidity, limited the commitment investors could make to mortgages. To resolve this dilemma, Congress took its first deliberate step toward the development of a secondary mortgage market with the creation of the Federal National Mortgage Association (FNMA), or "Fannie Mae," in 1938.

While the presence of FNMA had the desirable effect of providing liquidity for single-family mortgages, the lack of new home construction, resulting from a combination of the economic depression of the 1930s and the privations of World War II, crippled any strides toward a secondary mortgage market. In an effort to boost the number of homes being built and aid the five million military servicemen returning home after the war, the government passed the Servicemen's Adjustment Act which provided specialized loans under the direction of the Veteran's Administration (VA). The VA program, which is in existence today, grants mortgage funds at interest rates equal to or below the FHA rate, negates mortgage insurance payments, and most importantly, avoids any down-payment requirements. The end of the war also allowed financial institutions, which were by this time yield hungry and highly liquid, to trade their riskless war bonds for more profitable mortgage investments.

The combination of pent-up consumer demand, the availability of VA and FHA loans, and the highly liquid position of financial institutions propelled a massive expansion in new home construction. The importance of mortgage investments also grew as FNMA began purchasing VA loans in 1948. Beginning with the post-war era and continuing through today, mortgages have represented the largest sector of the long-term debt market.

In 1957, the Wisconsin commissioner of insurance licensed the Mortgage Guaranty Insurance Corporation (MGIC). The formation of MGIC marked the emergence of a private mortgage insurance industry; providing a vital component for the marketability of conventional loans (i.e., those not insured by any housing related agencies). Lacking a legitimate, financially viable guarantee, mortgages outside the auspices of FHA or VA approval would prove to be an impracticable investment. This is particularly true of financial intermediaries facing regulatory restrictions.

In the same year the secondary market received impetus from the Federal Home Loan Bank Board (FHLBB), when the regulatory agency decided to allow depository institutions insured by the Federal Savings and Loan Insurance Corporation (FSLIC) to hold or originate conventional mortgages. This ruling drastically reduced the obstacles faced by thrifts, who wished to participate in the secondary market and who had previously required authorization to buy loans that used properties outside the institution's designated area of operation as security.

Congress restructured FNMA as a government-sponsored, privately held corporation in 1968. At the same time, the Government National Mortgage Association (GNMA), or "Ginnie Mae" was chartered to carry out FNMA's prior duties in support of FHA, VA, and Farmers Home Administration (FmHA) loans. GNMA also was authorized to guarantee the timely payment of principal and interest on securities issued by lenders of FHA, VA, and FmHA-insured loans. The guarantee offered by GNMA, as a government-related agency, is backed by the full faith and credit of the U.S. Treasury, and proved to be the spark needed to stimulate the growth of the secondary market.

Mortgage-backed pass-through securities (MBS) were introduced to the market in February of 1970. The first issuance was guaranteed by GNMA and used VA and FHA loans as collateral. In its assigned capacity as a facilitator of the secondary market, GNMA's only purpose in the transaction was to supply a guarantee to the purchaser of the security, the State of New Jersey Pension Fund. Later in the year, Congress established the Federal Home Loan Mortgage Corporation ("Freddie Mac") in an effort to increase the liquidity and available credit for conventional mortgages and give FNMA some competition. It was hoped that Freddie Mac could dilute FNMA's monopoly in the market and create an environment for conventional loans similar to that for agency insured loans. To support Freddie Mac's efforts, FNMA also received authorization to purchase conventional mortgages in the same year.

The issuance of the first GNMA MBS marked the beginning of the modern secondary mortgage market. With the introduction of Freddie Mac Participation Certificates (PCs) in 1971, investors could also purchase securities backed by conventional home loans. The popularity of these instruments increased rapidly. By the time FNMA began issuing its own mortgage-backed securities in 1981, the mortgage market had become a multi-billion dollar industry.

The 1980s was a time of great flux in the mortgage lending sector. An increase in the level and volatility of interest rates in the early part of the decade caused severe problems for savings institutions, the majority suffering arduous drops in net worth that resulted from a mismatch in the durations of their assets and liabilities. Also, the federal government inadvertently promoted alternative methods of housing finance through the deregulation of most financial institutions. While ushering in the era of "creative financing," deregulation also forced ailing savings banks to compete for loanable funds by offering higher interest rates on their liabilities. However, the onset of "creative financing" allowed mortgage lenders to stimulate the sale (and construction) of new homes by offering below market rate home loans (Jaffee 1984a, b).

The Secondary Mortgage Market Enhancement Act (SMMEA) of 1984 vastly expanded the number of financial institutions that could hold mortgage related securities in their portfolios. Also, this legislation brought important issues relating to the regulation and definition of mortgage securities under federal jurisdiction, thereby further standardizing the market. The SMMEA allowed depository institutions to legally invest in mortgage securities; exempted the securities from "Blue Sky Laws"; and voided any state enacted restrictions on investment by pension funds and insurance companies. In this context, the term "mortgage-related" legally refers to securities that are backed by first mortgages on single or multi-family homes. The act also requires applicable securities to be registered with the Securities and Exchange Commission and, be rated by a nationally known bond-rating service in one of the top two quality categories.

Securitization, the process of pooling individual assets that are used as collateral for the issuance of securities, opened the mortgage market to investors who had previously been excluded because of concern about the liquidity and credit risk of whole loans. It is difficult to overstate the importance of government legislative efforts toward the establishment of a viable secondary market. Specifically, the creation of GNMA, FNMA, and Freddie Mac provided the implicit and explicit guarantees necessary to ease investor apprehension regarding mortgage debt. The next section examines the subtle differences between the three agencies, as well as the benefits attributed to the development of agency mortgage-backed securities.

11.2 The Agencies and Their Function

By the end of the 1960s, the U.S. had experienced an extended economic expansion, and the government was beginning to escalate its military effort in Vietnam and Southeast Asia. The combination of demand for consumer credit, government borrowing, and accommodating monetary policy produced a period of high and fluctuating inflation and interest rates. Commercial banks also began to compete vigorously with savings and loans for deposits, lowering the spread traditionally enjoyed by the mortgage lenders (Kaplan and Hartzog 1977). In this economic environment the mortgage market suffered from a lack of funds as depository institutions, which relied on short-term deposits, found long-term fixed rate mortgages a risky and undesirable investment. The situation was aggravated by government imposed interest rate ceilings on deposit accounts that resulted in the withdrawal of saving deposits and disintermediation.

Despite determined attempts to relieve mortgage credit droughts, agencies such as FNMA and the FHA were unable to originate loans. The bottleneck was the reliance of the housing finance system on depository institutions. It was correctly and generally believed that the development of a secondary mortgage market could circumvent this periodically, troubling dependence. The path to a viable secondary market relied on the creation of financial instruments that expanded the number and scope of investors and provide more reliable sources of funds. The answer to this problem, in the short-term at least, came in the form of agency guaranteed MBS.

Because of the guarantees attributed to agency MBS, the level of default risk associated with these securities is usually considered to be slight; nevertheless, it is to some extent related to the specific intermediary underwriting the security. Excluding the years 2005 and 2006, the majority of MBS on the market receive the guarantee of, or are issued by, one of the three housing-related federal "agencies".

GNMA is an instrumentality of the federal government and, as such, its guarantee is secured by the "full faith and credit of the United States Treasury". Consequently, GNMA MBS are generally considered to be free of default risk. To achieve its objective, GNMA does not underwrite or issue MBS. Rather, it guarantees securities issued by private entities such as commercial banks, thrifts, and mortgage bankers. Only MBS collateralized by pooled mortgages and federally insured by the FHA, VA or FmHA are eligible to bear a GNMA guarantee. A GNMA MBS is referred to as a 'fully modified pass-through,' because the agency's guarantee ensures the timely payment of both the interest and principal, regardless of whether or not the borrowers of the mortgage loans in the pool have made their payments.

The other two underwriting agencies, FNMA and Freddie Mac, are more appropriately described as Government-Sponsored Enterprises (GSEs). FNMA, Freddie Mac and GNMA are similar in that all three organizations were initiated by an act of Congress, but neither FNMA nor Freddie Mac receive government subsidies nor an appropriation, and the stock of these quasi-private corporations trade on the New York Stock Exchange. Furthermore, FNMA and Freddie Mac purchase conventional (i.e., non-government insured) mortgages, pool them, and subsequently issue securities using the pool of mortgage as collateral, whereas GNMA backs only federally guaranteed loans.

Freddie Mac offers both "modified" and "fully modified pass-throughs". All FNMA securities are "fully modified pass-throughs." A modified pass-through differs from its fully modified cousin in that the modified pass-through guarantees only the timely payment of interest; both types of securities offer assurance against default. Pass-throughs issued by the GSEs have slightly greater credit risk because they are backed by the respective agencies' corporate guarantees rather than the 'full faith and credit' guarantees of the U.S. Treasury provided by GNMA. Regardless of the differences in the guarantees offered by GNMA and the two GSEs, the yield differences are negligible, suggesting that investors believe the US Government will, in fact, step in to back FNMA and Freddie Mac's guarantees.

The introduction of mortgage-backed securities by GNMA, FNMA, and Freddie Mac has had a profound impact on the mortgage credit market. Securities backed by pools of individual mortgages and guaranteed by a federal agency, avoid the idiosyncrasies that typically hinder the transferability of a single mortgage. In its securitized form, mortgage debt is more comparable to other fixed-income assets of a similar maturity and risk.

Agency backed securities also have lowered the cost of home financing by decreasing transactions costs and increasing the liquidity of mortgage loans.

Standardization of mortgage securities by the agencies, as well as the guarantees, promoted the acceptability of the instrument. Mortgage rates have been favorably affected by the increased number of investors, and the accompanying growth in available mortgage funds. The acceptability of mortgage securities has allowed mortgage originators direct access to capital markets. The ability to market and rapidly sell large quantities of mortgages has reduced the liquidity risks faced by originators and consequently decreased the transactions costs associated with the loan origination process.

While most MBS are originated through one of the three housing-related agencies, private mortgage conduits have provided the market with an opportunity to access nonconforming mortgages. Traditionally, the agencies have provided no liquidity for mortgages that do not meet their specific underwriting requirements. For instance, "jumbo loans", mortgages too large to meet the maximum requirements for inclusion in a federally related agency pool, were not available to the secondary market until they were pooled and securitized by private conduits.

Because of the guarantees issued by the various housing agencies, the federal government exhibited an interest in the development of a market for private label MBS, because of the potential liability in the event of large numbers of homeowner defaults (Fabozzi and Modigliani 1992) – which, of course, began occurring in 2008. Similar to the other issuers of securitized mortgage debt, a private conduit buys loans from individual mortgage originators, forms collateral pools of the loans, and sells the securities backed by the pools in the secondary market. The market for private conduit issuances was nurtured by the federal government and was relatively slow to develop, but in 2005 and 2006 the number of non-agency, or Private-Label, issuances surpassed agency backed issues. Because Private-Label MBS are underwritten and issued by commercial banks, insurance companies, and savings institutions that are unaffiliated with the government created mortgage enterprises, they do not carry an explicit or implicit guarantee from the federal government. Nonetheless, through the use of private mortgage insurance and various other credit enhancement devices, such as third party letters of credit, bond insurance and subordinated structures and guarantees from the issuing firms, private label MBS and MBS backed by commercial mortgages have remained a viable credit risk for potential investors.

11.3 The Structure of Agency Guaranteed MBS

The creation of a residential, agency guaranteed mortgage-backed security (RMBS) is surprisingly straightforward. As shown in Fig. 11.1, the process begins when an originator issues a conforming loan to a residential borrower. If the loan conforms to established guidelines, a GSE (i.e., FNMA or Freddie Mac) will buy the loan in order to package it with other loans that pay a similar interest rate and a comparable term to maturity. The package, or pool, of mortgages is then sold as a mortgage-backed security to investors.

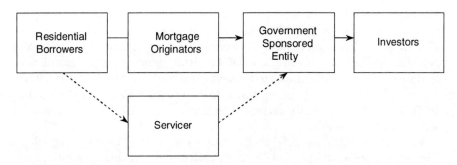

Fig. 11.1 Creating an agency MBS

Because FNMA and Freddie Mac guarantee the timely delivery of the cash flows, it is important that the borrowers pay as expected. Consequently, servicers are paid a fee by the GSE to monitor the performance of the loans and maintain contact with the individual borrowers. The servicers collect monthly mortgage payments, set aside taxes and insurance premiums in escrow, and forward interest and principal to the GSE, who then passes the residual onto investors. While the loan originator can also service the loan, typically firms that specialize in this administrative function get the business. Mortgage servicing rights can be bought and sold and the servicing industry itself has become a multi-billion dollar industry.

11.3.1 Mortgage Conformability

FNMA and Freddie Mac's conformability standards are established annually by the Office of Federal Housing Enterprise Oversight (OFHEO), the regulator of both GSEs. The OFHEO sets the size limit of conforming loans based on annual changes (October to October) in the average home price. The OFHEO also sets size limits for so called "Super Conforming Loans". Limits for super conforming loans are bumped up by 50% to reflect higher prevailing home prices in Alaska, Hawaii, Guam and the U.S. Virgin Islands. Table 11.1 provides the loan size limits for both FNMA and Freddie Mac for the year 2011.

The GSEs may only purchase conforming (and super conforming) loans. This significantly lowers the demand for non-conforming loans by removing the two biggest players from the market, making it more difficult for originators to sell them. Because of the lack of liquidity, borrowers whose loans fall outside the conformability standards can expect to pay an interest rate premium. Because non-agency/GSE participation in the secondary mortgage market has essentially disappeared since 2008, the yield spread on non-conforming mortgage loans has increased substantially (Guo 2011; Ambrose et al. 2004).

11.3 The Structure of Agency Guaranteed MBS

Table 11.1 2011 conforming loan size limits for FNMA and Freddie Mac

Property type	Conforming loan	Super conforming loan
1-unit	$417,000	$625,000
2-unit	$533,850	$800,775
3-unit	$645,300	$967,950
4-unit	$801,950	$1,202,925

Source: Office of Federal Housing Enterprise Oversight

11.3.2 Residential Mortgage Prepayments

RMBS are extremely complex financial instruments to value because of the uncertainty of the timing and magnitude of the cash flows they yield. The cash flows are unknown because a residential mortgagor (homeowner) has the option to prepay any part or all of the outstanding balance of a mortgage loan, without penalty, at any time. The uncertainty relating to the timing of the borrower's decision to exercise the option is called 'prepayment risk.' Prepayment risk accounts essentially for all of the risk faced by an agency RMBS investor and is the main reason investments in the securities perform differently than investments in Treasury bonds.

Borrowers can choose to prepay their mortgage for various reasons. The general level of interest rates can provide strong motivation for loan prepayments. For example, borrowers may find it prudent to refinance their homes when prevailing mortgage rates decline below the contract rate on their loan plus the transaction costs associated with the refinancing. Also, at a lower general level of interest rates, housing becomes more affordable and a bigger and better home becomes feasible.

Borrowers with accumulated savings might find it prudent to prepay only a portion of the total mortgage principal outstanding. Prepayments of this sort also are associated with the general level of interest rates. As the interest rate on long-term bonds declines relative to the rate on the mortgagor's loan, a higher return might be attained by paying down the outstanding balance of the mortgage relative to investing in an alternative, long-term security (Fabozzi and Modigliani 1992).

Homeowners may choose to sell their property, and prepay the balance of their mortgages, for a variety of reasons unrelated to the level of interest rates. For instance, a great job offer in another location might compel a borrower to abandon a very favorable mortgage rate in an otherwise high interest rate environment. Divorce or other life related discontinuities can be additional factors leading to the sale of a mortgaged property.

11.3.3 The Problem with Prepayments

The problem with prepayments is that they tend to amplify the interest rate risk of a fixed rate security by decreasing convexity on the upside (as interest rates fall) and the downside (as interest rates increase). Prepayment risk is similar to the call risk on a "straight" bond in that the option reduces convexity as increases in the market

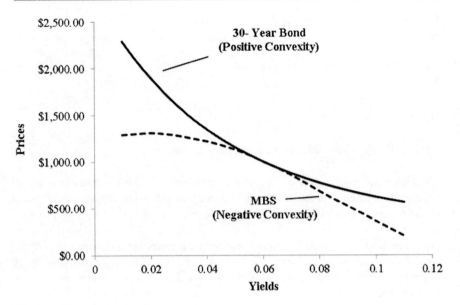

Fig. 11.2 Positive and negative convexity

value of the bond are limited by the possibility that the call will be exercised. Call risk can be even more significant for an RMBS because prepayments can intensify price decreases as prevailing mortgage rates increase – as prepayments begin to slow. So, prepayments truncate price advances and exaggerate price declines; for this reason RMBS securities exhibit "negative convexity" (see Figure 11.2).

In periods of falling interest rates, the value of an option-free (i.e., non-callable) bond rises because the future stream of contracted cash flows is discounted at a lower rate; this produces a capital gain for the bond investor. The probability of prepayment is inversely related to prevailing mortgage rates, so the increase in the value of an RMBS will be diminished. If prepayment occurs the RMBS investor will be repaid the par value of the debt rather than a price incorporating the premium normally dictated by the lower interest rates. Additionally, the investor will be receiving cash flows that cannot be reinvested at a rate comparable to the original rate on the mortgage.

Prepayments also produce a problem for investors when mortgage rates are rising. When interest rates increase the market value of existing debt securities decreases. RMBS values, however, fall further than a typical bond because the prepayment rates can be expected to decline. Prepayment rates decrease because the borrower's incentive to refinance or sell a home diminishes as mortgage rates rise. Of course, this implies that prepayments slow at a time when RMBS investors would like them to speed-up, so that the cash they receive can be reinvested at the higher market rate.

11.4 CMOs and REMICs

Prepayments are the lynchpin to pricing an RMBS. In fact, with an agency guarantee in place, prepayments are the only characteristic differentiating an option-free, Treasury security from a mortgage-backed security. The mortgages in a particular pool issued by a GSE are relatively homogenous with regard to the coupon rate, maturity and property-type. Nevertheless, to price an RMBS, an investor must form an expectation about the prepayment behavior of a particular pool of mortgage loans. If the pool prepays faster than expected (e.g., mortgage rates fall), the investor loses interest income and receives the principal in a low rate environment – reinvesting becomes a problem. Conversely, if the pool prepays slower than expected (e.g., mortgage rates rise), the investor ends up earning a relatively low rate for a longer period of time on the money remaining in the pool.

Despite efforts to devise strategies to deal with the prepayment problem (see Toevs and Hancock, 1988), the unpredictable and complex nature of prepayments caused RMBS prices to be volatile, and consequently, the securities were risky even with the agency guarantees in place. This severely stifled the growth of the market. In an attempt to attract a broader investor base, Freddie Mac issued the first Collateralized Mortgage Obligation (CMO) in 1983. A CMO is a multiclass, mortgage-backed security that allows the different classes to have different terms to maturity, interest rates, and most importantly, prepayment risks. CMO classes, referred to as tranches, accommodate the investors' preferences for varying prepayment speeds and risk. CMO investors can choose the risk characteristics that best fit their needs and risk tolerances – to earn a higher coupon rate, an investor must bear a larger portion of the prepayment risk.

The Tax Reform Act of 1986 included a provision for Real Estate Mortgage Investment Conduits (REMICs), which facilitated the issuance of CMOs. Unlike traditional pass-throughs, the principal and interest payments in REMICs are not passed through to investors pro rata; instead, they are divided into varying payment streams to create the different classes. A REMIC structure enjoys a tax advantage in that it avoids double taxation. Investors pay taxes on the income they earn from the REMIC, but the REMIC itself is exempt from income tax. As a result, almost all CMOs issued today are in issued in the form of a REMIC, accordingly, the terms "CMO" and "REMIC" are interchangeable.

The introduction of REMICs allowed investors to participate that otherwise had avoided the market. The ability to purchase securities with specific coupon, maturity and risk characteristics transformed mortgages into viable, liquid investment instruments.

11.4.1 Types of CMOs or REMICs

There are an assortment of ways the cash flows generated by the collateral backing a CMO can be allocated. Typically, the interest payments are initially distributed to all of the tranches. Then, the scheduled and unscheduled principal repayments are allocated to the various CMO classes based on a predetermined schedule that is presented in the security's prospectus.

Each of the tranches has an estimated *first payment date* on which investors can expect to receive their first principal payment and a *last payment date*, which identifies when they can expect their tranche to mature. Investors that purchase tranches that are initially insulated from principal repayments receive only interest payments during the lockout period. The window is the period of time when the principal repayments are scheduled to occur. Of course, CMO investors should understand that while the repayment dates are based on assumptions regarding the repayment performance of the mortgages in the pool, the actual repayments could occur at different times.

A CMO allocates payments to between 2 and 50 tranches; each tranche has its own expected average life. The average life of each tranche is based on the expected pattern of principal repayments.

11.4.1.1 Sequential Pay or "Plain Vanilla" CMOs

In the simplest CMO structure, the tranches are paid in a strict sequence with no overlap. Each of the tranches receives interest payments, but all of the principal payments are made to the first tranche alone. Only after the first tranche is completely retired does the second tranche begin to receive any principal. The principal is distributed in this manner until all of the tranches are paid off. So, with a five part CMO typical lifetimes might be: Tranche 1: 2–3 years; Tranche 2: 5–7 years; Tranche 3: 10–12 years; Tranche 4: 14–16 years; Tranche 5: 17–20 years. These CMOs are often referred to as "waterfalls," because the cash flows down to the lower rated tranches.

11.4.1.2 Planned Amortization Class (PAC) CMOs

A PAC CMO uses a support or "companion" tranche to provide protection for the PAC or "main" tranche and will do so as long as prepayments remain within a pre-specified range of prepayment speeds. The PAC structure allows investors in the main tranche to receive more certain, steadier cash flows; the PAC tranche's yield, average life and lockout periods adhere more closely to those estimated when the investment is purchased. The PAC tranche receives priority for receiving scheduled interest and principal, so that its payments are met first in accordance with the contract.

If prepayments are different from what was expected, then the support tranche absorbs the variable portion of the payments. If there is lower than expected prepayment of principal, most of the cash flow initially flows to the PAC tranche. If the prepayments are higher than expected, the PAC tranche receives the scheduled payment, with the excess flowing to the support tranche.

Some PAC CMOs offer more than one level of priority. "Type I PAC" tranches receive the highest priority, maintaining their scheduled payments over the widest range of prepayments, and have the most predictable income. "Type II" and "Type III PAC" tranches have lower priorities and their income is stable over a narrower range of unexpected prepayments.

Of course, to induce investment in the riskier, more variable, support tranches, investors are offered a higher yield.

11.4.1.3 Targeted Amortization Class (TAC) Tranches

TAC tranches are similar to PACs in that they provide more stable cash flows to investors, but TACs protect investors from a rise in the prepayment rate (usually associated with a lower interest rates), leaving them exposed to a lower than expected prepayment rate. If PAC and TAC tranches reside within the same CMO, PACs receive priority and the TAC tranche's payments will be more variable. Nonetheless, the companion tranches insulate both TAC and PAC tranches and receive the lowest priority in the CMO. TAC tranches are inferior to PACs, because they are protected only from unexpected increases in prepayments.

11.4.1.4 Z-Tranches (i.e., Z-Bonds, Accretion Bonds or Accrual Bonds)

As long-term investments, Z-tranches are usually used to support other tranches with higher priority such as PACs. The Z-tranche receives no interest until the lockout period ends and then finally begins to receive principal; the lockout period typically ends when all of the other tranches have been completely paid off. During the lockout period the tranche is credited with accrued interest that is taxable as income, even though investors receive no cash. As accrued interest accumulates the face value of the tranche is increased by an amount equal to the stated coupon rate on each payment date. The Z-tranche is the last in a series of sequential pay, PAC and companion tranches, and often have terms of 18–22 years. These types of investments have widely varying values throughout their lives and are best suited for tax-deferred retirement accounts.

11.4.1.5 Principal-Only (PO) Strips

PO Investors receive only principal payments and, accordingly, buy the securities at a deep discount from the face value, which is repaid through scheduled payments and potentially unscheduled prepayments. They are created from mortgage loans by separating or stripping the principal payments from the interest payments. Falling interest rates are associated with higher prepayment rates, shortening the effective term of the security, which means that the outstanding principal is paid more quickly – increasing the investor's yield. Conversely, rising interest rates tend to damage a PO investor's yield.

11.4.1.6 Interest-Only (IO) Strips

It takes an IO to create a PO, so any CMO offering a principal-only tranche will necessarily offer an interest-only tranche. IOs are sold at a deep discount to a notional principal, which is the amount used to calculate the interest paid to investors, but they have no par or face value. As the principal in the pool backing the CMO is paid down, the notional value of the tranche declines as do the interest payments. Note that IOs gain value as interest rates increase, because prepayments can be expected to decrease, lengthening the term of the underlying mortgages that are generating the payments. As interest rates fall and prepayments increase, however, these securities lose value rapidly and, in fact, may generate less cash than an investor initially paid for the security.

11.5 Securitization Without Agency Guarantees

Securitizations that lack agency backing comprise a vast array of assets that includes seemingly any asset that can be standardized, pooled and produces a stream of cash flows. Table 11.2 provides a partial list of assets that actually have been included in a securitization transaction.

Table 11.2 Assets used in securitization transactions

Trade receivables	Tax liens and credits	Public school facilities
Health care receivables	Auto leases	Insurance premium loans
Credit card receivables	Auto loans	Franchisee loans
Entertainment royalties	Equipment leases	Management fees
Nuclear fuel	Manufactured housing	Commercial mortgages
Natural gas	Student loans	Taxi medallions

The structural differences between agency securities and essentially any other type of securitization are directly related to the government guarantees offered by the GSEs, or more accurately, the lack of guarantees provided for anything else. Lacking the guarantees, non-agency CMOs must include additional participants into the structuring process. Figure 11.3 shows that the process of creating a non-agency, securitized asset begins when an originator makes loans.

A Special Purpose Vehicle (SPV), which enjoys an advantageous tax treatment, purchases them from the Originator and issues securities using the loans as collateral (Schneider 2004). The SPV is set up by a financial intermediary, such as a bank or insurance company, for the singular purpose of issuing the securities. In every sense, the SPV is an independent entity whose only assets are the loans backing the securitization. By removing the loans from the Originator's balance sheet, the SPV,

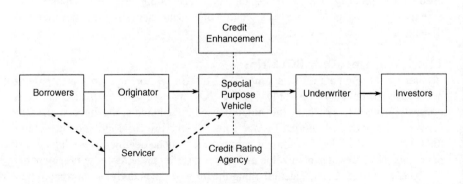

Fig. 11.3 Creating a private label MBS or asset-backed security

as a stand-alone entity, protects investors from deterioration of the Originator's credit quality or potential bankruptcy. The SPV also has the administrative role of controlling the collateral, collecting interest and principal payments and subsequently passing them through to the investors.

While the SPV structure protects investors from the bankruptcy risk of the Originator, for the securities to be competitively priced, the issuer must still provide some assurance to potential investors that the cash flows produced by the assets held in the pool will arrive. Furthermore, regulated investors such as money market funds, pension funds and life insurance companies cannot hold securities that are rated below investment grade. Consequently, the credit rating agencies, primarily Moody's, Standard and Poor's and Fitch, are a crucial element in the marketability of the securitization, because they provide the due-diligence that investors would otherwise have to obtain individually.

The credit rating agencies determine the size of the credit enhancements that are required, while the type and structure of the enhancement is left to the issuer. The factors that affect the size of the enhancement include the borrowers' credit quality and incentives to default, the size and variability of the potential loss facing investors, and the extent to which the loans in the pool are diversified (Fabozzi and Kothari 2008).

The Underwriter includes the banks, investment banks and brokers that sell the securities in a public offering or place them privately, often retaining a portion of the issuance for their own account. In this role, the underwriter is primarily responsible for pricing and marketing the securities to investors. Because of its ongoing relationship with institutional investors and knowledge of what the market will buy and the price at which it will buy it, the underwriter frequently seeks out originators for securitizations. It also often structures the classes of securities issued by the SPV, provides legal counsel and expertise on securities and tax issues.

11.6 Replacing an Agency Guarantee with Credit Enhancements

A critical factor in determining the demand for, and therefore success of, a securitization offering is the credit rating of the various bonds within it. Obviously, the credit quality is directly related to the yield that must be offered: the higher the credit quality, the lower the yield. Moreover, the credit quality impacts the demand for the issuance in that a better rating translates into a larger set of regulated investors that are permitted to hold the security, which means more demand, liquidity and a higher price. Banks, for example, may only hold securities with an investment-grade quality rating (i.e., at least: Moody's Baa3, S&P's BBB-, or Fitch BBB-) or better. The amount and type of credit enhancement will depend on the credit rating desired for each of the bonds.

Credit enhancement, or credit support, is the process used to reduce the risk of an entire security or specific bonds within a securitized asset such as a REMIC. It has been applied to an assortment of debt securities including public and corporate bond issues, but in the current context credit enhancement creates a financial buffer between investors and potential losses that could occur as a result of defaults on the pooled loans used as collateral. Without credit enhancement, the marketability of Private-Label securities would be difficult if not impossible.

Credit enhancements can be provided by external or internal sources, or are structural in nature. External Credit Enhancements rely on highly credit rated third parties, while Internal Credit Enhancements are based on injections of capital that are provided by the originator. As the name implies, Structural Credit Enhancements distribute risk among bonds so that some of them provide protection to those that enjoy a higher priority.

11.6.1 External Credit Enhancements

Third party or external enhancements invariably rely on the high credit quality of the guarantor. The most obvious form of external credit enhancements are provided by the GSE issuers FNMA and Freddie Mac. Similarly, GNMA, as a government agency, offers a guarantee secured by the "full faith and credit of the US Treasury" when it backs pools of Federal Housing Authority (FHA) and Veterans Administration (VA) mortgage loans. Non-agency securitizations, however, lack the explicit or implicit guarantee of the government and must rely on other methods.

11.6.1.1 Monoline Insurance
Some private insurers specialize in providing protection against loss on specific financial securities such as bonds or mortgages. These monoline insurers are employed by originators to provide particular guarantees, also known as credit wraps, to bolster the credit rating of specific bonds within a securitization structure – these guarantee the timely payment of both principal and interest to investors. Pool insurance policies are used for securitizations of residential mortgages to guarantee against defaults or foreclosures of the loans in the pool.

11.6.1.2 Letter of Credit (LOC)
A letter of credit is purchased from a financial institution, usually a bank, to provide credit enhancement for the senior bonds in a securitization. In this case, the bank assumes the default risk of the bonds to which the LOC applies. The LOC typically specifies the amount of cash that will reimburse the SPV for any shortfalls resulting from defaults of the collateral in the pool. LOCs have become less common, because the number of banks able and willing to offer them has declined substantially.

11.6.1.3 Liquidity Provider
A liquidity provider makes short-term, temporary payments when glitches in the system prevent the timely dispersal of funds from the servicer. Liquidity providers are most valuable and relevant in international securitizations, especially those originating in emerging markets.

11.6.2 Internal Credit Enhancements

With internal credit enhancements the originator itself provides protection in the form of financial support to cover potential losses. Here, the originator typically invests cash, assets or profits into the transaction, while simultaneously taking a position in a lower priority bond.

11.6.2.1 Excess Interest/Spread or Profit

The interest rate paid by borrowers on the loans used as collateral for a securitization transaction is not the coupon rate earned by the investors. Rather, the interest paid by the borrowers must cover any expenses faced by the trustee as well as servicing costs; the residual is then distributed to investors. When applying Excess Interest as a credit enhancement, an additional amount is deducted from the borrowers' interest payments and deposited into an account that accumulates over time to cover any losses that occur during a specified time period. For example, mortgages in a collateral pool might pay 8% interest, while the coupon of the MBS created from the pool may be 5%; the 300 basis point difference can be used for expenses and credit enhancement. If a loan defaults, the Retained Excess Interest can be used to make payments to investors. Once the Excess Interest has accumulated to a pre-specified target deemed to be adequate, any remaining excess interest is distributed to investors in residual bonds.

It is worth noting that the balance of the Retained Excess Interest is often unpredictable and is directly impacted by the performance of the underlying pool of loans. For this reason, the credit rating agencies consider it "soft credit enhancement," and therefore cannot be trusted as a definitive source of support (Kothari 2006).

11.6.2.2 Overcollateralization

With overcollateralization, the originator transfers a pool of collateral loans to the SPV that has a higher par value (usually 5–10%) than that of the issued securities. This means that the SPV holds a larger pool of assets than would be necessary if the loans in the pool pay as expected – more cash flows into the pool than is required to meet the coupon payments of the security. The idea is that even if some of the payments from the underlying loans are late or default, there are more than enough loans in the collateral pool to compensate. Overcollateralization is a very common if not ubiquitous form of credit enhancement.

11.6.2.3 Cash Collateral Account

Exactly as the name implies, a Cash Collateral Account can be created to offset losses to the collateral that threaten coupon payments to investors. As losses occur, cash is withdrawn from the account. To create a Cash Collateral Account an originator may choose to deposit funds when the securitization transaction is initiated, or excess interest (as explained above) can be deposited until an appropriate balance is acquired. When the securitization reaches termination, any proceeds in the account are returned to the originator.

11.6.3 Structural Credit Enhancements

Often referred to as Senior/Subordinated Structures, this type of credit enhancement is typical in CMO transactions (see Sect. 11.4.1). Here, subordinated, lower priority and lower credit rated classes of securities provide credit protection for the senior, higher rated securities in the transaction. If payment problems do occur, the subordinated bond investors, which often include the originator, are the first to experience the negative effects. Because the subordinated bonds carry a lower credit rating they offer a higher expected return.

11.7 Commercial Mortgage-Backed Securities

A commercial mortgage-backed security (CMBS) is backed by a pool of mortgage loans on commercial real estate property. CMBS differ from RMBS in that prepayments are not a significant issue and do not represent a substantial risk. Unlike residential borrowers, commercial borrowers lack the right to prepay the balance of their loans at any time without cost. As a result, lenders often apply a "lockout period," which usually lasts for the majority of the life of the loan, requiring the borrower to pay a penalty for early termination. This can severely discourage prepayment, but does not completely eliminate it. In certain situations, borrowers can substitute other income producing collateral in place of the real property securing the loan in a process known as defeasance. Typically, US Treasury obligations are the only acceptable form of collateral for this substitution.

Unlike the pool of loans usually backing an RMBS, CMBS are not standardized and the pool backing the security is often heterogeneous, comprising many mortgages of varying sizes and maturities, property types (i.e., office, retail, multifamily, industrial, etc.), location and quality of tenants. While heterogeneity enhances the diversification benefits to investors, it also adds to the complexity of rating and pricing the securities. Nonetheless, the CMBS market has proven to be an important source of capital for commercial lenders.

Like other securitized real estate assets lacking an agency guarantee, CMBS are typically set up as a REMIC in which a tax exempt trust (or SPV) distributes interest and principal payments to tranches (or bonds) that differ with regard to yield, duration and payment priority – similar to a sequential pay, senior/subordinated CMO or "waterfall".

Figure 11.4 illustrates the structure of a CMBS. In this example, $100 million in commercial mortgages generate cash flows to the various tranches. The bulk of the CMBS are issued as investment grade securities (Aaa/AAA to Baa2/BBB) and receive priority as interest payments are distributed. Each month the interest received from the loans in the pool is paid to the CMBS investors, beginning with those holding the highest priority tranches. Once all of the accrued interest is paid to the highest tranche, the next highest rated tranche is paid. The same process is applied to any principal payments received.

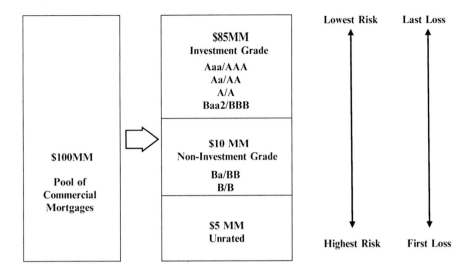

Fig. 11.4 Credit tranching a CMBS

Of course, the required yield on a tranche is inversely related to its priority – investors in low priority tranches rightfully expect to earn a higher return for the additional risk they bear. Here, risk resides in a higher probability of loss for the "B tranches" in the event of a borrower's default. For example, if a property is foreclosed upon and sold in a depressed market, it is likely that the proceeds will not be sufficient to pay off the remaining principal balance. For investors in a CMBS, the loss cascades upward, so that the lower priority tranches suffer the loss first and it is subsequently passed to the higher tranches. In this way, the "B tranches," or "B-Piece" provide protection for the "A tranches," or "A-Piece".

11.7.1 A Peculiar Arrangement: Servicing the Loans of a CMBS

In any securitization the servicer plays an important role in the viability of the transaction. Beyond passing payments onto the trustee, who then disperses them to investors, it is incumbent upon the servicer to maintain contact with borrowers in the pool backing the security and monitor changes in payment behavior. In the case of default, it is usually the servicer that contacts the borrower and, if necessary, pursues foreclosure and ultimately the sale of the property – passing the proceeds onto the trust.

Three separate roles are defined for CMBS servicers. The Master Servicer is ultimately responsible for servicing all of the mortgage loans owned by the CMBS trust *that are not in default*. The Master Servicer is responsible for collecting the interest, principal and escrow payments, approving borrower requests and compiling and maintaining property and borrower information. Because of its familiarity with the borrower and the property, the loan originator is sometimes hired by the Master Servicer to function as a Primary or Sub-servicer. In this capacity, the

Primary Servicer maintains direct contact with the borrower, essentially representing the Master Servicer as a sub-contractor to carry out specific loan administration duties.

Usually, the Master Servicer has little or no ability to modify a loan, or accept alternative collateral. In fact, accepting alternative collateral can negatively impact the tax exempt status of the REMIC (Butler and Steiner 2001). If a borrower defaults, the Master Servicer turns the servicing responsibility over to a Special Servicer. The Special Servicer is named at the issuance of the CMBS and traditionally this role has been performed by an affiliate of the B-piece investors.

It may seem odd that the low priority tranches take over once the loan defaults, but at its face, the logic is well-founded. As the holder of the first loss position, the B-piece faces the greatest risk and because of this the investors in the B-piece are likely most familiar with the deal and the loans in the pool. In essence, because of their risky position, the B-piece investors typically conduct a thorough review of the deal and the loans. This suggests that these investors are in the best position to determine the most prudent actions regarding a particular loan in default.

Nonetheless, the potential for a conflict of interest between the A-piece and B-piece exists. For example, suppose a borrower defaults in a market in which real estate values are depressed. The Master Servicer will turn the loan over to the Special Servicer who is hired by the B-piece investors. The Special Servicer may naturally determine that the best course of action is the one that most benefits the investors facing the greatest risk, the B-piece. In this case, Special Servicer has an incentive to grant an extension to the borrower in the hopes of generating additional interest income for the lower tranches while hoping that the borrower gets things under control. In fact, immediate foreclosure often leads to the greatest economic value for the A-piece investors. The conflict of interest arises because foreclosing in a depressed market will likely mean that the subsequent sale will not generate enough cash to pay off the remaining balance of the loan. The A-piece investors stand first in line for payment, leaving the B-piece with a loss of principal.

The perception of this conflict of interest has given some potential A-piece investors pause, especially in the current economic climate. In fact, some analysts have argued that this type of arrangement has contributed to the significant difficulties the CMBS market is currently experiencing.

11.7.2 Struggling to Recover: The CMBS Market

Once the market established a systematic approach to pricing CMBS, the demand for these relatively complex securities grew rapidly and became an important source of capital for commercial real estate lenders (Maxam and Fisher 2001). Unfortunately, the CMBS market was not immune to the ravages of the financial crisis experienced by the residential market in 2007/2008. Consequently, existing CMBS are struggling with delinquencies and the volume of issuances has plummeted.

Figure 11.5 shows the phenomenal growth of CMBS since 1995, peaking in the U.S. in 2007 at about $230 billion. In one year, however, issuances fell to $12 billion, hopefully bottoming out in 2009 at a mere $3 billion. Nonetheless, while it

11.7 Commercial Mortgage-Backed Securities

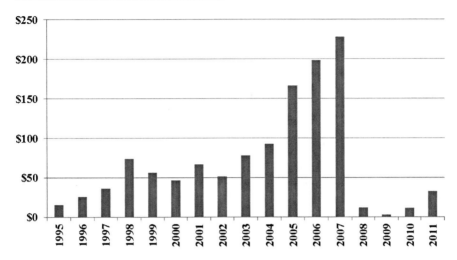

Fig. 11.5 Issuance of CMBS ($ billions)
Source: Commercial Mortgage Alert, WWW.CMAlert.com

is early and optimistic to find a trend, it may be that the CMBS market is trying to make a meager recovery. It is easy to blame the general malaise and over exuberance of financial and real estate markets leading up to 2007 for the trouble faced by CMBS, but many analysts argue that at least some of the difficulties can be attributed to the structure of the securities.

In an attempt to revitalize the market, issuers are overhauling the securitizations to address some of the structural issues with what is being termed "CMBS 2.0". In fact, "overhaul" is probably an overstatement; the changes are more akin to a "tweak". For example, in an attempt to simplify rating and pricing, the new issues have half the number of tranches of the original CMBS.

Some of the rights originally held by B-piece investors are shifting to the investment grade tranches. For example, the B-piece is less likely to appoint the Special Servicer (see Sect. 11.6.1) – that right will swing to A-piece investors. Hopefully, this simple change could eliminate the perception that the investment grade tranches give up economic value when a loan defaults and could induce more deep-pocketed investors to begin participating again. Furthermore, CMBS 2.0 grants more authority to the A-piece in defining the commitments made by the issuer of the security – previously the wording of the "representations and warranties" lay exclusively with the B-piece.

The relatively high yield associated with a non-investment grade security comes with a commensurate increase in risk. Typically, B-piece investors understand the relationship between risk and return and therefore dedicate an appropriate level of scrutiny before buying a CMBS. After all, before buying a security in which you hold a first loss position, you would be well advised to know what backs the security. The heightened scrutiny undertaken by a B-piece buyer provides economic discipline to the entire securitization structure. If the loans in the pool are questionable, the B-piece will sell at a significant discount, if at all, forcing the issuer to package loans with a reasonable level of quality.

But, what if the B-piece buyer does not intend to hold the security, but rather sells the investment into another pool, such as a Collateralized Debt Obligation (CDO)? In that case, the level of economic risk faced by the investor is reduced and the scrutiny need not be as stringent. By 2007, many B-pieces were, in fact, sold into CDOs and the quality of loans backing new issues fell accordingly. To alleviate this problem, the U.S. government, under the Dodd-Frank Act, has eliminated the ability of B-piece buyers from selling their part of a new CMBS issue.

All in all, these changes appear to be relatively minor adjustments to the original CMBS structure. Even so, it is clear that the non-investment grade tranches will have substantially less influence over the terms of the CMBS agreement, troubled loan workouts and less access to capital, given that they can no longer repackage and sell their investment. While this might make CMBS somewhat more palatable for A-piece investors, the B-piece will inevitably demand a higher yield.

11.7.3 Reviving the CMBS Market Through Quantity, Quality and Durability (QQD)

The structural changes being applied to new issuances of CMBS may induce some investors back to the market. Indeed, the additional authority and protections provided to A-piece investors will undoubtedly reduce the perception of a disadvantage they may have had. Ultimately, however, investors buy quality, and the quality of a CMBS relies on the pool of loans backing the security.

With the go-go days of real estate values behind us for at least quite a while, investors will be much more discerning about the assets backing their investment. In the end, what matters is the ability of the properties to generate the cash flows necessary to meet the borrowers' obligations. While loan-to-value (LTV) ratios essentially measure the skin a borrower has in the game, they provide little information about the quality of the income stream the property can produce. LTVs of 80% or higher at issuance, that were not uncommon before the crash (i.e., before 2008), have come down to about 65% on recent deals. While this is good sign, it is hardly an indicator of a property's quality.

To induce investors back to the CMBS market, issuers must provide them the transparency necessary to evaluate the properties on which the pool of loans is written. This means that the extent to which investors can be assured of the quantity, quality and durability (QQD) of the income produced by the property will determine the extent to which capital will return to the market. For example, the Debt-Service Coverage Ratio (DSCR) provides one bit of information that investors must be able to obtain on at least an annual basis. If a DSCR falls below 1.0, the property does not produce enough income to meet its debt service. This means, the borrower will either have to supplement the payments personally, eventually seek relief, or default.

Transparency will have the added benefit of reducing the price volatility of CMBS, because it will reduce investor uncertainty and the incidence of "market revelations". Consequently, once the capital begins to flow again, it will be less prone to feast or famine cycles, providing lenders a steady source of commercial real estate funding. And, this is important, because banks and insurance companies simply do not have the balance sheets to offer the kind of financing necessary to revive commercial real estate.

11.8 Financial Markets and the Problems of Securitization

There is no doubt that securitization has profoundly impacted the machinery, and some would argue the well-being, of financial markets. Certainly, the events leading to the "credit crisis" of 2007/2008 effectively took a great deal of the perceived bloom off of the securitization rose. Nonetheless, at its core securitization is no more than a process – it is not a thing in itself, and therefore, it has no characteristics outside of the purposes to which it is applied by individuals and institutions. Securitization unquestionably represents an innovation, but so did electricity and automobiles, both of which enhanced the quality of our lives and are simultaneously associated with the deaths of thousands of people every year.

11.8.1 Meltdown

The causes of the subprime mortgage crisis began long before the heat of the meltdown was first felt. Embedded deep in the psyche of most Americans is the idea that home ownership defines stability and success – it is, in fact, the American dream. It seems natural then that expanding home ownership is consistent with pursuing the national best interest. To that end, President Bill Clinton launched the National Home Ownership Strategy in 1993 and began the Federal government's concerted effort to include people in the American dream that had traditionally been excluded. At the same time, the U.S. economy boomed, interest rates remained low and home prices soared – buying a home had all the earmarks of a safe investment.

It seemed that everyone was participating in the boom. More people than ever were homeowners and their wealth was climbing, originators were booking more business than they could handle, and investment banks were creating complex mortgage derivatives feeding an insatiable market for the securities. Meanwhile, rating agencies (e.g., S&P, Moody's and Fitch) were hired as consultants by issuers to rate their securities. Furthermore, it did not seem important at the time that the rating agencies were inundated with securities they had little experience in evaluating.

The subprime market was not left behind; although it never exceeded 22% of the mortgage loans being originated, around 80% of the subprime loans were securitized. This represented a remarkable increase in the rate of participation by these loans. Essentially, both underwriters and investors were convinced that proper structuring of a securitization transaction, backed by relatively risky subprime loans that included appropriate support tranches and credit enhancements, could create viable investment grade securities. And, in fact, if default rates stayed within seemingly unimaginable limits, the higher priority tranches would remain safe.

Unfortunately for us all, the unimaginable sometimes becomes reality. What people did not notice, or chose to ignore, was the fact that the wealth being accumulated by American homeowners was being heavily leveraged in pursuit of bigger and better homes, more homes, and more consumption in general. Homes financed with more debt are simply more prone to default. In 1990 a typical loan-to-value ratio for a new home purchase was 80%, but by 2007 the ratio had risen to 90%. Also, originators were finding creative ways to increase the leverage of home buyers, so that the first rumblings of default were felt from mortgages with loan-to-value ratios of 100% – that is, home buyers were borrowing the entire value of the property.

The first rumblings of the disaster began in early 2007 as mortgage payments slowed or went missing, especially those on subprime loans. This essentially tells the story of the meltdown of 2008: borrowers simply failed to make the payments they had pledged to make. In fact, borrowers failed to such an extraordinary degree, that the default rate completely swamped expectations. Most of the faltering mortgages were bundled into CMOs; the support (i.e., low priority) tranches were immediately impacted receiving few if any payments at all.

As foreclosure rates climbed the value of the MBS on which they were based fell rapidly, creating a serious problem for the banks and the regulated financial institutions that held them. Mark-to-market or fair value accounting rules required the banks to recognize the loss on the value of the securities they held immediately, even though no cash loss had been suffered. Recognizing the lower value of the securities significantly lowered bank earnings, and in fact, in many cases generated substantial losses, decimating the equity on their balance sheets.

Regulators, concerned about the potential liabilities posed by the federally insured institutions, viewed the deteriorating capital positions as a real and present danger and threatened severe action, both implicitly and explicitly if matters were not dealt with posthaste. Counterparties, seeing the weakening position of the financial institutions, were reluctant to maintain their deposits or actually withdrew them, creating significant funding problems. Meanwhile traders heaped on short positions in an effort to profit from the inevitable decline in the banks' stock prices – adding to the downward pressure in the market.

The aptly named "credit crisis" resulted from the fact that banks simply had no access to funding. The interbank lending market that normally facilitates short-term loans between banks essentially froze, because it was virtually impossible to determine exactly what exposure another bank might have to mortgage-backed securities – banks simply would not loan to each other. In a desperate attempt to raise capital financial institutions resorted to: (1) trying to sell their mortgage securities, which only added downward pressure on the security values and (2) recalling loans that were scheduled for renewal, which essentially sent the crisis rippling through the economy.

Individuals, small businesses and major corporations shared the same experience in that there seemingly was no credit available. Banks with which they had long-standing relationships simply would not, or could not, continue to lend money. This had a profound and immediate impact on firms as they tried to honor short-term obligations (i.e., payroll and the power bill) while waiting for their customers to pay them. The lines of credit and short-term loans that had allowed them to ride temporary, predictable cash shortages just a few days earlier were gone and there did not seem to be any way to get them back. As investors realized the extent of the damage the Dow Jones Industrial Average (DJIA) began a slide that culminated in a 54% decline over the period extending from October, 2007 to March 2009.

In October, 2009, President George W. Bush signed the Troubled Asset Relief Program (TARP), which gave the U.S. Treasury the authority to purchase $700 billion in MBS from struggling financial institutions. At least in the short-term, TARP created the liquidity necessary to unfreeze money markets, if not jump start an economy.

11.8.2 Who's to Blame

At this point, the world is still trying to pull itself out of the morass that seemingly began in 2007/2008. We know that the groundwork for the disaster was laid years earlier, even if it was with the best of intentions. While abuses of the securitization technology can certainly be ascribed some of the blame, the technology itself is no more than a sophisticated financial tool. The blame for the financial crisis can be spread among a wide range of players including, but not limited to the government and its housing agencies, the mortgage originators, the underwriters, the rating agencies and the investors. So, pretty much everybody involved. Nonetheless, two players in the mortgage securitization process stand out as the proximate causes of the calamity: (1) Originators and (2) The rating agencies.

11.8.3 Originate to Distribute

In the old days, when an originator agreed to provide a mortgage loan to an individual, it represented the beginning of a relatively long, committed relationship. The bank or savings and loan kept the mortgage on its books as an income producing asset. So, depending on the movement of interest rates, an originator/lender could expect to receive payments from the borrower over a 3–10 year period. If things went bad, in that the borrower defaulted, the originator/lender took the hit. Because of the risks the lender faced, loan origination could be a grueling and humiliating experience as the lender verified income, assets, credit history, marital status, acquired a full appraisal of the property (as opposed to a curbside or drive-by inspection) and two or more years of tax returns. Furthermore, a lender usually required that the loan-to-value ratio did not exceed 90%, but 80% was more typical. Still, after all that, a fine, upstanding, bill paying individual could find himself denied a loan.

The originate-to-distribute model of securitization changed all that. As long as the mortgage fit within the agencies' (i.e., FNMA, Freddie Mac) purchasing guidelines, the originator could sell the mortgage and keep the upfront fees, plus any servicing rights. The moral hazard should seem obvious here: once the loan was sold it was off the originator's balance sheet. Any subsequent payment problems were borne by the MBS investor, not the originator. In too many instances, at least enough to generate a worldwide financial crisis, this led to a perverse process of credit analysis where the volume of loans was rewarded exclusive of any consideration of credit quality. The quality of a loan had absolutely no bearing on the decision to approve. In fact, originators became the conduit to successfully acquiring a loan rather than a viable safeguard against unnecessary risk. The question was no longer "Does this person qualify for a loan?" it became "How do I get this person qualified for a loan?"

11.8.4 The Rating Agencies

Historically, credit rating agencies [CRAs] have provided reliable information to fixed income investors, facilitating the growth of smoothly functioning financial markets. Nevertheless, the increasing complexity of securitized products that included the use of subprime mortgages, forced CRAs to provide ratings for financial instruments that did not fit traditional rating frameworks. In essence, Wall Street's innovations outpaced the CRAs ability to provide accurate information to the market. Consequently, ratings for complex securitized debt products were unreliable, and of little value to investors who had no reason to doubt the rating's validity. It appears that the probabilities of default were grossly understated, masking the risk actually confronted by investors (Mahon 2008).

Here also there is an element of moral hazard that cannot be ignored. The concept of corporate bond rating was introduced at the beginning of the twentieth century by John Moody, who provided bond ratings for U.S. railroads (Sylla 2002). The original revenue model required investors to purchase the information from the rating agencies. Once purchased, however, the potential for sharing had the rating agencies concerned about the long-run sustainability of their model. So, by the 1960s, the rating agencies began charging firms for rating their bonds. While there is no direct evidence to suggest that the ratings of MBS were affected by the demands of the firms issuing the bonds, the utter failure of the rating agencies to cast doubt on the quality of the issues leaves serious questions to be answered and has seriously damaged reputations.

11.9 The Upside of Securitization

The fundamental economic purpose of financial markets is to allocate resources to their most highly valued use. At this point in history, convincing people that the technology of securitization has enhanced the ability of financial markets to fulfill their fundamental purpose seems like a daunting task. Nevertheless, along with risk the world could hardly conceive just a few years ago, securitization also has brought benefits, the most obvious of which is that it has lowered the cost of moving capital from investors to borrowers. To a large extent, this has resulted from infringing on business of portfolio lenders, or banks.

Borrowers have benefited from the changes brought by securitization in that many firms tend to specialize in securitizing a specific type of asset as well as specific stages of the lending process, increasing the competition for the business. The competition has led to increased operational efficiencies and generated a lower cost of funds for borrowers. Furthermore, borrowing has become much more national in scope, so that the availability of funds tends to be more consistent.

Originators have benefited from the added liquidity that a ready secondary market offers. Also, for good or for ill, this has had a profound impact on the business model of many originators, because the fees for origination have become an important source of income. Furthermore, because they are typically most familiar with the borrower and the property or assets on which the loan is made, the originator often proves to be a natural servicer, especially early in the life of the loan. Once a payment schedule is established, the servicing rights can prove to have substantial value on the market.

Investors now have access to the cash flows provided by assets that were once out of reach, such as auto and student loans, equipment leases and credit card receivables (see Table 11.2). Obviously, these assets offer potential diversification benefits that have been absent in the past. Furthermore, given the proper structuring and credit enhancements, these relatively highly rated securities (or tranches) have achieved impressive yields.

Finally, investment bankers have reaped some impressive rewards from the introduction of entirely new product lines that also are conducive to financial engineering and innovation that leads to market expansion. Essentially, securitized assets have generated entirely new product lines and because of their expertise, investment banks benefit from trading activities in the secondary market for these securities.

Questions for Discussion

1. Why was the development of a secondary mortgage market important to the U.S. government?
2. What function did the "agencies" serve in developing the secondary market?
3. Why are credit enhancements necessary for non-agency MBS?
4. How did collateralized mortgage obligations (CMOs) contribute to the marketability of mortgage-backed securities?
5. As it relates to financial securities, what is convexity? What types of securities exhibit convexity?
6. Why do mortgage-backed securities exhibit negative convexity?
7. What risks do agency guaranteed residential mortgage-backed security investors face?
8. How does securitization provide an important funding source for lenders?
9. How are commercial mortgage-backed securities (CMBS) different from residential mortgage-backed securities (RMBS)?

10. In the past, who has appointed the CMBS Special servicer? Why has that been a problem?
11. What caused the credit crisis of 2007/2008?
12. Why do A-piece tranches earn a lower yield than B-piece tranches?

References

afme/esf Securitisation Data Report Q1: 2011.
Ambrose, B. W., LaCour-Little, M., & Sanders, A. (2004). The effect of conforming loan status on mortgage yield spreads: A loan level analysis. *Real Estate Economics, 32*, 541–569.
Bartlett, W. W. (1989). *Mortgage-backed securities: products, analysis, trading*. New York: New York Institute of Finance.
Butler, J. R., & Jeffrey E. Steiner (2001). New rules of engagement for workouts: REMICs & distressed real estate loans. *Real Estate Issues*, Winter.
Fabozzi, F. J., & Kothari, V. (2008). *Introduction to securitization*. Hoboken: Wiley.
Fabozzi, F. J., & Modigliani, F. (1992). *Mortgage and mortgage-backed securities markets*. Boston: Harvard Business School Press.
Guo, J. (2011). Fair valuation of residential whole loans. Interactive Data Corporation.
Hendershott, P. H., & Shilling, J. D. (1989). The impact of agencies on conventional fixed-rate mortgage yields. *Journal of Real Estate Finance and Economics, 2*, 101–115.
Jaffee, D. M. (1984a). Creative finance: Measures, sources, and tests. *Housing Finance Review, 3*, 1–18.
Jaffee, D. M. (1984b). House-price capitalization of creative finance: an introduction. *Housing Finance Review, 3*, 107–117.
Kaplan, D. M., & Hartzog, B. G., Jr. (1977). Residential mortgage markets: Current developments and future prospects. *Journal of the American Real Estate and Urban Economics Association, 5*, 302–12.
Kothari, V. (2006). *Securitization: The financial instrument of the future*. Singapore: John Wiley & Sons.
Mahon, P. M. (2008). Assessing credit risk, correlations and the collapse of the mortgage market. Honors thesis, Wake Forest University.
Maxam, C. L., & Fisher, J. (2001). Pricing commercial mortgage-backed securities. *Journal of Property Investment & Finance, 19*(6), 498–518.
Oberg, E. (2010). Securitization's role in the financial crisis. www.thestreet.com, May 21.
Sabry, F., & Chudozie O. (2009). Study of the impact of securitization on consumers, investors, financial institutions and the capital markets, for the American Securitization Forum by the National Economic Research Associates, Inc. Economic Consulting.
Schneider, W. H. (2004). US tax-specific points for issuers and SPVs. *Global securitisation and structured finance*. London: Globe White Page Ltd.
Shin, H. S. (2009). Securitisation and financial stability. *The Economic Journal, 119*(March), 309–332.
Sirmans, C. F., & Benjamin, J. D. (1990). Pricing fixed rate mortgages: some empirical evidence. *Journal of Financial Services Research, 4*, 191–202.
Sylla, R. (2002). An historical primer on the business of credit ratings. In R. M. Levich, G. Majnoni, & C. Reinhart (Eds.), *Ratings, rating agencies and the global financial system* (pp. 19–40). Boston: Kluwer Academic Publishers.
Toevs, A. L., & Hancock, M. R. (1988). Diversifying prepayment risk: Techniques to stabilize cash flows and returns from mortgage passthroughs. *Housing Finance Review, 7*, 267–94.

Real Estate Investment Trusts (REITs)   12

The best plan is to profit by the folly of others.
Pliny the Elder

Contents

12.1	The History of REITs	253
	12.1.1 Origins of REITs	254
	12.1.2 What is a REIT?	254
12.2	Various Forms of REITs	256
	12.2.1 Equity REITs	256
	12.2.2 Mortgage REITs	257
	12.2.3 Hybrid REITs	258
	12.2.4 Mutual Fund REITs	259
12.3	REIT Investment Strategy and Portfolio Diversification	260
	12.3.1 REIT Quantity Strategies	261
	12.3.2 REIT Quality Strategies	261
	12.3.3 REIT Durability Strategies	262
	12.3.4 REIT Portfolio Diversification	262
12.4	REIT Valuation Techniques	263
	12.4.1 Gordon Dividend Growth Model	263
	12.4.2 FFO Multiple	264
	12.4.3 Net Asset Value (NAV)	265
	12.4.4 REIT Valuation Issues	266
12.5	Internationalization of REIT Concept	267
12.6	The Sendoff!	270
References		271

12.1 The History of REITs

In this final chapter, the real estate investment trust (REIT) will be discussed. These special investment vehicles can serve as a portfolio diversification strategy for investors seeking an investment which provides return possibilities in a variety of property types and locations. The REIT, while of relatively recent creation, has roots which trace back to earlier times. Students of real estate finance should be aware of the functioning and the strategy of REITs, as these entities have become increasingly more involved in real estate investment. The chapter concludes with thoughts on international dimensions of real estate.

12.1.1 Origins of REITs

Before we discuss the characteristics of the modern REIT, an historical orientation is helpful. The first corporate organization somewhat similar to the modern REIT was the Massachusetts Trust, which began in nineteenth century New England in the United States (Chan et al. 2003). Massachusetts state law at the time prevented a corporation from owning real estate unless the property was crucial for the success of the business. Thus, there was not a permitted corporate entity for investment property. The Massachusetts Trust was created in order to fill this void in legal structure. The Massachusetts Trust eliminated federal taxation at the corporate level and allowed investors to receive distributions of rental income tax free at the individual level. After success of these entities in Boston, the trusts began to provide capital to develop real estate in other cities. The Massachusetts Trust remained a conduit for real estate investment until 1935, when the U.S. Supreme Court removed its favorable tax status.

Closed end mutual funds were created by the Investment Company Act of 1940, but advocates of the Massachusetts Trust continued to lobby the federal government for tax treatment equal to that approved for mutual funds by the 1940 legislation. In 1960, the tax law was amended, with the first REIT created in 1961 (Chan et al. 2003).

12.1.2 What is a REIT?

The real estate investment trust (REIT) is most simply viewed as a pool of properties or mortgages traded in the stock market. Modern REITs began in 1960 when the U.S. Congress enacted laws to create investment vehicles for small investors in real estate. The original intention was for REITs to be passive entities which held the various properties owned by the REIT in a corporate holding company. As per the 1960 legislation and continuing today, REITs are not subject to taxation at the federal level. Shareholders own shares in REITs and pay taxes based on dividends received as reported on Form 1099. These dividends are taxed at the individual tax rate of the investor. REIT losses are not passed through to shareholders, but are carried forward to offset income in future periods.

When compared with the Massachusetts Trust structure of prior years, the modern REIT does not allow for dividends to be received without tax being paid by the shareholder. In order to qualify for designation as a REIT under U.S. tax law, the following requirements are currently mandated (Brueggeman and Fisher 2010):
- There must be 100 shareholders at a minimum
- No more than 5% of shares can be held by five or fewer individuals

12.1 The History of REITs

- At least 75% of REIT assets must consist of real estate, cash, and government securities
- Not more than 5% of asset value may consist of securities from one issuer if those securities are not included under the 75% test
- At least 95% of REIT gross income must be received from dividends, interest, rents, or gains from real estate sales
- At least 90% of income must be distributed to the shareholders each year

As you can see from the characteristics listed above, the REIT is significantly restricted from an asset and profit distribution standpoint. Prior to 1986, REITs were also restricted as to being able to actively manage the properties under REIT ownership. The U.S. Tax Reform Act of 1986 effectively allowed REITs to provide services to tenants and to actively manage their portfolios. Prior to this act, REITs were mandated to use external advisors who often did not have a shareholder interest in the REIT. This external advisor scenario presented agency problems under the rationale that an internal advisor would have a more vested interest in the performance of the REIT than would someone external to the organization (Chan et al. 2003).

The Umbrella Partnership REIT (UPREIT) was established in 1992 to enable established real estate operating companies to bring properties already owned under the umbrella of a REIT structure. By so doing, capital gains taxes were avoided as the prior owned properties were not sold to the REIT. Instead, the REIT owns a controlling interest in a limited partnership that owns the real estate. The limited partnership could also be owned by management of the former real estate operating company, or by other private investors who owned the pre-REIT properties (Block 2002). The owners of the limited partnership have the right to convert their operating units into REIT shares, to vote like REIT shareholders, and to receive dividends as if they held publicly traded REIT shares.

Down REITs represent another variation of the REIT concept. Down REITs are similar to UPREITs but are formed after the REIT goes public. Down REITs generally do not have members of management among the limited partners of the controlled partnership (Block 2002). Down REITs can own numerous partnerships at the same time, leading to complicated organizational structures. Thus the Down REIT owns the properties directly in a REIT structure, but it holds some of the properties in a partnership having other partners. Partnership units are securities, and their issuance can have securities law implications. Owners of the Down REIT hold their interests as operating units in the partnership until they are either redeemed or exchanged for cash or REIT shares. In most cases, no tax liability is triggered until the partnership units are converted into stock or sold.

The limited partners of UPREITs and Down REITs can exchange operating partnership units for interests in other real estate partnerships that own properties that the REIT wants to acquire via a "like kind exchange". In the United States, IRC Section 721 deals with transferring a real estate investment into an UPREIT on a tax deferred basis. In order to contribute a property to an UPREIT, it typically must be considered of investment grade quality. In a 721 Exchange, rather than taking possession of another property, the investors receive operating units that carry the economic benefits of the REIT's entire portfolio, including any capital appreciation and distributions of operating income. These operating units may only be taxable when the owner elects to convert the units into REIT shares or when liquidation occurs. Thus, UPREITs and Down REITs can have competitive advantage relative to traditional REITs when dealing with tax sensitive sellers (Block 2002). As with any orchestrated management structure, there can be incentive related concerns when management does or does not own shares in the REIT. These concerns are similar to the internal versus external advisor issue noted earlier. If management owns only shares in UPREIT operating units rather than owning shares in the REIT, management will be subject to tax when properties are sold, unlike REIT shareholders. This could lead to management being reluctant to sell an underperforming property when prudent (Block 2002). Conversely, in a Down REIT, management who does not own any operating units could become "trigger happy" with property sales given the lack of tax consequences from property sale under the REIT ownership structure.

The REIT Modernization Act of 1999 allowed REITs to own taxable subsidiaries that can provide further services to tenants that REITs could not provide without disqualifying their rents from favorable tax treatment. This legislation took a further step toward movement of REITs to operating companies, and particularly benefitted REITs with hotel property investments (Brueggeman and Fisher 2010).

12.2 Various Forms of REITs

Generally, REITs can be either privately owned or publicly traded entities. Private REITs are not actively traded and are typically targeted toward institutional investors (Brueggeman and Fisher 2010). Sometimes various REITs can be syndicated and offered to investors, and sometimes REITs are originated as private entities with the intention of an initial public offering (IPO) once the REIT obtains certain size and performance objectives. Incubator REITs are typically funded by venture capitalists with the plan for an eventual IPO.

Publicly traded REITs offer the best potential for small investors given the amount of information available and the low levels of initial investment requirements in some cases. Whether public or private, REITs can generally be comprised of equity investments in properties, mortgages, or both.

12.2.1 Equity REITs

Equity REITs invest in specific properties in the form of equity investments. In order to be considered an equity REIT, at least 75% of the assets must be invested directly in real property (Chan et al. 2003). Some REITs have a stated time line when the various investments will be sold. These finite-life REITs have the goal of disposing of assets and of distributing net sales proceeds to investors by a specified date. This type of equity REIT was created in response to criticism of REIT share prices behaving more like stocks and not based on the inherent real estate values for properties owned by the REIT. The finite-life REIT thus was structured so that investors would know the date of disposition for all properties so that the terminal value, or reversionary cash flow, could be better estimated. While the known termination date is helpful for investors who are interested in estimating the value of properties held in a specific REIT, the specific date could prove problematic if it occurs during poor economic times. If the termination date of a finite-life (or self-liquidating) REIT occurs during poor economic times, often the decision is made to extend the date of disposition to allow for market recovery. Thus, most equity REITs do not have a stated termination date.

Equity REITs also differ in the composition of the properties held by the trust. The National Association of Real Estate Investment Trusts (NAREIT) classifies numerous forms of REIT investment diversification (NAREIT 2010). Many of these specializations will seem familiar based on the prior topics covered in this book. Specifically, NAREIT has categorized the following REIT classifications by property type: apartment communities, office properties, shopping centers, regional malls, storage centers, industrial parks and warehouses, lodging facilities (hotels and resorts), health care facilities, and natural resources (timber, farm land, etc.).

Some equity REITs focus on one category for investment, while other diversified REITs invest across multiple asset classes. The decision whether to invest in one property sector or many depends on the management philosophy of the REIT advisory board, and on the experience of the managers associated with the REIT. Some managers attempt to provide their clients with sector diversification while others attempt to achieve economies of scale and scope via specializing in one particular property sector. In some cases, REITs will limit the properties held to under five to further exhibit management's expertise in property selection. Equity REITs represent the majority of REITs today, but the early 1970s was dominated by mortgage REITs.

12.2.2 Mortgage REITs

Rather than owning equity interest in investment real estate, mortgage REITs obtain mortgage obligations, becoming a creditor to existing indebtedness for investment property. Mortgage REITs gained prominence in the early 1970s, when most major banks in the United States formed their own mortgage REITs. Given the nature of their investments, mortgage REITs are particularly sensitive to movements in interest rates. Mortgage REITs are essentially buying long term assets with short term funds on a highly leveraged basis. In the early 1970s, inflation increased substantially owing to the OPEC-led expansion in oil prices, and the effects that this had on the consumer price index. As interest rates rose, mortgage REITs sustained heavy losses during 1973 and 1974. The mortgage REIT concept fell out of favor after this time as the industry chose to invest in real property as an inflation hedge. Equity REITs fared better during the 1970s, achieving reasonable operating performance albeit associated with slow asset growth (Block 2002).

During the "Yes era", mortgage REITs again came to prominence in the form of collateralized debt obligations (CDO) and commercial mortgage backed securities (CMBS), which were discussed in Chap. 11. If mortgage REITs are to have a sustained presence, there is a need for transparency concerning the strength of those who are in fact repaying the loans in the investor pool.

As with other forms of REITs, the composition of management can pose incentive issues for mortgage REITs. If mortgage companies, thrifts, and banks are sponsoring the REIT by selling to the REIT the loans that they originate, similar problems as evidenced in the "originate to distribute" financing model described by Kolb (2011) could emerge. Lenders that are not envisioning holding the mortgages that they originate have more of an incentive to sell riskier loans to REITs and other investors as the default risk will be borne by another.

Generally, mortgage REITs invest in commercial mortgage backed securities (CMBS) issuances, residential mortgages, construction and development loans, and mezzanine loans. Mezzanine loans are typically bridge loans between the first mortgage debt on the property and the equity investment. These loans are like second mortgages, only the mezzanine loans are secured with an investor's equity stake in the property. This type of debt is attractive to mortgage REITs since mezzanine loans typically have a conversion option, which when exercised gives the lender an equity interest in the property. Thus, the lender is in more control during workout situations which allows for a quicker foreclosure process than for traditional debt and equity scenarios. Mezzanine loans also contain an inter-creditor agreement whereby the holder can take over the first mortgage in the event of default (Brueggeman and Fisher 2010).

12.2.3 Hybrid REITs

A third type of REIT is the hybrid REIT. As the name implies, hybrid REITs invest in both equity and debt instruments in order to obtain a return for their investors. Traditionally, the dividend payout ratios for hybrid REITs have trailed those of equity and mortgage REITs (Chan 2003). This could be owing to the mortgage REIT assets burdening the equity REIT assets with lower returns when interest rates are high, and by the opposite scenario during times of low interest rates.

Over the last 40 years, REITs have generally provided returns similar to the S&P 500, and have periodically been subject to down years every 10 years or so. Fig. 12.1 reveals the annual the annual returns of the FTSE NAREIT U.S. Real Estate Index Series from 1972 to 2011 (NAREIT 2012) and compares the total U.S. NAREIT Index with U.S. Equity REIT returns (not including timber REITs) and with U.S. Mortgage REIT returns over the same period. Figure 12.1 reveals that mortgage REIT returns have been much more volatile over the last 40 years than have been the equity REIT returns.

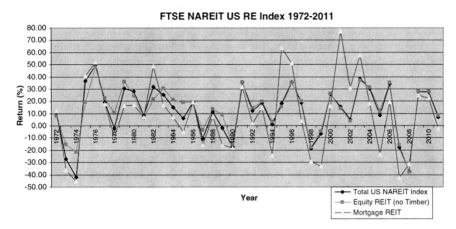

Fig. 12.1 Comparison of U.S. REIT returns from 1972 to 2011

12.2.4 Mutual Fund REITs

A final class of REITs is mutual funds which invest in REIT stocks. Mutual fund REITs may represent the lowest cost investment alternative for investment real estate. Investors seeking portfolio diversification can often invest in mutual fund REITs for as low as $2,500. A mutual fund, or open end investment company, is a group of related investment companies owned by their shareholders (or trusts effectively owned by their beneficiaries) and governed by their directors (Bogle 1999).

The U.S. mutual fund industry began with the formation of the Massachusetts Investor's Trust in 1924. This first organization was truly a *mutual* fund as it was organized, operated, and managed by its own trustees, and not by a separate external management company (Bogle 2005). John Bogle refers to the original internally managed mutual fund as the "alpha fund", contrasted today by the dominance of externally managed "omega funds". In most cases, each "omega" fund in the group contracts with an external management company to manage its affairs in return for a fee. The management company provides all activities necessary for the funds existence: investment advisory services, distribution and marketing services, and operational, legal and financial services. The "omega funds" of today are subject to higher costs given the competing goals of shareholder wealth creation as well as profit generation for the external management company. Modern mutual funds were created by the Investment Company Act of 1940, which states that the aim of mutual funds is to serve in the shareholder interest.

The 1940 legislation allows for both internal and external management, which is an issue of importance in the REIT context as well. Under the "alpha fund" model, mutual fund shareholders own mutual funds which own the management company. In the "omega fund" structure, management company shareholders own the management company which controls the mutual funds owned by the mutual fund shareholders. Under the "omega fund" format, management company shareholder interests are introduced to the operating model and strategy of the mutual fund. Since every dollar paid to the external management company is one less dollar paid to the mutual fund shareholders, the benefits of internally managed mutual funds under the "alpha" model appears to better serve the interests of mutual fund shareholders.

Mutual funds with internal management teams are more likely to produce lower operating expenses when compared with externally managed funds. Internal managers answerable to mutual fund shareholders help ensure that expense control is maintained. Investors in mutual fund REITs are encouraged to determine if the "alpha" or the "omega" is at play, and are encouraged to consider funds with lower operating expenses and fees relative to peer funds. Some mutual funds charge redemption fees for shares sold after being held for less than a year, and some funds charge their investors 12B-1 fees, which essentially reimburse the mutual fund for marketing expenses incurred by the external management firms. In the words of John Bogle, founder of the Vanguard Group,

> the easiest and surest way for a fund to achieve the top quartile in investment performance among peer funds is to achieve the bottom quartile in expenses. (Bogle 1999).

Firms with external management companies are essentially serving two masters: shareholders of the mutual fund and shareholders of the management company. Since the external management company is paid prior to dividends being paid to the mutual fund shareholders, one might question why management company shareholders are entitled to profit from the equity investments of mutual fund shareholders. In the final analysis, mutual fund REITs can provide low costs, high dividends, and portfolio diversification benefits, but it does pay to shop around. Now that we have introduced REITs as an investment class, we will next turn to REIT investment strategy and portfolio diversification.

12.3 REIT Investment Strategy and Portfolio Diversification

Based on our prior discussion, the advantages of investing in REITs are as follows: REITs serve as an inflation hedge, REITs have low correlation to other investment classes, and REITs allow for the avoidance of double taxation for investors. Depending on the sophistication of a given REIT, a prospective investor can review the annual report in order to ascertain the investment strategy of the advisory team. Some REITs focus on one particular property type, while others pursue a more varied investment strategy. Additionally, some REITs prefer to invest in properties in a limited geographic area, while others prefer geographical distribution so that the general economies will not necessarily move in tandem.

In the following section, we will review various methods for managing a portfolio of real estate investments as seen in light of the quality, quantity, and durability (QQD) framework utilized in earlier chapters of this book. None of the strategies that comprise the QQD framework are considered mutually exclusive, but it does help to shed light on how REIT strategy can fit our investment model.

12.3.1 REIT Quantity Strategies

Some REIT strategies can be categorized under the "quantity" heading as REIT managers attempt to invest in an appropriate array of properties so that the REIT shareholders can receive strong dividends over the holding period of their investment in the REIT. Whether the REIT has a focused or a diversified portfolio of properties, the goal of the "quantity" strategy is to maximize gross potential income by keeping overall vacancy rates as low as possible.

Investment managers preferring a focused strategy typically will cite a better understanding of the subject markets and the presumed lower operating costs associated with following a limited number of properties and or markets as a basis for their investment strategy (Ali et al. 2008). Possible negative outcomes associated with a focus strategy are higher risk given the lack of property and or market diversification, and a higher volatility of earnings associated with placing all of the REITs investment "eggs" in one proverbial basket.

Investment managers preferring a diversification strategy typically will cite more stable income streams, less risk given a higher level of property and or market diversity, and general psychological well-being for the prospective buyers of REIT shares given the lower fear concerning volatile earnings (Chan et al. 2003). Possible negative outcomes include higher operating costs associated with following more property types and more markets than firms focusing their efforts.

While maintaining high portfolio occupancy rates is important, just as crucial to the quantity strategy is to discover properties that are poised for growth in the future. The search for future growth might involve consideration of "turnaround" properties sold at a discount relative to their projected future values. What is also crucial for the quantity focus are understanding the market and general economic trends so that the selection of prospective investments does not detract from the overall portfolio occupancy rates.

12.3.2 REIT Quality Strategies

Another REIT investment strategy is the concentration on properties that have high quality, nationally known companies as tenants. The "quality" strategy can be gleaned as part of a REIT investment strategy when annual reports provide detailed information concerning the top five to ten tenants by square footage for a given REIT portfolio. Some REITs concentrate on "triple net" lease properties to well known tenants. This approach is prevalent enough for the category of "NNN REITs" to appear in many industry publications. Prospective investors can gain a general understanding of whether the quality strategy is being employed based on the amount of discussion concerning tenants in annual reports and other REIT generated publications.

Other quality-minded REITs can be considered Blue Chip REITs. These REITs have a long track record of value creation, strong management, established access to capital in order to fund growth, and a high level of insider stock ownership. Blue Chip

REITs are also characterized by strong balance sheets, no conflicts of interest, and relatively low dividend payout ratios in order to retain cash for external growth (Block 2002).

Blue Chip REIT managers pride themselves on achieving high returns for their shareholders and are a strong example of the "quality" investment strategy. The quality strategy is thus highly correlated to value investment strategy.

12.3.3 REIT Durability Strategies

The issue of durability primarily deals with tenant rollover risk, which is a concern of any real estate investor. Other than attempting to purchase investment properties occupied by well-known anchor tenants, the durability strategy also focuses on the length of the leases in place. REIT annual reports which disclose average tenant remaining lease terms, or the percentage of properties where tenant rollover is possible in a given future period provide some evidence that the durability strategy is being employed.

Another durability strategy is geographic dispersion, where properties in a given REIT portfolio are purchased in varying domestic and international markets. By widening the scope of locations invested by the REIT, the less likely the REIT will be subject to earnings volatility based on the poor economic metrics of one particular market. REITs which disclose the percentage of properties held that are located in a given metropolitan statistical area (MSA) provide some evidence of a durability orientation to portfolio management. Another example of durability strategy via geographic dispersion is when the annual report provides a listing of the percentage of properties located in specific regions of the domestic or foreign markets. In terms of foreign markets, most U.S. based REITs have yet to truly penetrate foreign markets, outside of well-known, upscale locations such as London, Paris, Frankfurt, and Tokyo. In recent years, some effort has been made to include emerging market economies in REIT portfolios (Lynn and Wang 2010).

12.3.4 REIT Portfolio Diversification

All of the aforementioned REIT investment strategies have the goal of diversifying unsystematic risk. This form of risk can be minimized through portfolio diversification strategies which invest in a portfolio of properties that are less than perfectly correlated. Rather than focusing efforts on finding the few perfect investments for a REIT portfolio, investment managers are better served by compiling a portfolio of assets which together produces a reasonable return for the shareholders while undertaking a medium amount of risk. Risk management via portfolio diversification should not have the goal of completely extinguishing risk. Since risk and return are related, a completely risk free portfolio, if this is possible, would more than likely not meet shareholder return requirements. Systematic risk will exist even within a diversified portfolio, as market risk always remains since it is non-diversifiable. The goal of risk management should be in reducing the variability of returns in a portfolio of investments.

When considering whether or not to purchase an investment property as an additional holding in an existing portfolio, the QQD framework can be utilized to assess the inherent risk of the property experiencing return variability owing to lease rollover risk, or to general business risk given the property sector and market location. Historical property operating performance and historical market metrics should also be evaluated prior to the property being acquired. Historical metrics are helpful in assessing possible default risk owing to property leverage, and liquidity risk based on prior sales of comparable properties. As discussed in Sect. 6.3, statistical measures such as the coefficient of variation and the standard deviation should be calculated for individual properties being considered as additions to an existing asset pool. The prospective purchase can then be viewed in relation to the existing risk profile of the entire portfolio so that the investment manager can estimate whether adding the asset to the portfolio adds to returns while lowers or at least maintains portfolio risk.

Portfolio risk is viewed in terms of the standard deviation of possible returns and by the covariance of the portfolio. Covariance is an absolute measure of how two data series, or assets, move together over time. Correlation is the relative measure of movement between the series or assets. Traditionally REIT stocks have experienced low correlation when compared to other asset classes (NAREIT 2011). This bodes well for REITs adding value to investor portfolios. What remains is a discussion of methods for evaluating the current REIT stock price relative to the composition of assets held by the REIT.

12.4 REIT Valuation Techniques

As we now turn to the various methods utilized by prospective REIT investors in determining the market value of a REIT share price, it is helpful to remember our discussion in Sect. 4.2 concerning the three forms of real estate property valuation: income, sales, and cost. While REIT valuation techniques are not perfectly correlated with investment property valuation methods, elements of all three of these approaches are present in the techniques which follow. REIT share price valuation can best be seen as a blend of real estate and general stock market share price valuations, which considering the subject matter makes intuitive sense. The general tendency is to "buy" when the intrinsic value as determined by the prospective investor is higher than the current stock price, and to "sell" when the intrinsic value is less than the current price of the REIT stock.

12.4.1 Gordon Dividend Growth Model

What should be clear from the discussion in this chapter is that REITs are required to pay out the majority of their income in the form of dividends to shareholders.

Fig. 12.2 Gordon dividend growth model

Div 1	$ 3.00
K	10.00%
g	4.00%
V=	$ 50.00

The Gordon Dividend Growth Model utilizes the future dividend per share expected to be paid out next year, and calculates the value of the stock as the present value of expected future dividends. The model assumes a constant dividend growth, and is similar to the discounted cash flow analysis in formulation.

Prospective investors begin the analysis with a construction of a discounted cash flow based on the income statement of the company. Projections are then created utilizing assumed revenue and expense growth rates by line item. The net operating income, or the fund from operations (FFO) for a REIT, is calculated for an assuming holding period, with the final period ending with a reversionary cash flow. This leads to the present value of the firm, which is then converted to a per share basis by dividing by the number of shares outstanding. The expected future dividend per share forms the basis for the Gordon Dividend Growth model as is shown below:

$$V = \frac{D_1}{(k - g)}$$

Value (V) is equal to the expected future dividend (D_1) divided by the required rate of return (k) minus the dividend growth rate (g). Assuming an expected future dividend of $3.00 per share, an investor required rate of return of 10%, and a dividend growth rate of 4%, the intrinsic value of a hypothetical REIT stock is calculated at $50.00 per share as is shown in Fig. 12.2.

While the determination of the expected future dividend is based on the investor's expectations of the future operating performance of the REIT, the intrinsic value of the REIT is also based on the investor's required rate of return and the constant dividend growth rate assumed.

12.4.2 FFO Multiple

The FFO multiple approach to valuing REIT stocks is one of the more popular methods utilized by Wall Street analysts. The FFO multiple approach is similar to the price to earnings multiple utilized by analysts when valuing traditional equity investments. Funds from operations (FFO) is defined as net income, computed in accordance with generally accepted accounting principles, excluding gains or losses from sales of property or debt restructuring, and adding back real estate depreciation (NAREIT 2010). Once FFO is estimated, it is then divided by the number of shares outstanding, or the projected number of future shares outstanding, to arrive at FFO per share.

Fig. 12.3 FFO multiple model

FFO/share	$ 4.50
FFO Multiple/share	$ 11.00
Share Value	**$ 49.50**

The intrinsic value of the REIT stock is estimated by multiplying the FFO per share by the FFO multiple. FFO multiples can be estimated by researching the REIT's historical multiple and via peer group and industry multiples. The peer group FFO multiple consideration is similar to the use of comparables in the sales approach valuation for investment property. Since investment real estate is better viewed when compared to similar properties, REIT stock valuation offers no exception to this general rule.

As an example of FFO multiple value construction, the intrinsic value of our hypothetical REIT stock is estimated at $49.50 per share as is shown in Fig. 12.3.

As similar to the sales approach valuation for a specific investment property, the FFO multiple valuation approach for REITs is only as valid as the accuracy of the REIT stocks deemed comparable by the analyst or prospective investor.

12.4.3 Net Asset Value (NAV)

A third method for valuing the intrinsic value of a REIT stock is the net asset value (NAV) approach. The NAV is calculated by aggregating the stabilized net operating income (or FFO) for the entire company and dividing by an appropriate blended cap rate for the company's real estate assets. As was discussed in Sect. 4.3, estimating a cap rate for a specific real estate investment can become complicated once the prospective investor attempts to include comparables into the analysis. Thus, the determination of a blended cap rate has similar issues, especially when the investment portfolio is highly diversified.

A more appropriate method of calculating NAV is to estimate the net operating income for each property in the portfolio, and to divide by a specific cap rate for each property. Then the aggregate value of the portfolio can be determined. NAV is the result of debt being subtracted from the total property value held in the REIT. A simple example of NAV estimation is shown in Fig. 12.4.

The gross revenue for our hypothetical REIT is $35 million, which is offset by $15 million in operating expenses. Since REITs report the value of their real estate holdings at book value in their financial statements, there is a need to estimate the market value of the investment properties by dividing NOI by a cap rate. Once debt is subtracted from the aggregate market value of the properties, the estimate of NAV is achieved. Once divided by the number of outstanding shares, the NAV model estimates the share price of our hypothetical REIT at $49.28. The NAV model can be seen as a blend of the cost and income approaches to value given the desire to obtain an estimate of the current market value of the assets held by the REIT rather than simply focusing on the book value as per the REIT financial statements.

Fig. 12.4 Net Asset Value (NAV) model

			Per Share
Revenue	$	35,000,000	$ 12.73
Op Ex	$	15,000,000	$ 5.45
NOI	$	20,000,000	$ 7.27
Cap Rate		9.50%	
Value		210,526,316	$ 76.56
Debt		75,000,000	$ 27.27
NAV		135,526,316	**$ 49.28**
# of Shares		2,750,000	

Obviously, subjective factors are also included in the valuation of REIT shares. Our three hypothetical REIT valuation models returned similar results, but this is not always the case. When significant variation exists among the valuation alternatives, the method chosen by an analyst or a prospective investor will invariably include considerations such as the strength of management, the reputation of the REIT, future growth prospects for the REIT, and opportunity costs associated with giving up competing investment alternatives.

12.4.4 REIT Valuation Issues

As with any investment analysis, REIT share valuation is a mixture of objective and subjective components. One significant complication is that while the definition of FFO is standardized, the interpretation of that definition can lead to differing pathways to the end share price estimate. The classification of recurring expenses will impact the estimation of FFO. If an incurred cost is classified as an expense, it is subtracted out of FFO. Alternatively, if an expense item is classified as a capital improvement on the balance sheet, the expense is not deducted from cash flow as the capital improvement is amortized over time rather than treated as an expense item incurred in a given fiscal year.

Some tenant improvements are capitalized and are depreciated over the life of the lease. Other tenant improvements may be deducted in the year in which the expense was incurred. Leasing commissions may also be capitalized with depreciation and amortization occurring over the life of the lease, but there is not uniform treatment of these expenses across the REIT industry.

Figure 12.5 below provides a simple illustration of how the income reported on a REIT financial statement can differ from how FFO is calculated. In our hypothetical example, the earnings per share based on the income statement is $2 less than the FFO per share. The difference in this example is owing to depreciation being subtracted from cash flow in the income statement, and owing to the gain on sale being added back.

For REIT valuation purposes, gains on sale of investment property are not typically considered recurring sources of income, as the gain was based on a one-time event.

Fig. 12.5 Differences in earnings per share

		REIT Income Statement	REIT FFO
	Gross Rev	$ 1,000	$ 1,000
-	Op Ex	$ 400	$ 400
	NOI	$ 600	$ 600
-	Deprec	$ 400	$ -
+	Gain on Sale	$ 200	$ -
	Cash Flow	$ 400	$ 600
	EPS	$ 4	
	FFO/Share		$ 6
	# of shares	100	

In order to circumvent the problem of different implementations of FFO calculation, many analysts prefer to calculate **adjusted funds from operations (AFFO)**. AFFO adjusts for expenses, while capitalized, which do not enhance property value and further adjusts by eliminating the straight-lining of rents (Block 2002). In order to arrive at AFFO, the FFO is further reduced by recurring capital expenditures, the amortization of tenant improvements and leasing commissions, and adjustments for straight-lined rents. FASB 13 requires that free rent or rental increases must be equalized, or straight-lined, over the term of the lease for accounting purposes (FASB 2011). Depending on the length of the leases and on the general economic environment, rent increases and rent concessions may not be entirely representative of the future revenue potential of the properties held by the REIT.

The AFFO concept helps to reveal that all FFO growth is not the same. Some REITs may have experienced historical FFO growth externally via the acquisition and development of the portfolio, while other REITs may have grown internally via improved profitability in the existing portfolio of investment properties. Internal FFO growth via rent increases, expense reductions, and occupancy rate improvement is inherently more stable than external "capital recycling" approaches to growth. What is clear from the preceding discussion is that the better a prospective investor understands the composition of REIT income, the better their intrinsic valuation of the REIT stock price will reflect the future expected market movements.

12.5 Internationalization of REIT Concept

After the creation of the modern REIT in the United States, similar designs were established throughout the world. In 1971, Australia established the REIT concept, followed by Canada and Turkey in 1985. Other important REIT markets were established in Singapore in 1999, Japan in 2000, Hong Kong and France in 2003, and Germany in 2007. While the concept of REIT was established in the years denoted, typically there was an open end mutual fund or some other investment vehicle available prior to REIT issuance. For example, while G-REITs in Germany were established in 2007, open end real estate funds (so called "Offene Immobilienfonds") were permissible since the German Investment Company Act of 1969 (Maurer, et al. 2004).

While tax considerations for international REITs are of similar design to U.S. legislation, European REITs have some additional requirements. For example, the overall REIT leverage has historically been limited to 50% in Germany, France, and Spain, while Austrian REITs have been limited to 20% leverage. In the United Kingdom, REITs have an additional requirement of maintaining a minimum of $1.25\times$ coverage of earnings before interest and taxes relative to interest expense (Suarez 2009). UK REITs have also been limited to 40% of assets being allocated to one property type, while G-REITs have historically been limited to a 15% concentration. Asian REITs began to rise in prominence after 2002, but the ground work was provided much earlier as Malaysia was the first Asian economy to create a market for REITs in 1986. M-REITs came to market officially in 2005, with the first Islamic REIT being created in 2006. Islamic finance has proven receptive to real estate investment trusts given the verifiable nature of the asset being financed (Visser 2010). Islamic REITs are required to maintain Shariah compliant portfolios which are managed by restrictions on business activities conducted by tenants, and via requirements for interest-free and speculation-free financing and insurance (Newell and Osmadi 2009).

Thus the REIT concept has proven flexible depending on the level of government intervention and the specific market requirements. The move toward internal advisory capability allowed for more customization in the international REIT alternatives. Fig. 12.6 provides a table which itemizes the predominance of internally managed versus externally managed REITs in various markets across the globe (Brueggeman and Fisher 2010).

It should be noted that while the United States allows for both internal and external management of REIT portfolios, after the 1986 tax law change, most U.S. REITs follow the internal management paradigm. The prevalence of internally managed REIT options is on an upward trend given the agency issues noted earlier with the external structure.

As U.S. based investors consider equity investment in domestically owned REITs with a foreign diversification strategy, or equity investments in foreign owned REITs, a number of things may influence investment performance. Differences in legal structures, political stability, currency exchange rates, and government market intervention can all effect the performance of a given property investment. Additionally, differences in accounting standards may lead to a

Internal	External	Both
France	Australia	United States
Turkey	Japan	Canada
	Singapore	Netherlands
	Hong Kong	Belgium
	Malaysia	Germany

Fig. 12.6 Predominance of REIT management structure

12.5 Internationalization of REIT Concept

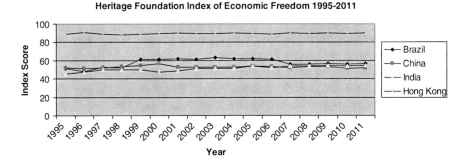

Fig. 12.7 Emerging market property rights scores

difference in transparency abroad, especially in developing countries. These issues not withstanding, international property diversification has significant advantages especially as developing countries improve property rights legislation and enforcement in the future. Some foreign markets, such as Hong Kong and Singapore, have strengthened property rights enforcement to equal that of western economies, while other emerging markets have still a road to travel.

If REIT investment managers are to continue to expand their portfolios into second tier cities of developing countries, property rights and transparency must be improved. The Heritage Foundation (2011) specifically ranks property rights as part of their composite index for economic freedom. For 2011, China received a score of 20, while India and Brazil received scores of 50. A score of 20 indicates that private property is weakly protected, with property rights very difficult to enforce. A score of 50 indicates that the court system is inefficient and is subject to delays. Thus, property rights enforcement still has a long way to go in these developing economies. Hong Kong has traditionally scored much higher that the other three nations ranked in Fig. 12.7.

Hong Kong has seen strong levels of foreign direct investment and REIT IPO activity owing to their market friendly approach to doing business. As was noted in Sect. 1.5, international real estate investment is an exciting, relatively new, branch of business study. Given the increasing integration among international markets, the geographic reach of real estate has been extended, weakening the nexus between "local" and "location" (LeComte et al. 2010). In Sect. 10.3, we discussed the impact of the outsourcing of services to foreign markets and the potential impact on the domestic property operating performance. Focusing on the metropolitan statistical area (MSA) economic base analysis may not be sufficient in the future. Issues such as differential wage rates among competing areas of the globe, the percentage of space in a given property which is utilized for import or export activity, the percentage of foreign tenants in a given property, and the percentage of foreign direct investment in a given property may allow an investor to assess the impact of globalization on their investment portfolios. The authors expect much progress and interest in this area of study in years to come.

12.6 The Sendoff!

Our motivation for writing this book stem from the comments of our students over the years. Existing real estate finance books are either too short with a lack of relevant information or too long and loaded with academic jargon. We have attempted to keep the length of this book relatively short without sacrificing important concepts that students of investment real estate must master in order to be successful students as well as investors. Our goal is that students utilize the concepts discussed in this book as a basis for forming their own personal investment philosophy.

Exposure to the investment real estate valuation techniques covered in this book should help in framing a sustainable investment strategy for long term investment success. While such success is seldom guaranteed, students with a value orientation should find that their investment decisions are less geared toward the flavor of the month, toward speculative, or toward overly aggressive project proposals. While the real estate investment arena is prone to periods of excess, students should not interpret this as a reason not to invest. In the words of the German philosopher Friedrich Nietzsche (1995), *"one repays a teacher badly if one always remains a pupil"*. Now that the "Yes era" is behind us, investors with a value orientation may find that happy days are here again. Go forth and invest in real estate!

Questions for Discussion

1. Describe the characteristics which make up a REIT. Differentiate the UPREIT and the Down REIT in your answer.
2. Describe the various forms of REITs. Why do mortgage REITs experience higher return volatility than do their equity counterparts?
3. Elaborate on the benefits and weaknesses of internal and external management of REITs and REIT mutual funds.
4. Utilize the QQD framework to classify the various investment strategies for REITs.
5. Describe the three REIT share valuation techniques. Discuss the strengths and weaknesses of each approach in your answer.
6. Elaborate on the benefits and potential pitfalls of international real estate investment. What factors contribute to earnings volatility for international property and how might these factors best be averted?

Problems

1. Nelson REIT recently reported an NOI of $12 per share, and is currently paying a dividend of $9 per share. The dividend is projected to increase by 3% next year, and continue to increase by 4% each year thereafter. Assuming

a blended cap rate of 9% and a required rate of return of 11%, what share value would the Gordon Dividend Discount Model provide?
2. Using the same information as in number 1 above, what would the net asset value be for the Nelson REIT?
3. A REIT with 1,000 shares outstanding earns $15,000 in gross revenue and incurs operating expenses of $6,000. In addition, the REIT owns property with an historic cost of $45,000 and depreciates it over a 30 year period using straight line depreciation. What are the FFO and EPS for this REIT?
4. Considering the REIT in question 3 above, what is the least dividend payment per share that it must make to maintain its tax exempt status?
5. Consider the following financial information for the Partridge Family REIT:

Net revenue	$50,000,000
Less:	
Operating expenses	$32,000,000
Depreciation/amortization	$11,000,000
Income from operations	$7,000,000
Less:	
Interest expense	$2,500,000
Net income	$4,500,000

Price multiples for comparable REITs are about ten times the current FFO. What price does this suggest for Partridge Family REIT shares if 1,000,000 shares were issued?

References

Ali, Z., McGreal, S., Adair, A., Webb, J. R., & Roulac, S. E. (2008). Corporate real estate strategies and financial performance of companies. *Journal of Property Research, 25*(3), 241–267. Routledge, Taylor & Francis Group.
Block, R. L. (2002). *Investing in REITs* (2nd ed.). Princeton: Bloomberg Press.
Bogle, J. C. (1999). *Common sense on mutual funds*. New York/Chichester/Weinheim/Brisbane/Singapore/Toronto: John Wiley & Sons, Inc. p. 92.
Bogle, J. C. (2005). *The battle for the soul of capitalism*. New Haven/London: Yale University Press.
Brueggeman, W. B., & Fisher, J. (2010). *Real estate finance & investments* (14th ed.). New York: McGraw-Hill.
Chan, S., Erickson, J., & Wang, K. (2003). *Real estate investment trusts: Structure, performance, & investment opportunities*. Oxford/New York: Oxford University Press.
Financial Accounting Standards Board (FASB) (2011). FASB 13 website. http://www.fasb.org/pdf/fas13.pdf. Accessed June 28, 2011.
Heritage Foundation (2011). Index of economic freedom. Retrieved June 28, 2011, from http://www.heritage.org/index/Ranking
Kolb, R. W. (2011). *The financial crisis of our time*. Oxford/New York: Oxford University Press.
Lecomte, P., Whitaker, W., & McIntosh, W. (2010). Butterfly spotting: An industrial real estate perspective on globalization and transnational economies. *Journal of Real Estate Literature, 18*(1), 55–75.
Lynn, D. J., & Wang, T. (2010). *Emerging market real estate investment: Investing in China, India, and Brazil*. Hoboken: John Wiley & Sons.

Maurer, R., Reiner, F., & Rogalle, R. (2004). Return and risk of German open-end real estate funds. *Journal of Property Research, 21*(3), 209–233. Spon Press, Taylor & Francis Group.

NAREIT (2010). The investor's guide to REITs. National Association of Real Estate Investment Trusts website, www.reit.com/AboutREITs/~/media/.../2011_InvestorsGuideToREITs.ashx. Accessed June 15, 2011.

NAREIT (2011). REIT watch, a monthly statistical report on the Real Estate Investment Trust Industry, May 2011. Accessed on June 14, 2011 via website: http://www.reit.com/IndustryData-Performance/NAREITStatisticalPublications/REITWatch.aspx

NAREIT (2012). REIT watch, a monthly statistical report on the real estate investment trust industry, February 2012. Accessed on February 11, 2012, http://www.reit.com/IndustryData/FNUS-DailyReturns.aspx

Newell, G., & Osmadi, A. (2009). The development and preliminary performance analysis of Islamic REITs in Malaysia. *Journal of Property Research, 26*(4), 329–347. Routledge, Taylor & Francis Group.

Nietzsche, F. (1995). Thus Spoke Zarathustra. Modern Library. New York: Random House, p. 48.

Suarez, J. L. (2009). *European real estate markets*. New York/Hampshire: Palgrave Macmillan.

Visser, H. (2010). *Islamic finance: Principles and practice*. Cheltenham/Northampton: Edward Elgar.

Glossary

95% rule 1031 exchange requirement where 95% of the aggregate fair market value of properties indentified for replacement properties are actually selected.

200% rule 1031 exchange requirement where the aggregate fair market value of replacement properties cannot exceed 200% of the fair market value of exchanged properties.

Absorption When a property is leased or purchased after construction or renovation.

Absentee management Manager of a given property is located far from the physical location of the property.

Abstract of title Legal opinion disclosing prior sales or prior claims on the subject property by other individuals, the present seller, or other various market participants.

Acquired property clause Loan documentation requirement that anything after the documents are executed that becomes part of the real estate is included in the collateral securing the bank loan.

Active income Government classification of income where the investor actively participates in its creation.

Adjustable rate mortgages (ARM) Mortgage loan that allows the interest rate to be changed at specific intervals over the maturity of the loan.

Adjusted funds from operations (AFFO) A financial performance measure primarily used in the analysis of REITs. AFFO adjusts for expenses, which while capitalized, do not enhance property value and further adjusts by eliminating the straight-lining of rents.

Adverse possession Seller does not have sufficient proof that they own the property being sold.

Adjusted basis Value used as a starting point in computing depreciation or gain on the disposition of fixed assets tax purposes. This is similar to the concept of book value.

Agglomeration Tendency of one type of industrial property being concentrated in one place. This is also known as clustering tendency.

Air rights Right to use, control, or occupy the space above a designated property.

Amenities Amenities can include specific features in a particular unit, such as vaulted ceilings and fireplaces, or could consist of things offered in the complex as a whole, such as tennis courts, swimming pools, or laundry facilities.

Americans with disabilities act (ADA) US law enacted in 1990 requiring handicap accessible buildings and parking spaces.

Anchor ratio This is the percentage of the center's total square footage that is attributable to the anchor tenants.

Anchor tenant Tenant in an investment property which serves as the primary draw for customers. The anchor tenant occupies the largest square foot in a given property, and the leases for in-line tenants may contain domino clauses.

Annual depreciation The depreciable basis of the property divided by the cost recovery period.

Apartment property Investment property that contains at least five third party residential tenants. This is also known as multi-family property.

Approved use The approved use allows the landlord to restrict the use of the tenant during the lease term.

Arbitrage investing Form of investing that is motivated by the desire to capitalize on differences in prices paid in different geographic areas for similar investment properties by purchasing a large property in the private market.

Arm's length transaction Where the buyer and seller are not related.

Assignment Method by which a right or contract is transferred from one party to another.

Attornment Tenant's formal agreement to be a tenant of a new landlord.

Balloon payment Final payment on a loan, when that payment is greater than the preceding installment payments and pays the loan in full.

Band of investment Income property appraisal technique where the overall capitalization rate is derived from weighting mortgage and equity rates.

Bank for international settlements (BIS) An international organization for cooperation among central banks, with headquarters in Basel, Switzerland.

Basel accord International banking supervision accords administered by the BIS.

Basis points One one-hundredth of one percentage point.

Blue chip property investing Investing in the best quality properties with well-known tenants in well-known locations, and with a strong historical earnings record.

Blue chip REIT REIT with a long track record of value creation, strong management, established access to capital in order to fund growth, and a high level of insider stock ownership.

Blue sky laws State legislated restrictions on the sale of corporate securities through investment companies. These laws are usually enacted to protect citizen-investors from investing in fraudulent firms. Most blue sky laws require the registration of new issues of securities with a state agency that reviews the selling documents for accuracy and completeness.

Boot Additional proceeds used to balance the equities in a 1031 exchange; an old English term meaning "an addition".

Breakage fee Fee required in terminating a fixed rate loan or interest rate swap.

Break even analysis Sensitivity analysis to determine the required interest rate or vacancy factor required in order for the net operating income to equal the annual debt service.

Break even interest rate The maximum interest rate paid on debt before leverage becomes negative is known as the break even interest rate.

Break-even occupancy rate Where the gross revenue is equal to the sum of the annual operating expenses and annual debt service requirements.

Business parks Master planned developments which encompass a group of predominantly industrial buildings on large acreage tracts.

Build-to-suit Properties constructed for a particular use or tenant.

Bulk warehouses Commercial property whose primary use is the storage of goods and wares.

Business risk Risk of a general decline in operating performance given economic conditions.

Cap rate The ratio of projected net income relative to sales price.

Capital call Lender required principal curtailment by investor's in a given property.

Capitalization effects Increased property value owing to proximity to certain favorable factors.

Cash collateral account An account created to offset losses to the collateral that threaten coupon payments to asset-backed security investors.

C-corporation Corporate structure reserved for larger corporations. It does not have a shareholder limitation requirement and taxation occurs at both the corporate and personal level.

CDO Collateralized Debt Obligation.

Certificate of occupancy (C/O) A document issued by a local government to a developer permitting the structure to be occupied by members of the public.

Chattel Personal property; anything owned and tangible other than real estate.

CMBS Commercial Mortgage Backed Securities.

Coefficient of variation Statistical measure of the dispersion of data points in a data series around the mean.

Collateralized mortgage obligation (CMO) A mortgage-backed security that generates separate cash flows for different classes of securities called tranches. Tranches have varying maturities and prepayment risks and offer varying expected returns for investors.

Collection loss Reduction in expenses due to trouble collecting rent, or possibly due to lease rate concessions during the year.

Commencement date When a commercial lease takes effect.

Common area maintenance (CAM) An agreement for tenants to reimburse the landlord for their pro-rata share of certain expenses.

Community centers Retail shopping centers which provide general merchandise, encompassing from 100,000 to 350,000 square feet of GLA.

Competitive buildings Those that offer space to the public for lease.

Concessions Benefits granted by a seller or lessor to induce a sale or lease.

Condominium A system of ownership of individual units in a multi-unit structure, combined with joint ownership of the commonly used property.

Consideration A legal classification involving something of value being exchanged by parties to an agreement, thus making the agreement legally binding.

Constant Payment mortgage (CPM) Another name for a level payment, amortizing mortgage. The payment made by the borrower remains constant, while the portion representing principal repayment increases as the interest portion declines throughout the life of the loan.

Constant amortization mortgage (CAM) Commercial mortgage payment structure also known as "straight line" repayment as the client is paying the same amount of principal each month, with interest varying depending on the outstanding balance.

Consumer price index (CPI) A typical benchmark for lease rate increases, with the goal of the lease payment received at least keeping up with expected price increases over the term of the lease.

Contrarian investing Investing in properties that are currently out of favor.

Convenience center Also known as a strip center; it may be configured in a straight line or it may have an "L" or "U" shape.

Conventional home loan A home loan that is eligible for purchase by FNMA and Freddie Mac, but are not insured by the Federal Housing Authority (FHA) or Veterans Administration (VA).

Convexity Non linearities in a financial model; the relationship of bond price with respect to interest rates.

Corporate veil Legal decision to treat the rights or duties of a corporation as the rights or liabilities of its shareholders or directors.

Correlation Statistical measure of how two securities move in relation to each other.

Cost approach A method of appraising property based on the depreciated reproduction or replacement cost (new) of improvements, plus the market value of the site.

Counter party A party to a contract. In an interest rate swap, it may be an investor seeking to profit given movement of interest rates relative to the contract rate of interest.

Covariance A measure of the degree to which returns on two risky assets move in tandem. A positive covariance means that asset returns move together. A negative covariance means that asset returns move inversely.

Credit enhancement Techniques used to improve the creditworthiness of a borrower.

Credit wrap A form of financial guarantee that does not cover all of the debts of the borrower, but a specific loan, debt issuance, or other financial transaction.

Curtesy rights The right of a husband to all or part of his deceased wife's realty regardless of the provisions of her will.

Debt coverage ratio The relationship between net operating income (NOI) and annual debt service (ADS). The formula is NOI divided by ADS.

Deed A written instrument used to convey the title of real estate from one entity to another.

Deed of trust Means of conveyance where the property is transferred to a trustee (neutral party) by a borrower (trustor) in favor of a lender (beneficiary) and is re-conveyed upon payment in full.

Default provision Any particular event that can trigger the balance of the loan to be due to the lender in its entirety prior to the stated maturity date.

Default risk The risk that the loan will not be repaid.

Defeasance Process in commercial real estate finance whereby a borrower substitutes other income producing property (typically US Treasury obligations) for real estate collateral to facilitate the removal (defeat) of an existing lien without paying out the existing note.

Deglomeration Factors that detract from an otherwise favorable location.

Delayed or deferred exchange 1031 exchange where transactions are not consummated on the same day.

Derivative An asset whose value is derived, or based on, another asset.

Discount rate Required rate of return (or rate of interest) for the investor taking into account opportunity costs, inflation, and the certainty of payment.

Discounted cash flow (DCF) model Income approach valuation methodology utilizing a multi-year view of the revenues and the expenses for a property which are received during the defined holding period of the investment.

Disintermediation In general it means to remove the middleman. In banking and finance, it refers to investors removing their funds from banks and other financial institutions to pursue higher yielding opportunities.

Distribution warehouses Storage buildings designed to promote the logistical movement of goods.

Domino clause Provision in a retail lease whereby if the anchor tenant vacates the subject property, the other tenants have a provision to break their leases. This is also known as a "Go Dark" or "Co-tenancy" clause.

Dower rights The legal right of a wife or child to part of a deceased husband's or father's property.

Down REIT Similar to UPREITs but are formed after the REIT goes public. The Down REIT directly owns the properties directly in a REIT structure, but it holds some of the properties in a partnership having other partners.

Due on sale clause Provision in a mortgage that states the loan is due upon the sale of the property.

Duplexes Two dwelling units under one roof.

Efficiency (or studio) apartment A small dwelling unit, often consisting of a single room, within a multi-family structure.

Effective age The effective age of a property is the remaining economic useful life of the property given the quality of construction and improvements made to the property since it was originally constructed.

Effective tax rate Tax divided by taxable income. Net income becomes taxable income via reductions for interest expense, depreciation, and any other allowable deductions.

Effective gross income (EGI) Stabilized income a property is expected to generate after a vacancy allowance.

Eminent domain The right of a government or a public utility to acquire property for necessary public use by condemnation. The owner must be fairly compensated.

Easement Is a non-possessory interest in land by an entity other than the owner of the property.
Economic base Industry within a geographic market area that provides employment opportunities which are essential to support the community.
Employment multiplier Determined by dividing total employment by the base employment in a given market.
Environmental risk Concerns the specific property and its effect on land, water, air, sewage, and aesthetics of the surrounding area and of the community at large.
Escalator lease Is typically structured where the lessor pays expenses for the first year, and the lessee would pay any increase in expenses over the established level as additional rent over the subsequent years of the lease.
Escape clauses Whereby a tenant can end their lease prior to the stated lease expiration, as long as they notify the landlord by a specified date.
Escrow account Account where a portion of the net cash flow from the investment property is retained in case an unforeseen expense item materializes.
Estate at sufferance The same tenant occupies the property without the owner's knowledge.
Estate for years Lease with an exact duration of tenancy.
Estate from year to year This form of estate involves periodic tenancy, where the tenant can remain in the property until the tenant provides a notice of termination of tenancy.
Estate at will Tenant occupies a property with the knowledge of the owner, but without an official lease.
Estoppel certificate Document that certifies that all of the parties to the lease (lender, borrower, and tenant) understand and accept the bank's position as mortgagor and subsequent landlord in the event of foreclosure.
Excess interest/spread or profit An internal credit enhancement technique where remaining net interest payments from the underlying assets of an asset-backed security (after all payables and expenses are covered) are deposited into an account to provide assurance of performance of senior securities.
Expansion phase Phase of real estate cycle that exhibits increased new construction and declining vacancy.
Expense stop Lease agreement provision whereby a property owner agrees to pay operating expenses up to a certain point; beyond that is paid by the tenant.
External credit enhancements Techniques used to enhance a borrower's, or securitization structures creditworthiness that involves third party guarantees such as a Letter of Credit or Monoline Insurance.
Fannie mae Federal National Mortgage Association (FANNIE MAE). A government-sponsored entity created by an act of Congress in 1938 to purchase residential mortgage loans. Fannie Mae does not receive an explicit government subsidy or an appropriation, and the stock of this quasi-private corporation trades on the New York Stock Exchange.
Fashion or specialty centers Retail properties which cater to higher end and often fashion-oriented tenants and anchors.
Federal funds rate Rate charged in the interbank lending market.

Fee simple Absolute ownership of real property; owner is entitled to the entire property with unconditional power of disposition.

FDICIA Federal Deposit Insurance Corporation Improvement Act of 1991 was passed during the savings and loan crisis and allows the FDIC to assess insurance premiums according to risk.

Financial risk Risk associated with debt financing and leverage.

Finite life REITs REITs that have a stated duration when the various investments will be sold.

FIRREA The Financial Institutions Reform, Recovery and Enforcement Act (FIRREA) of 1989 dramatically changed the savings and loan industry and dealt with ensuring that real estate appraisals met certain standards.

Fixed rate A traditional fixed rate loan eliminates the upside risk of interest rate movements during the term of the loan for an investor.

Flex space Generally contains glass on three of the outer walls and has additional parking relative to other warehouse properties.

Flood zone A level land area subject to periodic flooding from a contiguous body of water; also known as a flood plain.

Flood way Designated river channel and adjacent land which is built near waterways that experience periodic flooding. A floodway is designed to absorb and safely contain excess flood water.

Forbearance period Period of time when a Lender's decision not to exercise a legally enforceable right against a borrower in default, in exchange for a promise to make regular payments in the future; a temporary relief period for a borrower in default.

Foreclosure Termination of all rights of the mortgagor or the grantee in the property covered by the mortgage.

Four food groups of investment property Defined as: 1) Apartments 2) Retail 3) Office 4) Industrial and Warehouse.

Fractured ownership Numerous owners for the units under one roof. See condominium.

Freddie mac Federal Home Loan Mortgage Corporation (FHLMC). A government-sponsored entity created by an act of Congress in 1970 to purchase conventional, residential mortgage loans. Freddie Mac does not receive an explicit government subsidy or an appropriate, and the stock of this quasi-private corporation trades on the New York Stock Exchange.

Freehold estate This is the most complete form of ownership, as the owner can sell, lease, divide up, or otherwise exploit the property as they see fit, as long as the intended act is legal.

Frictional vacancy The small amount of vacancy that exists in even the best markets given the movement in and out of tenants in investment properties.

Friendly foreclosure When the borrower agrees to cooperate with the lender to achieve a quick foreclosure.

Future estate The property rights are not conveyed until some date in the future.

Functional obsolescence Loss of value from all causes within the property, except those due to physical deterioration. It involves weaknesses in design due to technology, style or current tastes and preferences.

Funds from operations (FFO) Net income excluding gains or losses from sales of property or debt restructuring, and adding back real estate depreciation.

Garden apartments A housing complex whereby some or all tenants have access to a lawn area.

General warranty deed Deed in which the grantor agrees to protect the grantee against any other claim to title of the property; it is the most complete form of title assurance.

Glass-steagall act 1933 US legislation that established the FDIC and separated commercial and investment banking activities. The act was repealed in 1999.

Gordon dividend growth model Method of calculation of the intrinsic value of a share price based on a future series of dividends that grows at a constant rate.

Graduated payment mortgage (GPM) Another common repayment structure. While these types of mortgages are more common in residential real estate, the repayment structure can be applied in a commercial situation.

Grantee The party to whom the title to real property is conveyed; the buyer.

Grantor Anyone who gives a deed.

Gross income multiplier (GIM) A blend of the sales and income approaches, and is typically utilized in appraisals for single family residences or for generally smaller properties with stable income streams.

Graduated rental lease This lease form allows for the increasing or decreasing of rent or reimbursed expenses over the term of the lease.

Gross building area An aggregate measure of the building according to exterior dimensions.

Gross lease A lease of property whereby the landlord (lessor) is responsible for paying all property expenses, such as taxes, insurance, utilities, and repairs.

Gross potential income (GPI) The total income that the investment property can produce based on full occupancy at market rental rates.

Gross rent multiplier (GRM) Same as the gross income multiplier.

Growth strategy Investment strategy geared toward long term asset value appreciation.

Ground lease Lease agreement concerning renting the land only.

High-rise Generally a building that exceeds six stories in height and is equipped with elevators.

Highest and best use Appraisal term meaning the legally and physically possible use that, at the time of appraisal, is most likely to produce the greatest net return to the land and or buildings over a given period.

Holding period of the investment The time span of ownership.

Homeowners association (HOA) Organization of the homeowners in a particular subdivision, or condominium; generally for the purpose of enforcing deed restrictions or managing the common elements of the development.

Hybrid REIT REIT that invests in both equity and debt instruments.

Hyper-supply phase Phase of the real estate cycle commencing after the peak or equilibrium point.

Improvement exchange Type of 1031 exchange where the investor desires to acquire a property and arrange for the construction of improvements on the property before it is received as the replacement property.

Income approach A method of appraising real estate based on the property's anticipated future income.

Income-producing property Real Estate held as investment where the tenants have different ownership than the owner of the property.

Incubator REIT REITs funded by venture capitalists with the plan for an eventual initial public offering.

Industrial density The projected area required per worker.

Inflation A loss in purchasing power of money; an increase in the general price level.

Inflation risk When the income increase experienced during the investment holding period does not keep pace with overall price level or operating expense increases.

In-line tenants Supporting players relative to the anchor tenants.

Industrial property Property used for industrial purposes, such as manufacturing, warehousing, and storage.

Interest rate swap Contract in which two counter-parties agree to exchange interest payments of differing character based on an underlying notional principal amount that is never exchanged.

Interbank lending market The market in which banks extend short-term loans to one another. Most interbank loans are for a term that is one week or less. Most loans do not extend past one night.

Interest-only (IO) strips IO investors receive only the interest payments from a fixed income security. IOs are sold at a deep discount to a notional principal, which is the amount used to calculate the interest paid to investors, but they have no par or face value. If interest rates fall and the prepayment rate increases, it is possible that IO investors could receive less cash than they pay.

Internal credit enhancements Techniques used to enhance securitization structure's creditworthiness that involves injections of capital by the securities' issuer or originator.

Internal rate of return (IRR) The true annual rate of earnings on an investment. It is the rate that equates the present value of the cash inflows with the present value of the cash outflows.

Internal rate of return on equity (IRR_e) The time adjusted yield that represents an expected rate of return to the equity investor.

Investment holding period The time span that the investor owns the property. The holding period begins once the property is purchased and ends once the property is sold. The length of the holding period will vary according to the investment objectives and investment horizon of the investor.

Investment horizon The going-in assumption as to how long a property will be owned after purchase.

Junior mortgage A second lien of indebtedness on a property.

Kyoto protocol International agreement to reduce the collective emissions of greenhouse gases to achieve cleaner air in all parts of the world.

Lease A contract in which, for a payment called rent, the one entitled to the possession of real property transfers those rights to another for a specified period of time.

Leasehold estate Lessee (tenant) of real estate has a lease.

Leasehold improvements Fixtures attached to real estate that are generally acquired or installed by the tenant.

Leased fee Landlord's ownership interest of a property that is under a lease.

Leasing commissions Commissions paid by investor to a broker in order to obtain a tenant in an investment property.

Lease rollover risk Risk that tenants will not renew their leases.

Legal description Legally acceptable identification of real estate.

Legislative risk This particular risk is defined as a change in the regulatory environment which raises costs or makes certain types of lending unattractive for lenders.

Lesse Tenant.

Lessor Owner.

Life estate A freehold interest in real property that expires upon the death of the owner or some other specified person.

Lifestyle center Shopping center or mixed-used commercial development that combines the traditional retail functions of a shopping mall with leisure amenities oriented towards upscale consumers.

Like kind exchange Tax deferred exchange of properties held for investment or business purpose; also known as 1031 Exchange.

Liquidity provider A bank, insurance company or corporation that makes short-term, temporary payments when glitches in the system prevent the timely dispersal of funds from the servicer.

Liquidity risk This risk is associated with the ease of converting assets to cash.

Line of credit (LOC) A loan that can be accessed at the borrower's discretion. The borrower may take any amount up to a maximum that is established by the bank. Typically interest is paid only on the outstanding balance, although annual fees to maintain the line are common.

Load factor Percentage difference between the useable versus rentable square footage in a building.

Loan points Amount paid to the lender at the time of the origination of the loan.

Lockout period The time frame during which a CMO tranche receives only interest payments.

Make whole provision A provision that allows a borrower to prepay the remaining fixed rate term debt. However, the borrower has to make an additional payment that is derived from a formula based on the net present value of the future debt payments. The bank is entitled to obtain this additional fee which allows the bank to recoup the costs associated with the early payout.

Management risk Risk of a variance from expected returns given inadequate administration of the property.

Marginal tax rate The rate paid on the last dollar of taxable income. This is typically viewed as the tax bracket of the investor including all sources of income.

Maturity date Date on which the principal balance of a loan is due and payable to the holder.

Market comparison approach See Sales Approach.

Market equilibrium The currently experienced level of occupancy, lease rates, and property expenses per property type in the region.

Market feasibility study Research to prove whether a property investment is viable.

Market timing strategy Investment strategy that seeks to time the trough of a business cycle, in order to profit by investing at the bottom point.

Mark-to-market or fair value accounting An accounting system that requires recording an account to reflect its current market value rather than its historical book value. Difficulties can occur when the market-based measurement does not accurately reflect the underlying asset's true value, such as when a company is forced to determine the selling price of its assets or liabilities during an unfavorable or volatile period.

Mechanics lien A lien given by law upon a building or other improvement upon land, and upon the land itself, as security for the payment for labor done and materials furnished for improvement.

Medical office buildings Office buildings primarily or exclusively leased to tenants engaged in healthcare services.

Mid-rise apartments Apartments typically consisting of from six to nine stories in height.

Millage rate The rate of tax assessment per dollar of value. Each mill represents $1 of tax assessment per $1,000 of assessed property value.

Mineral rights Subsurface rights, pertaining to the landlord's rights regarding mineral, oil, and gas deposits on the property.

Mini-storage A building separated into relatively small lockable individual units, typically with a garage-door styled opening, that provides storage. Also known as Mini-warehouse and self-storage.

Modified internal rate of return (MIRR) The MIRR assumes a reinvestment rate equal to the required rate of return, or some other realistic rate rather than at the IRR.

Modified gross lease Lease agreement whereby the expenses are shared between the lessor and the lessee.

Money market A segment of the financial market in which securities with high liquidity and short maturities are traded. The money market is used for borrowing and lending in the short term, defined as one year or less.

Money multiplier effect Relationship between the monetary base and the money supply. The multiplier explains the money supply has expanded through the banking system by distribution of excess reserves.

Monoline insurance An insurance company that specializes in providing guarantees to issuers that enhance the credit of the issuer.

Moral hazard The risk that the party to a contract has an incentive to behave in a manner is detrimental to a counterparty.

Mortgage A written instrument that creates a lien upon real estate as security for the payment of a specific debt. The purpose of this document is to provide evidence that the investor (borrower) has pledged real property (the investment being purchased) to another party (the lender) as security for the loan.

Mortgage REIT REITs that own mortgage obligations rather than equities.

Mortgagee One who holds a lien on a property or title to property, as security for a debt; the lender.

Mortgagor One who pledges property as security for a loan; the borrower.

MSA Metropolitan Statistical Area

Mutual fund A group of related investment companies owned by their shareholders and governed by their directors.

Negative absorption When the market occupancy rate falls over a given time period.

Negative amortization An increase in the outstanding balance of a loan resulting from the failure of period debt service payments to cover required interest charged on the loan.

Negative leverage When the cost of debt is higher than the expected rate of return on the total funds invested.

Neighborhood center Retail properties typically catering to convenience, and typically having an anchor tenant, usually a supermarket, as well as other nationally or regionally known tenants.

Net asset value (NAV) A method for valuing the intrinsic value of a REIT stock. The NAV is calculated by aggregating the stabilized net operating income for the entire company, dividing by an appropriate cap rate for the company's assets, and subtracting total company indebtedness.

Net lease Where the tenant pays such expenses as taxes, insurance, and maintenance.

Net operating income (NOI) Income from property or business, after operating expenses have been deducted, but before deducting income taxes and financing expenses.

Net present value A method of determining whether expected performance of a proposed investment promises to be adequate. It is a technique that discounts the expected future cash flows at the minimum required rate of return.

Net rentable area The total area available for occupancy by tenants excluding stairwells, common areas, and mechanical areas.

Net sales proceeds Sales price minus selling costs.

NOIe Pre-tax annual cash flow to the investor. NOI after debt service has been subtracted.

Non-compete clause Where one party pledges not to pursue a similar trade or profession in competition against another party. Typically requested by tenant so that competing businesses are not located within a certain radius of a given property.

Non-dilution/radius clause Lease provision whereby the tenant is restricted from leasing another location within a certain radius.

Non-disturbance An agreement in mortgage contracts that provides for the continuation of leases in the event of loan foreclosure.

Non-recourse financing Where no personal liability is available to the lender for a loan.

Note A document which serves as evidence of debt and a promise to pay between a borrower and the lender.

Notional principal For an Interest Only Strip, it is the predetermined dollar amount on which the interest payments are based.

Off-campus student housing Multi-family housing property catering to college students located in proximity to college campuses.

Off shoring Repositioning of operations in global markets.

Office warehouses Warehouse buildings with a percentage of the property designed as office space.

Open end mortgage Clause in a mortgage allowing mortgaged property to be used for additional advances on the same note, up to a preset amount. Subsequent advances are dated back to the recording of the original mortgage deed.

Open market operations Purchase or sale of government securities by the Open Market desk at the Federal Reserve Bank of New York, as directed by the Federal Open Market Committee.

Opportunity costs Present value of the income that could be earned (or saved) by investing in the most attractive alternative to the one being considered.

Opportunistic investing Investment strategy characterized by targeting under-performing and/or under-managed properties, or properties that are temporarily depressed, and then using high degrees of leverage (borrowed funds) to acquire the property, hold it for a short period of time, and then sell it at an expected profit.

Originate to distribute Contributed to the financial crisis of 2007/8 because originators could immediately sell (i.e., distribute) their conforming mortgages to one of the government housing agencies or a private conduit, they were less concerned with credit quality than volume.

Outlet centers Retail properties catering to manufacturer's outlet stores.

Outparcel Retail pads which are typically free-standing buildings with one to a few tenants located in the front of the primary property.

Outsourcing Engaging a third party company to operate and administer a specific set of business processes.

Overage rent See percentage rent.

Overcollateralization An internal credit enhancement technique in which the originator transfers a pool of collateral loans to the SPV that has a higher par value (usually 5 to 10%) than that of the issued securities. This means that the SPV holds a larger pool of assets than would be necessary if the loans in the pool pay as expected.

Passive income Any income or loss where the investor does not actively participate in the management of the company or activity.

Percentage rent Rent payable under a percentage lease where the percentage applies to sales in excess of a pre-established base amount of the dollar sales volume; is also known as overage rent.

Phase one environmental report This report is prepared by a certified expert in environmental risk assessment.

Phase two environmental report This environmental risk assessment includes soil testing, as well as the testing of groundwater and air, if necessary.

Physical depreciation Includes any deficiencies in structural quality or issues associated with the aesthetic appeal of the property.

Planned amortization class (PAC) A class of tranche or bond within a CMO that receives a primary payment schedule as long as the actual prepayment rate remains within a pre-specified range of prepayment speeds.

Portfolio income This includes interest and dividends where the investor does not control majority ownership in the property or company.

Portable on-demand storage (PODS) Transportable self-storage units.

Portfolio lender A financial institution that holds a portfolio of loans and earns profit from origination and the difference (spread) between the interest they charge on the loans and the interest they pay on deposits.

Positive leverage Is achieved when the investor is borrowing at an interest rate lower than the expected rate of return on the total funds invested in a property.

Power centers Retail properties that typically have three or more anchors.

Public storage A facility that includes an operating business that provides warehousing storage and service for multiple customers.

Prepayment risk Risk associated with the desired rate of return being impacted by the early repayment of loan principal.

Principal-only (PO) strips PO Investors receive only the principal payments from a fixed income security and, accordingly, buy the securities at a deep discount from the face value, which is repaid through scheduled payments and potentially unscheduled prepayments. They are created from mortgage loans by separating or stripping the principal payments from the interest payments.

Primary market Also called the "new issue market," it is the market in which securities or assets can be purchased for the first time. The primary market is where companies raise money to expand their operations. Once the asset has been purchased, subsequent transactions occur in the secondary market.

Private-label securities A mortgage-backed security or other bond created and sold by a company other than a Government Sponsored Enterprise (GSE). The security frequently is collateralized by loans that are ineligible for purchase by Freddie Mac or Fannie Mae.

Professional building Office building leased primarily or exclusively to professionals.

Profitability index The present value of the cash flows divided by the investment cost.

Pro-forma statement Projection of the income and expense statement for a property for the current year.

Property rights The rights to control, occupy, develop, improve, exploit, pledge, lease, or sell a given property.

Property sector investment Strategy involving investing in one particular type of property (i.e. multi-family units).

Pro-rata guarantees Form of guarantee is considered a conditional guarantee as the individual is only obligated on the loan balance up to the percentage of ownership.

Primary trading area The measurement for the potential revenue of the tenants in a given property from the geographic boundary where sixty to eighty percent of the sales in a given property originate.

Qualified intermediary (QI) A qualified intermediary is a person or entity who is not the tax payer or a disqualified party such as an attorney, accountant, realtor, or related party, and who enters into a written exchange agreement with the taxpayer desiring to facilitate the 1031 exchange.

QQD framework Framework which includes analysis of the quantity, quality, and durability of the income stream for an investment property.

Quit claim deed A deed that conveys only the grantor's rights or interest in real estate, without stating the nature of the rights and with no warranties of ownership.

Real estate investment trust (REIT) Specially designated corporations whose primary aim is investment in income-producing properties.

Real estate mortgage investment conduit (REMIC) A type of special purpose vehicle, created by the Tax Reform Act of 1986, that holds commercial and residential mortgages in trust and issues interests in these mortgages in the form of securities to investors.

Real rate of interest The minimum rate where savers will agree to forego current consumption and to save.

Recession phase The phase of the real estate cycle where rental rates drop, vacancy rates rise, and new construction ceases.

Recovery phase The phase of the real estate cycle where vacancy rates fall and new construction begins.

Regional center A type of shopping center containing 300,000 to 900,000 square feet of shopping space and at least one major department store.

Revaluation lease Lease structure allowing for periodic rent adjustments based on the revaluation of the real estate, at prevailing market conditions, during specific intervals during the lease term.

Reverse exchange In this type of 1031 exchange, the replacement property is purchased and closed before the relinquished property is sold.

Replacement cost The cost of erecting a building to replace or serve the functions of a previous structure.

Replacement reserve An amount set aside from net operating income to pay for the eventual wearing out of short-lived assets.

Reproduction cost The cost of exact duplication of a property as of a certain date.

Renewal option The right, but not the obligation, of a tenant to continue a lease at a specified term and rent.

Rent roll Document which lists all of the current tenants in the subject property, the amount of square footage for each tenant, the annual lease rate paid per square foot for each tenant, and the expiration dates of the current leases.

Remainder An estate that takes effect after the termination of a prior estate.

Reserve requirements Portion of their deposits banks and savings institutions are required to maintain as legal reserves. A method of controlling the money supply and interest rates is via reserve requirements.

Reversionary interest The interest a person has in property upon the termination of the preceding estate.

Right to cure Period of time that a borrower has to correct a default to a mortgage.

Riparian rights Rights pertaining to the use of water on, under, or adjacent to one's land.

Risk absorption ratio The annualized net present value divided by the initial equity investment, where annualized NPV is the maximum amount by which cash flow each year could be reduced without reducing the NPV to below zero.

Risk of non-payment Risk involving the uncertainty of receiving payment in the future.

Sales approach One of the three appraisal approaches to value. Value is estimated by analyzing sales prices of similar properties recently sold. Also known as the market comparison approach.

S-corporation A "small corporation" whereby the number of shareholders is limited to one hundred.

Schedule A US personal tax return schedule that itemizes state and local taxes as well as itemized deductions.

Schedule B US personal tax return schedule where interest and dividend income is reported.

Schedule C US personal tax return schedule which is essentially an income statement for sole proprietorships.

Schedule E US personal tax return schedule where on the first part of the schedule E, income from rental property and royalties are reported; and on the second part pass-through net income from closely held corporations are reported.

Schedule F US personal tax return schedule where the profit and loss of a personal farming operation is reported.

Schedule K-1 US Corporate tax return schedule that is prepared for each shareholder and serves as the basis of taxation at the personal level. The financial performance of the corporation is pro-rated via the K-1 schedule.

Schedule K US Corporate tax return schedule that aggregates the net income, equity contributions, equity distributions, interest and dividends, and capital gains for all shareholders during the calendar year.

Secondary market A market in which investors buy assets, or securities, from other investors rather than the issuing entity. A secondary market is delineated by the fact that the traded assets have been sold at least once before.

Secondary mortgage market enhancement act (SMMEA) of 1984 An act passed in the US allowing federally chartered and regulated financial institutions to invest in mortgage-backed securities. The SMMEA was an overt move by the US government to increase liquidity in the residential mortgage market by increasing the number of investors who could participate in the secondary market.

Seisin and warranty Validates that the owner of the property does in fact own the title conveyed, and that there are no significant issues concerning the title of the property.

Self-storage See mini-storage.

Senior/subordinated structures A structural credit enhancement technique that induces investors to accept higher levels of risk in exchange for higher returns. Risk is layered into subordinate bonds or tranches that pay higher returns than do the senior classes with higher payment priority. Losses caused by homeowner delinquencies and defaults first are allocated to the junior classes of investors, thereby protecting the senior classes from losses.

Sensitivity analysis A technique of investment analysis whereby different values of certain key variables are tested to see how sensitive investment results are to possible change in assumptions. It is a method of evaluating the riskiness of an investment.

Sequential pay or "Plain Vanilla" CMO A CMO comprising several tranches. Tranche investors receive interest payments as long as the tranche's assigned principal amount has not been completely paid. The first tranche receives all of the initial principle payment until it is completely paid off, at which point the second tranche receives all of the principle, and so on.

Service warehouses Warehouse properties that typically contain more office space than in distribution or storage warehouses.

Simultaneous exchange 1031 exchange scenario where there is no time interval between the closings of the relinquished and replacement property.

Single family residence A type of residential structure designed to include one dwelling.

Size strategy Investment strategies focused on investing in properties of a particular price point.

Special purpose property When the use and design of a particular property makes it difficult for a tenant in a different line of business to conduct operations at the sight without significant renovation.

Special purpose vehicle (SPV) An entity created for the purposes of providing bankruptcy remoteness from the originator and issuing securities.

Special economic zones Zones that have been created near airports and other key transportation hubs to help encourage the export of products at favorable customs rates.

Special warranty deed A deed in which the grantor limits the title warranty given to the grantee to anyone claiming by, from, through, or under him, the grantor. The grantor does not warrant against title defects arising from conditions that existed prior to his ownership in the property.

Speculative investment Investment where the profit potential comes solely from the eventual resale of the property. This would be the case for investment properties where there is no tenant, or for various owner occupied properties such as single family residences. An alternative definition would be where income from the property exists, but where there is a shortfall of monthly income relative to the debt service for the loan that is secured by the property. In either case, the investor has purchased the property for reasons other than the annual cash flow produced after debt service is paid.

Standard deviation Is a measure of dispersion about the mean.

Strip center Retail property consisting of an attached row of stores or service outlets which are managed as a unified retail entity.

Structural credit enhancements Techniques used by securitizations to distribute risk among bonds so that some of them provide protection to those that enjoy a higher priority.

Storage warehouses Warehouse properties which are designed for the storage of wares, goods, and merchandise where obsolescence has occurred such that the characteristics of the building are no longer considered competitive with distribution warehouses.

Subordination clause A clause or document that permits a mortgage recorded at a later date to take priority over an existing mortgage.

Subordination Moving to a lower priority, as a lien would if it changes from a first mortgage to a second mortgage.

Subrogation The substitution of one person in the place of another with reference to a lawful claim, demand, or right so that he or she that is substituted succeeds to the rights of the other.

Substitute basis The basis in a property acquired via a 1031 exchange.

Super-regional centers Retail properties similar to regional centers, but are larger and have more variety in tenant offerings; also known as a mall.

Survey exception A typical encumbrance of a title where the acreage surrounding the subject property is measured and it is verified that a portion of the land parcel encroaches onto the land of an adjacent property.

Swap risk Risk that the borrower has to pay a breakage fee to terminate an interest rate swap.

Systematic risk The risk inherent to the entire market or market segment. Risk that exists even within a diversified portfolio.

Targeted amortization class (TAC) A class of tranche within a CMO that is protected from unexpected increases in the prepayment rate.

Tax shelter Exists whenever the depreciation expense is greater than the amortization of principal.

Tenant strategy Investment strategy involving purchasing investment properties that have specific tenants deemed desirable by the investor.

Terminal cap rate The cap rate based on the eventual sale of the investment at the end of the holding period.

Third party tenants Tenants in income producing property that have ownership other than the ownership of the real estate. These tenants could be unrelated parties in an apartment complex, or operating companies located in an office, retail, or warehouse property.

Title insurance A one time insurance payment to ensure that the policy holder is protected from loss due to title defects.

Total leasable area Calculated from outside wall to outside wall.

Townhouse A dwelling unit, generally having two or more floors and attached to other similar units via party walls.

Triple net leases (NNN) Commercial lease arrangement requiring the tenant to pay all property operating expenses.

Troubled asset relief program (TARP) A government program created to curb the ongoing financial crisis of 2007-2008. The TARP gave the U.S. Treasury purchasing power of $700 billion to buy mortgage backed securities (MBS) from financial institutions in an attempt to create liquidity and lubricate the money markets.

Umbrella partnership REIT (UPREIT) REIT established in 1992 to enable established real estate operating companies to bring properties already owned under the umbrella of a REIT structure.

Underwriter Banks, investment banks and brokers that sell the securities in a public offering or place them privately, often retaining a portion of the issuance for their own account.

Unsubordinated No subordination agreement exists. For example, when ground rent is not paid, the landowner can foreclose and terminate the leasehold rights.

Unsystematic risk Company or industry specific risk that is inherent in each investment. The amount of unsystematic risk can be reduced through appropriate diversification.

Value investment Investment where the monthly receipts from third party tenants at least equal the monthly property operating expenses and monthly debt service requirements on the mortgage.

Variable rate Non-fixed rate loans which are typically priced relative to a market index.

Voluntary conveyance Transfer of title from a delinquent property owner to a lender to satisfy the balance on a loan in default. The delinquent borrower transfers title on a voluntary basis, in order to avoid foreclosure.

Walk-up apartment Multi-family residential unit typically consist of from three to five floors.

Warehouse Industrial property that is designed and utilized for the storage of wares, goods, and merchandise.

Window The estimated time period during which a tranche receives payment of principal.

Yes era Period from 2002-2007 when low interest rates, vibrant economic growth, and a strong appetite for lending created a period of mass speculation in real estate markets and generally loose bank underwriting standards.

Zero amortization mortgage (ZAM) Interest only loan; repayment structure does not reduce the principal on the loan during the loan term.

Zoning restrictions Limitations on ownership owing to government mandated uses of a property given its location or other distinguishing characteristics.

Z-tranche The Z-tranche receives no interest until the lockout period ends and then finally begins to receive principal; the lockout period typically ends when all of the other tranches have been completely paid off. During the lockout period the tranche is credited with accrued interest that is taxable as income, even though investors receive no cash.

Index

A
Absentee management, 121
Absorbed, 17, 136, 138, 202
Abstract of title, 12
Acquired property clause, 28
Active income, 150
ADA *See* Americans with Disabilities Act (ADA)
Adjustable rate mortgages (ARM), 36, 40, 41, 124
Adjusted basis, 147, 154, 160
Adjusted funds from operations (AFFO), 265
Adverse possession, 12
AFFO *See* Adjusted funds from operations (AFFO)
Agglomeration, 208
Air rights, 128
Amenities, 70, 71, 166–168, 173, 178, 179, 188, 190
Americans with Disabilities Act (ADA), 166–168
Anchor ratio, 186, 187
Anchor tenants, 6, 99, 107, 109–111, 186, 187, 260
Annual depreciation, 142–144, 154
Apartment property, 6, 7, 168, 176
Approved use, 106
ARM *See* Adjustable rate mortgages (ARM)
Arm's length transaction, 68, 69, 89
Assignment and sublease section, 107
Attornment, 107

B
Balloon payment, 30, 224–225
Band of investments, 84–87, 89
Bank for International Settlements (BIS), 42, 123
Basel accord, 42, 123, 125
Basis points, 31, 40, 80, 82, 124, 125, 160, 239

BIS *See* Bank for International Settlements (BIS)
Blue Chip REITs, 259, 260
Blue Sky Laws, 227
Bond insurance, 229
Boot, 155–157
Breakage fee, 32, 33, 127, 130
Break even analysis, 133
Break even interest rate, 132, 137
Break-even occupancy rate, 165, 179, 180, 196, 199, 217, 221
Build-to-suit, 186
Bulk warehouses, 207, 208
Business parks, 191, 206
Business risk, 120, 261

C
CAM *See* Common area maintenance (CAM); Constant amortization mortgage (CAM)
Capital call, 32, 149
Capitalization effects, 15, 167, 208
Capitalization rate, 7, 76, 80, 86
Cap rates, 26, 50, 76, 77, 79–90, 96, 133, 146, 160, 179, 180, 196, 198–200, 202, 217, 219, 221, 263, 269
C-Corporation, 149–151
CDO *See* Collateralized debt obligations (CDO)
Certificate of occupancy (C/O), 164
Chattel, 28
CMBS *See* Commercial mortgage backed securities (CMBS)
CMO *See* Collateralized mortgage obligation (CMO)
C/O *See* Certificate of occupancy (C/O)
Coefficient of variation, 261
Collateralized debt obligations (CDO), 243–244, 256

Collateralized mortgage obligation (CMO), 233–236, 240, 246, 249
Collection loss, 76, 77, 171–172
Commencement date, 106, 112–114
Commercial mortgage backed securities (CMBS), 240–244, 249, 256
Common area maintenance (CAM), 36, 38–40, 44, 45, 100–102, 110, 111, 196
Community centers, 6, 186–187
Competitive buildings, 191
Concessions, 76, 107, 122, 175, 265
Condominiums, 166, 184, 193, 195
Conforming loan, 229–231
Consideration, 15, 28, 34, 47, 52, 55, 62, 63, 78, 80, 91, 92, 96, 100, 102, 106, 112, 122, 145, 154, 157, 163, 167–171, 173, 183–184, 187–192, 195, 196, 205–211, 213, 216, 259, 263, 264, 266
Constant amortization mortgage (CAM), 36, 38–40, 44, 45, 100–102, 110, 111, 196
Constant payment mortgage (CPM), 36–39, 44, 45, 143
Consumer price index (CPI), 78–79, 99, 105, 109, 110, 112, 113, 160, 255
Convenience center, 186
Conventional home loans, 226
Corporate veil, 150
Correlation, 258, 261
Cost approach, 5, 69, 71–75, 89, 90, 142
Counter party, 33, 126, 246, 268
Covariance, 261
CPI *See* Consumer price index (CPI)
CPM *See* Constant payment mortgage (CPM)
Credit enhancements, 229, 237–240, 245, 249
Credit wraps, 238
Curtesy rights, 28

D

DCF *See* Discounted cash flow (DCF)
DCR *See* Debt coverage ratio (DCR)
Debt coverage ratio (DCR), 31, 43, 44, 47, 83, 84, 133, 157, 159, 169, 170, 200, 214, 219
Debt-service coverage ratio (DSCR), 31, 90, 179, 180, 196, 198, 199, 217, 221, 244
Deed, 12, 13, 29, 135, 142
Deed of trust, 29
Default provision, 31, 35
Default risk, 125, 126, 148, 228, 238, 256, 261
Deglomeration, 208
Depreciable basis, 142, 144
Derivative, 33, 126, 245

Discounted cash flow (DCF), 74–79, 96, 168–171, 180, 194, 195, 198, 201–202, 217, 219, 220
Discount rate, 50, 55–57, 62, 63, 78, 82, 110, 124, 160, 202, 217, 219
Disintermediation, 227
Distribution warehouses, 207, 210, 217
Dodd-Frank Act, 244
Domino clause, 107, 109
Dower rights, 28
Down REITs, 253, 254, 268
DSCR *See* Debt-service coverage ratio (DSCR)
Due on sale clause, 28
Duplexes, 23, 165

E

Easement, 13, 28
Economic base analysis, 16, 267
Effective age, 71
Effective gross income (EGI), 76, 77, 79, 92, 97, 102, 111, 121, 159, 160, 168, 171, 180, 194–196, 198, 202, 217, 219
Effective tax rate, 149
Efficiency apartment, 166
EGI *See* Effective gross income (EGI)
Eminent domain, 10
Employment multiplier, 16
Environmental database report, 130
Environmental risk, 128–130, 137, 206
Escalator lease, 105
Escape clauses, 106
Escrow account, 76, 168–169
Estate at sufferance, 12
Estate at will, 12, 51
Estate for years, 11, 102
Estate from year to year, 11, 102
Estoppel certificate, 107–108
Expansion phase, 18
Expense stop, 101, 110, 160, 191
External credit enhancements, 238

F

Fashion, 36, 39, 77, 83, 103, 107, 121, 149, 186, 187, 189, 215
FDICIA *See* Federal Deposit Insurance Corporation Improvement Act (FDICIA)
Feasibility analysis, 164
Feasibility study, 15, 16, 136
Federal Deposit Insurance Corporation Improvement Act (FDICIA), 122–123
Federal funds rate, 124

Fee simple, 10, 11, 79
FFO *See* Funds from operations (FFO)
Financial Institutions Reform, Recovery and Enforcement Act (FIRREA), 122
Financial risk, 130–135, 144
Finite-life REITs, 254, 255
FIRREA *See* Financial Institutions Reform, Recovery and Enforcement Act (FIRREA)
Fixed rate, 32, 33, 36–41, 125–126, 131, 169, 227, 231
Flex space, 207
Flood way, 129
Flood zone, 107, 128–129, 135
Forbearance period, 32, 42
Fractured ownership, 193
Freehold estate, 10
Frictional vacancy, 76, 77, 79
Friendly foreclosure, 42
Fully modified, 228
Functional obsolescence, 72–74, 142
Funds from operations (FFO), 262–265, 269
Future estate, 11, 20

G
Garden apartments, 166
Garden unit, 166
General warranty deed, 12, 13
GIM *See* Gross income multiplier (GIM)
Glass-Steagall Act, 122
Go-dark clause, 107
Gordon dividend growth model, 261–262, 269
Graduated payment mortgage (GPM), 36, 39, 40, 44
Graduated rental lease, 105
Grantee, 11, 13
Grantor, 11–13
GRM *See* Gross rent multiplier (GRM)
Gross building area, 190
Gross income multiplier (GIM), 74–75
Gross leases, 104–105, 110, 186, 189
Gross rent multiplier (GRM), 75
Ground leases, 11, 103–104, 185, 200

H
Highest and best use, 5, 79, 111, 165
High-rise apartments, 167
HOA *See* Homeowners Association (HOA)
Holding period of the investment, 52, 53, 59, 77, 78, 100, 103, 123, 166, 168, 171, 176, 193, 206, 209

Homeowners Association (HOA), 166–167
Hybrid REITs, 256–257
Hyper-supply phase, 19

I
Improvement exchange, 154
Income approach, 5, 69, 74–80, 88, 100, 193, 263–264
Incubator REITs, 254
Index lease, 105
Industrial density, 209
Industrial property, 6, 15, 70, 72, 76–79, 99, 194, 195, 205–221
Inflation risk, 123
In-line tenants, 107, 110, 111, 187
Interbank lending market, 122, 246
Interest rate risk, 33, 124–128, 130, 231
Interest rate swap, 29–30, 32, 33, 126, 127, 130
Internal credit enhancements, 238, 239

J
Jumbo loans, 229
Junior mortgage, 28

K
Kyoto protocol, 128

L
Leased fee, 79
Leasehold estates, 11, 102
Leasehold improvements, 106, 115, 191, 206
Lease rollover risk, 108–109, 202, 261
Leasing commissions, 78, 96, 108, 111, 194, 195, 199, 264, 265
Legal description, 27, 68, 106
Legislative risk, 122–123, 144
Lessee, 10, 11, 95, 97, 102, 104, 105, 107, 112–116
Lessor, 10–11, 95–97, 102, 104, 105, 107, 112–116
Life estates, 11
Lifestyle centers, 172–173
Like kind exchanges, 141, 144, 152, 153, 158, 253
Lines of credit, 28, 46, 47, 246
Liquidity risk, 122, 229, 261
Load factor, 106
Loan constant, 37, 38

Loan points, 143–144, 147
Lockout period, 160, 234, 235, 240

M
Make whole, 126
Management risk, 109, 120–121, 130, 133, 260, 261
Marginal tax rate, 148, 149, 151
Market equilibrium, 16, 17, 174, 176, 213
Market feasibility study, 15
Market study, 90, 136
Mark-to-market or fair value accounting, 246
Maturity date, 31, 37, 61, 106
Mechanics lien, 13
Medical office buildings, 46, 47, 191
Metropolitan statistical area (MSA), 21, 260, 267
Mezzanine loans, 256
Mid-rise apartments, 167
Millage rate, 142
Mineral rights, 10, 128
Mini-storage, 205, 211, 220
Mixed use properties, 172–173, 184, 193, 201
Modified, 36, 60, 62, 107, 114, 228
Modified gross lease, 105, 109
Money markets, 237, 247
Money multiplier effect, 125
Moral hazard, 247, 248
Mortgage, 3, 11, 14, 16, 27–30, 32, 34–42, 44–46, 54, 56, 61, 76, 86, 104, 107, 108, 122, 124, 143, 146, 200, 214, 224–236, 238–249, 252, 254, 256
  originators, 229, 247
  REITs, 255–256, 268
  servicing rights, 230, 247
Mortgagee, 11, 27, 28, 42
Mortgagor, 27, 28, 104, 107–108, 231
MSA *See* Metropolitan statistical area (MSA)
Mutual fund, 53, 252, 257–258, 266, 268

N
NAV *See* Net asset value
Negative absorption, 17, 19
Negative amortization, 39, 40
Negative leverage, 132
Neighborhood center, 6, 186
Net asset value (NAV), 263–264, 269
Net operating income (NOI), 7, 31, 34, 41–43, 52, 69, 76, 77, 79, 80, 82, 84–90, 96, 100, 101, 104, 137–139, 141–142, 145–147, 149, 158, 160, 179, 180, 196, 198, 199, 217, 219, 221, 262, 263, 269

Net rentable area, 190, 198
Net sales proceeds, 147, 254
NOI *See* Net operating income (NOI)
NOIe, 84–88
Non-compete clause, 107
Nonconforming mortgages, 229
Non-dilution/radius clause, 107
Non-disturbance, 107
Non-recourse, 136
Note, 4, 5, 25, 27–33, 35, 36, 44, 48, 53, 74, 83, 101–102, 113, 125, 130, 165, 168, 198, 199, 210, 216, 217, 235, 254, 261, 266, 267
Notional principal, 235

O
Occupancy date, 106, 135
Off-campus student housing, 10, 20, 163, 167, 173–178
Office park, 191
Office showrooms, 207
Office warehouses, 208
Off shoring, 14, 124–125, 195, 205, 209–211, 216
Open ended mortgage, 28
Open market operations, 124
Outlet centers, 186
Outparcel, 185–186, 188
Outsourcing, 96, 121, 124–125, 195, 205, 209–211, 216, 267

P
Passive income, 121, 150
Percentage (or overage) rent, 99
Phase one environmental report, 130
Phase two environmental report, 130
Physical depreciation, 72–73
PODS *See* Portable on-demand storage (PODS)
Pool insurance, 238
Portable on-demand storage (PODS), 212
Portfolio income, 150
Portfolio lenders, 248
Positive leverage, 131
Power centers, 187
Prepayment risk, 125, 231–233
Primary trading area, 185
Private-label, 229, 236
Private-label securities, 237
Private mortgage conduits, 229
Private mortgage insurance, 225, 229
Professional building, 191

Pro-forma statement, 31
Pro-rata guarantees, 34
Public storage, 211

Q
QI *See* Qualified intermediary
QQD *See* Quantity, quality, and durability (QQD)
Qualified intermediary (QI), 153, 158
Quantity, quality, and durability (QQD), 7, 19, 20, 52, 62, 95, 133, 135, 163, 193, 220, 244, 258, 261, 268
Quit claim, 42
Quit claim deed, 13

R
Real estate investment trust (REIT), 4, 26, 52, 53, 187, 216, 251–269
Real Estate Mortgage Investment Conduits (REMICs), 233–235, 237, 240, 242, 255
Real rate of interest, 125
Recession phase, 19
Recovery phase, 17
REIT *See* Real estate investment trust (REIT)
Remainder, 11, 53, 110, 145, 179
REMICs *See* Real Estate Mortgage Investment Conduits (REMICs)
Renewal options, 103, 105, 106
Rent roll, 31, 64, 135, 136, 169, 170, 179, 180, 193, 196–201
Replacement reserve, 76, 77, 79, 91–92, 104, 168, 180, 202, 217, 219, 221
Reproduction cost, 73, 74
Reserve requirements, 125
Retained excess interest, 239
Revaluation lease, 105
Reverse exchange, 153
Reversionary interest, 11
Right to cure, 104
Ring study, 69–70
Riparian rights, 128
Risk absorption ratio, 135
95% rule, 157
200% rule, 157

S
Sales (or market) approach, 5, 69–71, 73, 79, 89, 91, 263
Schedule A, 151
Schedule B, 150, 151
Schedule C, 150
Schedule D, 150
Schedule E, 150, 151
Schedule F, 150
Schedule K, 151
Schedule K-1, 151
S-Corporation, 149, 151
Seisin, 27
Self-storage, 6, 7, 98–100, 104, 121, 159, 195, 205, 211–216, 220
Senior/Subordinated structures, 240
Sensitivity analysis, 16, 133, 139, 141, 158, 159
Servicers, 230, 236, 238, 241–243, 249
Service warehouses, 207, 208
Simultaneous exchange, 152
Single family residence, 2, 6, 23, 74, 165
SNDA, 107, 109
Special economic zones, 208
Special purpose property, 206
Special purpose vehicle (SPV), 236–240
Specialty centers, 186
Special warranty deed, 12
SPV *See* Special purpose vehicle (SPV)
Standard deviation, 132, 134, 137, 138, 261
Storage warehouses, 207
Strip center, 6, 89, 186, 196
Structural credit enhancements, 238, 240
Subordinated, 104, 240
Subordinated structures, 229
Subordination, 28, 104, 107
Subordination clause, 28
Subrogation, 107, 115
Substitute basis, 154
Super-regional centers, 187
Support or "companion" tranche, 234, 235
Survey exception, 13
Swap risk, 128
Systematic risk, 260

T
Tax shelter, 144
Tenant improvements, 17, 78, 97, 98, 108, 111, 171, 194, 195, 264, 265
Tenant mix, 61, 64, 107, 108, 168, 188, 189, 195, 196, 199
Terminal cap rate, 79, 80, 90, 146, 160, 171, 202, 219
Third party, 3, 6, 7, 11, 14, 28, 33, 50, 74, 85, 165, 168, 176, 185, 190, 191, 195, 206, 209, 213, 214, 229, 238
Third party letters of credit, 229
Title insurance, 12, 13, 20, 29

Total leasable area, 99
Townhouse, 166
Tranches, 233–235, 240–246, 249
Triple net leases (NNN), 104, 105, 121, 259

U
Umbrella Partnership REIT (UPREIT), 253
Underwriter, 237, 245, 247
Unsubordinated, 104
Unsystematic risk, 261
UPREIT *See* Umbrella Partnership REIT (UPREIT)

V
Value investment, 2–3, 5, 7, 20, 49, 123, 260
Variable rate, 29, 30, 40–41, 46, 124, 125
Voluntary conveyance, 42

W
Walk-up apartments, 167
Warehouse, 5, 6, 15, 82, 90, 92, 104, 163, 194, 195, 205–221, 255
Warranty, 12, 13, 27, 42, 243
Window, 124, 168, 234

Y
Yes Era, 1–3, 5, 6, 11, 17, 19, 26, 29, 32, 33, 42–44, 49, 53, 60, 68–70, 79, 81, 88, 98, 107, 120, 122, 124, 125, 127, 128, 134, 148, 157, 165, 183, 214, 256, 268

Z
ZAM *See* Zero amortization mortgage (ZAM)
Zero amortization mortgage (ZAM), 36, 41
Zoning restrictions, 10

CPSIA information can be obtained at www.ICGtesting.com
Printed in the USA
LVOW102137280313

326601LV00007B/45/P

9 783642 235269